FOOD SCIENCE AND TECHNOLOGY

SEEDS AS FUNCTIONAL FOODS AND NUTRACEUTICALS

NEW FRONTIERS IN FOOD SCIENCE

Food Science and Technology

Additional books in this series can be found on Nova's website under the Series tab.

Additional e-books in this series can be found on Nova's website under the e-book tab.

FOOD SCIENCE AND TECHNOLOGY

SEEDS AS FUNCTIONAL FOODS AND NUTRACEUTICALS

NEW FRONTIERS IN FOOD SCIENCE

ROSALVA MORA-ESCOBEDO
JOSE DE J. BERRIOS
AND
GUSTAVO FIDEL GUTIERREZ LOPEZ
EDITORS

nova publishers
New York

Additional color graphics may be available in the e-book version of this book.

Library of Congress Cataloging-in-Publication Data

Seeds as functional foods and nutraceuticals : new frontiers in food science / editors, Rosalva Mora-Escobedo, Jose De J. Berrios and Gustavo Fidel Gutierrez-Lopez (National Polytechnic Institute, National School of Biological Sciences, Prol. Carpio y Plan de Ayala, S/N Col. Sto. Tomas, Mexico, and others).
pages cm
Includes bibliographical references and index.
ISBN 978-1-62808-489-4 (hardcover)
1. Functional foods. 2. Seeds--Therapeutic use. 3. Nuts--Health aspects. I. Mora-Escobedo, Rosalva, editor of compilation. II. Berrios, Jose de J., 1950- editor of compilation. III. Gutiirrez-Lspez, Gustavo F., editor of compilation.
QP144.F85S44 2014
613.2'6--dc23
2013043712

Published by Nova Science Publishers, Inc. † New York

Contents

PREFACE

Fundamental aspects in human development and the different cultures that have formed are closely linked to the seeds. The cause or reason that prompted the speed development of man was closely related when men started the domestication of plants. With this cultural development of man, came a social transformation that provided a structure and organization that brought about a change in lifestyle, as man nomad became sedentary. Most fruits and seeds of plants have provided man with vital source of foods. Seeds in particular, have played an important role in the human diet. They have served as a means of spreading crops that man requires. They have been eaten as part of the human diet, providing a complex plant food rich in macro and micronutrients, dietary fibers, as well as bioactive components with healthy and disease-prevention benefits, way beyond their basic nutritional role.

Seeds as legumes, oilseeds, and other seeds such as grape seed, avocado, and mango, have excellent nutritional benefits, including high levels of dietary fiber, complex carbohydrates, bioactive peptides, and phytochemicals: chemicals found as a natural components of foods or other ingestible forms that have been determined to be beneficial to the human body in preventing or treating one or more diseases, or improving physiological performance. Therefore, these phytochemicals are excellent contributors to good health. Seeds as Functional Foods and Nutraceuticals: New Frontiers in Food Science bring together the contribution of experts in food science, biological science, nutritionists, and food engineers, to provide the reader with the latest information of fundamental and applied research in the role of seeds and their products as functional foods and nutraceuticals, and discuss the benefits of consuming them. In this book you will find relevant information regarding origin and taxonomy of seeds, global markets, physicochemical composition, and the effect of phytochemicals in seeds components on chronic degenerative diseases, such as obesity, diabetes, cancer, cardiovascular disease, inflammation and arthritis. Giving the importance and challenges, derived from environmental concern, regarding the effective utilization of residues of industrial by-products and agroindustrial wastes, this book also report the inclusion of seeds and certain fruits' by-products in foods and the presence of phytochemicals with potential medicinal benefits.

While promising advancement have been made in the area of functional foods and nutraceuticals in last decade, there are many challenges ahead that need to be address regarding bioavailability, elucidation of pathways and mechanism of action, innovation in formulations of functional ingredients and delivery, functional ingredient synergism, the effect of processing on the biological activity of bioactive compounds, and the clinical outcomes associated with consumption of functional foods and nutraceuticals. This book is intended to placing the concept of seeds as functional foods and nutraceuticals in a broad context for readers, stimulates increased interests in research regarding this topic, and the application of these findings to beneficial alternatives.

ABOUT THE EDITORS

Dr. Rosalva Mora-Escobedo is a full professor at the National Polytechnic Institute at the National School of Biological Sciences, Mexico DF, Mexico. She received her BSc degree in Chemistry from the University of San Luis Potosí, México, in 1978; her MSc degree at the University of Querétaro in 1978; and her Ph.D. at CINVESTAV Irapuato Unit in 1994. Dr. Mora-Escobedo completed a Sabbatical leave at CINVESTAV Irapuato Unit (1987-1988) and a postdoctoral training at the University of Reading, UK (1998-1999). She has been a member of the Mexican Academy of Sciences since 2004. She belongs to the National Research System level III, which is the highest scientific level in Mexico. She received National awards in 1999 and 2011 for her leading work in Food Science and Technology from the Coca Cola Company and CONACyT-Mexico, respectively. She also received an award in 2004 for Best Paper award-Runner Up, by the Tortilla Industry Association. Dr. Mora-Escobedo has conducted extensive research work in the areas of nutrition and nutraceutical properties of foods, with special emphasis on carbohydrates and proteins of cereals and legumes. She is an author and co-author of 56 peer reviewed journal publications and 5 book chapters. She has given 50 presentations at society meetings, universities, special interest groups, and commodity group meetings.

Dr. Jose De J. Berrios is currently a research food scientist with the Processed Foods Research Unit at the USDA-ARS-Western Regional Research Center (WRRC), in Albany, California. He received his Ph.D. degree in Food Science from Washington State University in 1995. Dr. Berrios is the leader of food extrusion at WRRC conducting experiments and developing novel processes and formulations to increase the utilization of legume pulses and their fractions, cereal grains, and fruits and vegetables, through the manufacture of new value-added extruded food products, such as pasta products and beverages; high protein and dietary fiber and gluten-free products with optimized nutritional, functional properties, and sensory attributes. Dr. Berrios has been a renowned leader and internationally recognized authority in innovative food processing research for over 20 years. During this period, he has authored and co-authored more than 52 publications, including 2 patents, 4 invention disclosures, 5 book chapters, and 80 abstracts, and has delivered more than 100 invited talks at national and international meetings. He has been a Scientific Adviser to the United Nations' World Food Program in Ecuador, Canadian Pulse Innovation Project, USA Dry Pea and Lentil Council, Washington State Potato Commission, PepsiCo-Frito Lay Co., and Pulse America. Dr. Berrios has mentored more than 15 professors and visiting scientists and 25 graduate students from around the world, who have requested to work under his guidance. He has collaborative research projects with scientists from 5 national and 8 International Universities. Dr. Berrios was selected Chair of the II International Symposium on Food Extrusion in Brazil, 2010; keynote speaker at International meetings in Mexico, Brazil, Spain, Germany, France, and Peru. His research has been widely reported by the electronic and print media. He was selected Chair of the California Section of the AACC; member of the AACC Pulse/Legume Technical Committee; Jury for the AACC International Travel Award, and nominated Federal Employee of the Year by the San Francisco Bay Area Federal Executive Board 2012. He acts as a reviewer of numerous scientific journals in the areas of food science, food technology, food engineering, and post-harvesting of fruits and vegetables.

Dr. Gustavo Fidel Gutierrez-López is a Full Professor of Food Engineering at the National Polytechnic Institute of Mexico (IPN-Mexico) and Chairman of the Food Science and Technoloby PhD Program of The National school of Biological Sciences of IPN-Mexico . He received his bachelor of Science in Biochemical Engineering and Master of Science in Food Science and Technology from IPN and his Master of Science in Food Process Engineering and PhD in Food Engineering of The University of Reading, U.K., where he spent two years as a postdoctoral researcher. He was chairman of the Postgraduate School of the National School of Biological Sciences and received the APV Seligman Fellow Award (1995) of the Society of Chemical Industry, UK and was distinguished by the Committee of Principals and Vice-chancellors of the Universities of the UK in 1987 and 1988. In Mexico, he received the Food Science and Technology National Award in 1991 and 1992 as well as the prize for the best research in Food Science at IPN-Mexico and supervisor of the most distinguished PhD thesis at this institution. He was President of the Mexican Society of Biotechnology and Bioengineering and former President of the Jure of the National Prize of Food Science and Technology. He has been invited as a key-note speaker to ISOPOW, ICEF, EUoFST, IUoFST, Food Innova, etc. Is the leading editor of two books by CRC and Springer. He is author/co-author of more than 100 scientific publications in the fields of food engineering and structure-function-processing relationships.

Part 1. Seeds

In: Seeds as Functional Foods and Nutraceuticals
Editors: R. Mora-Escobedo, J. De J. Berrios et al.
ISBN: 978-1-62808-489-4

Chapter 1

SOYBEAN: BIOACTIVE COMPOUNDS IMPROVED BY GERMINATION, FERMENTATION AND ENZYMATIC HYDROLYSIS

Marcio Schmiele[1*], Maria Gabriela Vernaza[2], Leomar Hackbart da Silva[3], Luz Maria Paucar-Menacho[4] and Yoon Kil Chang[1]

[1]University of Campinas – Brazil
[2]University of Equinoctial Technology – Ecuador
[3]Federal University of Pampa – Brazil
[4]National University of Santa – Peru

ABSTRACT

The growing of soybean as a crop began by the year 2800 BC, but in recent years has been gaining attention since it is considered a functional food. Its consumption has been linked to the prevention of cancer, type 2 diabetes, cardiovascular and gastrointestinal diseases, cholesterol reduction and obesity. The main components responsible for the protection of health are the bioactive compounds such as peptides, lunasin, lectins, Bowman-Birk inhibitors, isoflavones, saponins and others. The germination and fermentation processes can promote the increase in nutritional value, by improving the protein digestibility and the value of the protein efficiency ratio, decreasing the antinutritional factors in legumes, such as the proteolytic and lectin inhibitors and causing the hydrolysis of oligosaccharides (stachyose and raffinose) present in soybeans, which are the cause of flatulence. The fermentation processes promote the conversion of isomers of the glycosylated isoflavones into aglycones and the formation of low molecular weight peptides (<20 kDa), improving the nutritional and functional properties. The germination process in combination with alcalase hydrolysis presented a notable antioxidant capacity and a significant inhibition on inflammatory markers.

* Corresponding author: marciosc@fea.unicamp.br.

INTRODUCTION

The oldest data on the cultivation and use of soybean (*Glycine max* L. Merrill) were reported in the 11th century BC. China is the largest producer, and the grain was considered sacred as was rice, wheat, barley and millet [1]. The expansion of this culture was slow in Asia (China, Korea and Japan). In the early 19th century soybeans were introduced in the United States, but only in 1880 did this grain acquire great economic importance in that country. In Brazil, soybean was introduced in the late 19th century in the State of Bahia and in the early 20th century in the state of São Paulo. However, large-scale production occurred only after 1914, when soybean cultivation began in the state of Rio Grande do Sul, in Santa Rosa city, a region of increased domestic production until mid-1930s [2].

In 1975, Embrapa Soja was established as one of the 45 units of the Brazilian Agricultural Research Corporation (EMBRAPA - Empresa Brasileira de Pesquisa Agropecuária), aiming to expand the cultivation of this grain, develop new cultivars and increase domestic production [3]. As a result, Brazil is the world's second largest soybean producer with 66.5 million tons in 2012 (approximately 28 % of world production), while the U.S. is the largest producer (84 million tons). Other major soybean producers are Argentina, China, India, Paraguay and Canada [4].

The soybean belongs to the family *Fabaceae*, a large family of legumes composed of dicots, which include beans, peas, lentils, chick-peas and alfalfa. Its major components are approximately 40 % protein, 35 % carbohydrate, 20 % fat and 5 % ash (in dry basis) [5]. It is an agricultural commodity with great importance in the world economy due to both new soybean byproducts on the market, such as biodiesel, and functional properties evidenced by the scientific studies [6].

As a rich source of protein for food and animal feed, soybean is the most protein-rich plant when compared to any other cereal or legume. Soy protein has high nutritive value [7, 8, 9]. The main components are the storage proteins known as β-conglycinin and glycinin, which represent around 50-85 % of the total proteins of the grain [10]. Other minor proteins and glycoproteins include lipoxygenases, lectins, trypsin inhibitors and α-amylases [5]. Soy glycinin contains no carbohydrate, whereas conglycinins and glycoproteins contain about 4 % carbohydrate.

As minor components, soybean contains vitamins such as thiamine, riboflavin and vitamin E as well as, several bioactive compounds such as isoflavones, saponins, phytates, phytosterols, trypsin inhibitors, lectin and lunasin Bowman Birk inhibitor [12]. The soy intake is related to the prevention of heart disease, kidney and intestine disorders, cholesterol reduction, prevention of cancer, type 2 diabetes and obesity [12, 13, 14, 15, 16, 17, 18, 19, 20].

In 1999, the FDA (Food and Drug Administration) has issued a document to formalize the therapeutic potential of soy in preventing heart disease. The FDA has determined that a daily intake of 25 grams of soy protein significantly reduce total blood cholesterol and low density lipoprotein (bad cholesterol) and increase high density lipoprotein (good cholesterol), thereby reducing the risk of cardiovascular diseases such as myocardial infarction, thrombosis and atherosclerosis [21].

In addition to the health benefits provided by soybeans, some techniques allow increasing concentrations of bioactive components and/or decreasing the anti-nutritional factors of the grains or flours, such as germination and fermentation followed or not by enzymatic hydrolysis [22] due to the size reduction of the protein chains, converted them into peptides.

Bioactive Compounds

In this chapter we discuss the protein (lunasin, Bowman-Birk inhibitor and lectins) and non-protein (isoflavones and saponins) bioactive compounds. The peptides from soy beans have antioxidant [23] and anti-obesity [24] activities. Moreover, they also contain bioactive proteins with anticancer activity, including lectins [25] and the more recently discovered peptide, the lunasin [26].

Lunasin

Lunasin is a single peptide originally isolated from soybean and more recently from barley, which may prevent cancer [27, 28]. It contains 43 amino acid residues and nine aspartic acid residues at its extremity with function of cellular adhesion due to the presence of arginine-glycine-aspartic acid peptide [29], as shown below:

SKWQHQQDSCRKQLQGVNLTPCEKHIMEKIQGRGDDDDDDDDD

Where: C = Cysteine, D = Aspartic Acid, E = Glutamic acid, G = Glycine, H = Histidine, I = Isoleucine, K = Lysine, L = Leucine, M = Metionin, N = Asparagine, P = Proline, Q = glutamine, R = Arginine, S = Serine, T = Threonine, V = Valine, W = Tryptophan.

Lunasin was discovered by accident by Dr. Alfredo Galvez in 1996 as a result of his research to improve the nutritional profile of soy protein. In the laboratory of Dr. De Lumen, at UC Brekeley, Dr. Galvez discovered that this amazing soy peptide found in small amounts in soybeans and soy-based foods, blocks cell division by binding to specific chromosomal proteins called "deacetylated histones" [30].

The efficacy of lunasin against chemical carcinogens and oncogenes has been demonstrated in cell cultures and in a mouse skin cancer model. Researchers at UC Berkeley [31] have isolated and cloned a cDNA encoding a post-translationally processed 2S albumin (Gm2S-1) from mid-maturation soybean seeds. Lunasin is a predicted helix with structural homology to a conserved region of chromatin-binding proteins [32, 33]. Lunasin appears to be a novel peptide that demonstrates potential anticancer activity against cancer induced by chemical agents [28]. Clearly, research is fundamental to confirm these preliminary observations and possible health uses, including its role in the prevention of chronic diseases.

In 2004, a study of lunasin concentration in different soybean genotypes from the U.S. was performed in the laboratory of Food Science and Human Nutrition, University of Illinois at Urbana-Champaign [34]. This research described a methodology for quantifying lunasin by enzyme-linked immunosorbent assay (ELISA) aimed to identify and quantify variations in lunasin concentration in 144 soybean genotypes selected by USDA Soybean Germplasm Collection, whose method is commercially available to evaluate soy protein fractions and isoflavone-enriched products. With synthetic lunasin and monoclonal antibodies, the ELISA method showed good reproducibility with a linear concentration range of 24-72 ng / mL, a detection limit of 8 ng / mL, and a recovery of 90 % on soybean samples. Lunasin concentration in the tested materials ranged from 0.10 to 1.33 g/100 g of flour. Differences that exceeded 100 % were observed among accessions of similar maturity that were grown in the same environment, indicating that genetic differences in soybeans exist for lunasin. The average of 23 major ancestral U.S. cultivars was similar to the average of 16 modern cultivars selected to represent the diversity of this crop. Soy protein concentrate, isolate, and hydrolyzate contain 2.81, 3.75 and 4.43 g lunasin/100 sample, respectively, whereas soy flour and soy flakes contain 1.24 g lunasin/100 g flour. Isoflavone-enriched products contain very little lunasin, and sometimes the presence of this bioactive compound may be null. The relative mass of lunasin in the samples was 5.45 kDa. The wide range of lunasin concentration within the *Glycine max* species indicates that the levels of this important bioactive peptide can be genetically manipulated. In addition, soy protein isolates and hydrolysates contain the highest concentrations of lunasin.

Bowman-Birk Inhibitor

The Bowman-Birk inhibitor has 71 amino acid residues, molecular weight of 7.975 kDa, and seven disulfide bonds in the molecule. It inhibits the activity of trypsin and chymotrypsin stoichiometrically at 1 molar ratio, in an independent and simultaneous manner [34]. The amino acid sequence containing the disulfide bonds and the centers of complexation with enzymes can be seen in Figure 1.

The isoelectric point of Bowman-Birk inhibitor ranges from pH 4.0 to 4.2, whereas the isoelectric point of various vegetables and fruits range from pH 4.0 to 9.77 [35]. This variation may be due to the differences in amino acid residues. The Bowman-Birk inhibitor is rich in cysteine amino acid (about 20 %).

Bowman-Birk inhibitor is stable within the pH range found in most foods, it withstands the temperature of boiling water for 10 min, resistant to acidic pH and proteolytic enzymes in the gastrointestinal tract, is bioavailable and non-allergenic. Previous studies have shown that protease inhibitors are not considered bioactive ingredients for fortifying food due to its low specificity. The FDA considers Bowman-Birk inhibitor as a drug.

The most bioactive compounds act in prevention of diseases by inhibiting the enzyme that catalyzes the pathological process. For this reason, Bowman-Birk inhibitor fits the definition of a functional food with bioactive capability. Whether the positive results of the research on Bowman-Birk inhibitor could be grounded in large-scale population studies, Bowman-Birk inhibitor would be a good economic alternative to public health as a functional food [36].

In particular, the role of protease inhibitors of a food source such as Bowman-Birk inhibitor is recognized by biomedical researchers as potential chemopreventive agents [37, 38, 39, 40], specifically in cases of breast cancer.

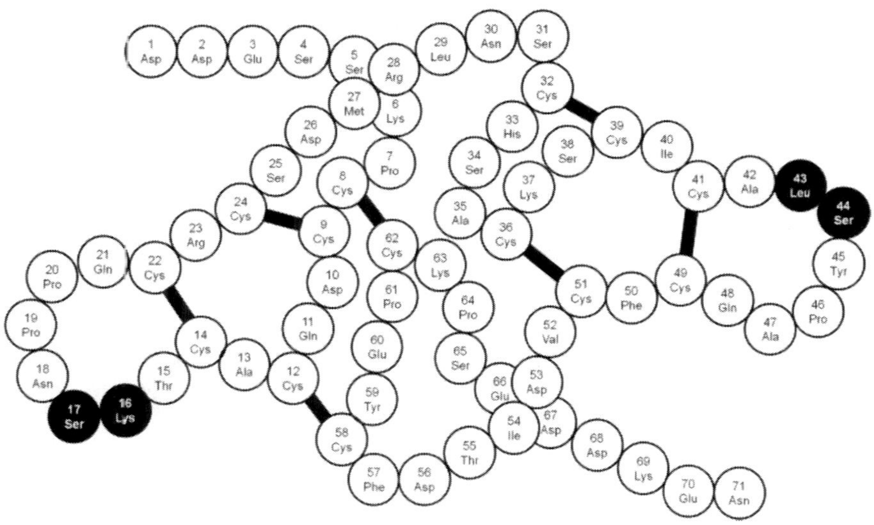

Figure 1. Amino acid sequence for Bowman-Birk inhibitor, disulfide bonds between Cys-Cys and centers of complexation with enzymes: trypsin (Ser, Lys) and chymotrypsin (Ser, Leu).

Lectin

Lectins, also called as hemaglutinins, are a group of proteins that bind specifically to sugars of glycoproteins and glycolipids on the surfaces of cells. The antinutritional factors limit the application of soybean in animal and human nutrition as they interfere with the utilization of proteins and minerals in the digestive tract [41].

Lectin in the diet of mice showed a significant reduction in weight gain as compared to controls, and also reduced nitrogen digestibility and nitrogen retention, increasing fecal and urinary nitrogen loss [42]. Antibodies specific for soybean lectin were detected in the serum of poults, which implies that the soybean lectin remained active in the gastrointestinal tract [43].

Lectins are accumulated in the seed protein storage vacuoles and cotyledons which are degraded during germination of the seeds [44, 45]. Lectins have several anti-nutritional properties, but also anticancer properties [46, 47].

Various lectins with anticancer properties were found in in vitro and in vivo studies (case studies in humans) and are used as therapeutic agents, binding to cancer cell membranes or their receptors preferentially, causing cytotoxicity, apoptosis and / or growth inhibition of tumors. These compounds can become internalized into the cells, leading to cancer cell agglutination and / or aggregation [47]. Some lectins may be related to the growth of several cancer cell lines, and some dietary lectins can cause a chemopreventive effect in breast cancer of humans by inhibiting cell growth and proliferation in vitro [48].

Isoflavones

Soybeans and its derivatives have great potential in the functional food market due to the presence of bioactive compounds, such as isoflavones. Which have been widely studied for their beneficial physiological effects on human health, such as antifungal activity, estrogenic and anti-estrogenic properties (especially on climacteric symptoms and osteoporosis), hypocholesterolemic and anti-carcinogenic effects and reducing the risk factors associated with arteriosclerosis [49, 50, 51, 52, 53]. These beneficial effects may be most evident in Asian populations because of their high intake of soy and derivatives.

In general, isoflavones are present in soybeans especially in the glycosylated form and in small quantity in the aglycone form. The concentration of isoflavones in soy products depends on the variety of soy foods and processing techniques, such as application of heat treatment, fermentation or germination [54, 55, 56, 57, 58, 59].

Isoflavones are a subclass of flavonoids, which belong to the group called phytochemicals, compounds not included as nutrients but have been getting attention from researchers due to their estrogenic properties and prevention of cancer and other chronic diseases.

Flavonoids include all the phenolic compounds from a plant and its basic structure consists of two benzene rings linked by a heterocyclic pyran ring. Other examples of flavonoids are anthocyanins, flavones, flavonols, flavanols, chalcones and aurones.

The bioavailability of soy isoflavones differs between the glycosylated forms and the aglycones, although the mechanism of absorption and bioavailability is not elucidated. Studies have shown that isoflavone aglycones are absorbed faster and in greater amounts than the glycosylated forms [49, 60, 61, 62]. This sparked interest in increasing the amount of aglycone isoflavones in soy products and their derivatives.

Soy isoflavones contain 12 forms of isomers, as shown in Figure 2. Soybeans are the only natural source that contains large amounts of isoflavones, above 3 mg / g on a dry basis. The original isoflavones present in soybeans are genistein, daidzein and their β-glycoside conjugates. The glycitin and glycitein also appear in smaller amounts.

Table 1. Total isoflavones levels (mg/100g) in soybean seeds sown in different months, harvested in 2004/2005

Cultivars	October	November	December
BRS 133	219.09	160.81	171.55
BRS 185	215.92	168.84	145.91
BRS 260	206.08	181.69	108.52
BRS 214	195.27	152.20	159.36
BRS 261	193.22	118.02	79. 37
BRS 213	191.63	145.18	104.90
BRS 184	169.31	144. 88	129.05
BRS 259	150.03	97.74	109.00
BRS 215	146.86	90.60	123.36
EMBRAPA 48	143.95	73.50	91.89
BRS 257	140.35	123.24	93.33
BRS 232	120.87	66.13	72.31
BRS 231	120.61	113.49	90.06
BRS 262	110.64	139.65	165.87
BRS 230	81.56	62.34	56.47
BRS 258	56.37	47.95	37.86

Source: Mandarino and co-workers [63].

Saponins 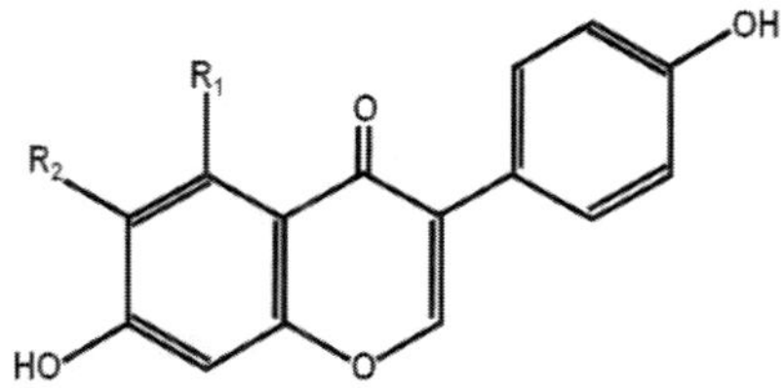

R_1	R_2	Isomers
H	H	Daidzein
OH	H	Genistein
H	OCH_3	Glycitin

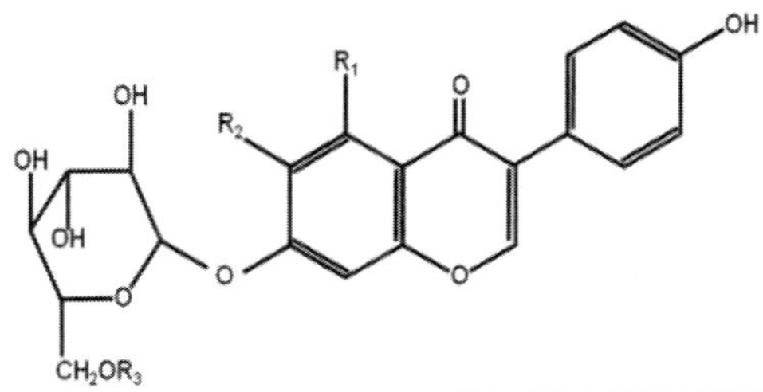

R_1	R_2	R_3	Isomers
H	H	H	Daidzin
OH	H	H	Genistin
H	OCH_3	H	Glicitin
H	H	OCH_3	6”-*O*-Acetyldaidzin
OH	H	OCH_3	6”-*O*-Acetylgenistin
H	OCH_3	OCH_3	6”-*O*-Acetylglycitin
H	H	$COCH_3COOH$	6”-*O*-Malonyldaidzin
OH	H	COCH3COOH	6”-OMalonylgenistin
H	OCH_3	COCH3COOH	6”-*O*-Malonylglycitin

Figure 2. Chemical structures of 12 soybean isoflavone isomers.

In non-fermented soybean products, isoflavones are present in their conjugated form, whereas in fermented aglycones are predominant, due to the *β*-glucosidase enzyme produced by microorganisms responsible for fermentation, which breaks the glycosidic bond, resulting in aglycone.

The isoflavone content in soybean cultivars of Embrapa Soja was presented at the III Congress of Soy of MERCOSUL - MERCOSOJA 2006, occurred in Rosario, Argentina, ACSOJA, described by Mandarino and co-workers [63] and are shown in Table 1.

Saponins

Saponins are triterpenoids naturally found in different plant foods. They are secondary metabolites of plants containing a steroid or triterpenoid aglycone with a number of carbohydrate molecules linked by ethers and esters connecting to one or more glycosylation sites. Seeds of soybeans contain from 0.6 % to 6.5 % (db) triterpenoid saponins depending on the variety, year of cultivation, place of culture, and degree of maturity. The saponins have antifungal, antiviral, spermicidal, expectorant, diuretic, and anti-inflammatory activities [64].

During the maturation of soybean, the soy saponins have been divided into group B saponins, showing a basic structure of aglycone. Group B saponins appear to exist in the intact plant tissue as conjugates of 2,3-dihydro-2,5-dihydroxy-6-methyl-4H-pyran-4-one (DDMP), at the 22-hydroxyl position [65]. The DDMP conjugates are relatively labile and easily degraded, resulting in formation of non-DDMP group B saponins. The other various forms of group B soy saponins can arise from alternate sugars in the oligosaccharide attached to the 3-hydroxyl position of the aglycone. Group A saponins are didesmosidic with alternative sugar compositions in both sets of oligosaccharides attached to the aglycone at the 3- and 21-hydroxyl positions [66]. Several other forms of saponins, including the Group E, had shorter sugar chains saponins, probably resulting from extraction and processing steps [67].

Berhow and co-workers [64] published a summary of 20 different forms of soy saponins and their processed products. The structures and names of the nomenclature of the soy saponins are shown in Figure 3.

SOYBEAN FERMENTATION PROCESS

Studies have shown that isoflavone glycosides are hydrolyzed into their corresponding aglycones during fermentation of soybean foods such as tempeh, soy extract, miso and natto [56, 58, 68, 69]. Furthermore, the fermentation can promote protein hydrolysis, thus forming peptides and reducing the anti-nutritional factors, which increases the beneficial health effects of soybeans [70, 71].

During thermal and fermentation processes, the transformation of isoflavone glycosides to aglycones takes place, and the latter are accumulated in the final product [72]. This conversion in glycosylated forms occurs in the presence of the enzyme β-glucosidase secreted by the fungus *Aspergillus oryzae* [73]. The fermented soy products have a higher content of peptides when compared with products produced with non-fermented soy [69].

The fermentation process is used to prepare traditional soy foods in Asia, commonly known as "dou-bain-jiang" in China, "miso" in Japan and "doenjang" in Korea [74, 75]. These fermented soy foods are easily digestible and nutritious, contributing to important nutrients such as calcium and vitamin A and vitamin B complex, as well as presenting functional properties with laxative effect and anticancer properties [76].

Obtaining fermented soy products may be by liquid, semi-solid or solid fermentation, according to the method and the characteristics desired in final products [77]. The fermentation may be performed in either whole grain or whole or defatted flour, which may be raw, cooked or toasted. This process can occur through natural fermentation, in which the microorganisms are already present in soybeans, or by controlled fermentation. In the last case, strains of microorganisms selected for obtaining certain technological sensory and functional characteristics desired in the final product are inoculated in soybeans, and the conditions of fermentation, such as humidity, temperature and incubation time are controlled to obtain better quality products [75, 78]. The main microorganisms involved in the production of fermented soybean foods are *Bacillus subtilis*, *Rhizopus oligosporus* and species of *Aspergillus oryzae* and *Aspergillus sojae* [79].

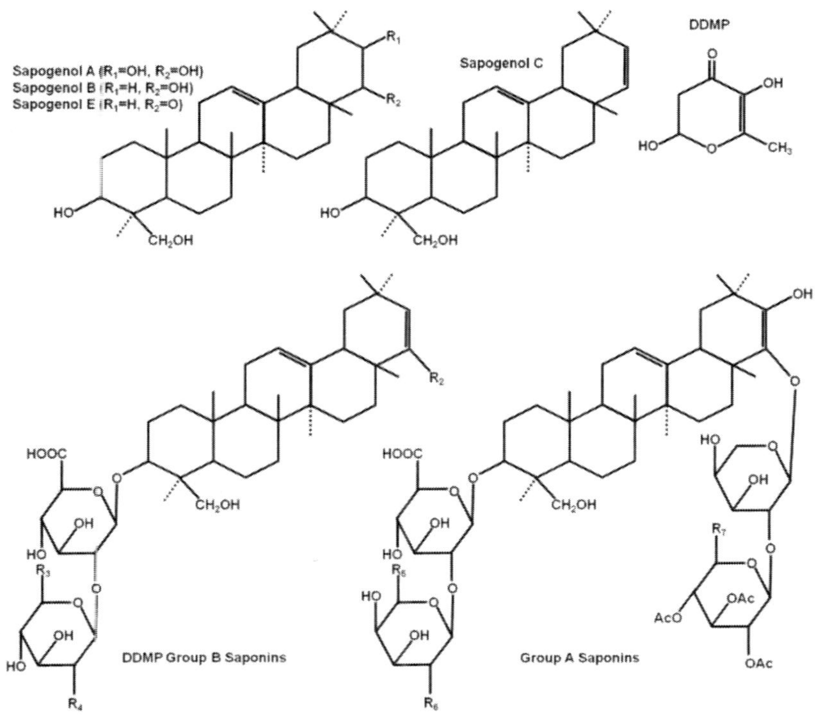

DDMP and group B saponins	MW	R_2	R_3	R_4
Soyasaponin I	942	OH	CH_2OH	*O*-*β*-D-Glucose
Soyasaponin II	912	OH	H	*O*-*α*-L-Rhamnose
Soyasaponin III	796	OH	CH_2OH	OH
Soyasaponin IV	766	OH	H	OH
Soyasaponin V	958	OH	CH_2OH	*O*-*α*-L-Rhamnose
Soyasaponin	940	O	CH_2OH	*O*-*β*-D-Glucose
Soyasaponin	956	O	CH_2OH	*O*-*α*-L-Rhamnose
Soyasaponin *β*g	1068	*O*-DDMP	CH_2OH	*O*-*β*-D-Glucose
Soyasaponin *β*a	1038	*O*-DDMP	H	*O*-*α*-L-Rhamnose
Soyasaponin *γ*g	922	*O*-DDMP	CH_2OH	OH
Soyasaponin *γ*a	892	*O*-DDMP	H	OH
Soyasaponin *α*g	1084	*O*-DDMP	CH_2OH	*O*-*α*-L-Rhamnose
Group A acetyl-saponins	**MW**	R_5	R_6	R_7
Soyasaponin aA_1	1436	CH_2OH	*O*-*β*-D-Glucose	CH_2OAc
Soyasaponin aA_2	1274	CH_2OH	OH	CH_2OAc
Soyasaponin aA_3	1202	H	OH	CH_2OAc
Soyasaponin aA_4	1364	CH_2OH	*O*-*β*-D-Glucose	H
Soyasaponin aA_5	1202	CH_2OH	OH	H
Soyasaponin aA_6	1172	H	OH	H
Soyasaponin aA_7	1420	CH_2OH	*O*-*α*-L-Glucose	CH_2OAc
Soyasaponin aA_8	1406	H	*O*-*β*-D-Glucose	CH_2OAc

Note: Nomenclature source obtained from Berhow and co-workers [64].

Figure 3. Structure and nomenclature of soyasaponins.

Most studies on the fermentation of grain or defatted soy flour perform the fermentation process with the raw material previously autoclaved or subjected to heat treatment at 65 °C for 1 hour, followed by incubation at 30 °C for 48 hours, with the fungus *Aspergillus oryzae*. Under these conditions, the fungal growth takes place as well as the production of enzymes, such as amylases, proteases, *β*-glycosidases, among others, which cause several changes in nutritional and functional properties of the final product [73, 75].

Esaki and co-workers [78] isolated an antioxidant complex of soybeans fermented with *Aspergillus oryzae* and *Aspergillus saitoi*, that is, 6-hydroxy daidzein (6-OHD), 8-hydroxy daidzein (8-OHD) and 8-hydroxy genistein (8-OHG). These isoflavones, which have O-dihydroxy structure at positions 7 and 8 or 7 and 6, were obtained from daidzein and genistein, respectively, by hydroxylation reaction during the fermentation process.

The microbial *β*-glucosidase, which is involved in the hydrolysis of soy isoflavone glucosides to aglucone was obtained by using several strains of microorganisms participating to the fermentation process, and *Aspergillus oryzae* showed high activity in the conversion of isoflavone glucosides to aglucone. The enzyme efficiently converted isoflavone glycosides from defatted soy flour in their aglycone forms, and an increased concentration of aglycones showed high antioxidant activity [78, 80].

Silva [81] evaluated the effect of fermentation time and temperature of autoclaved whole soybean flour on the conversion of isoflavone isomers from glycosides to aglycones. The author noticed that the fermentation using *Aspergillus oryzae* promoted the transformation of isoflavone glycosides into aglycones, whose contents increased from 35.13 μg / g in the non-fermented flour to the range of 245.23 to 535.94 μg / g in the fermented flour. According to the conditions studied, the author observed an increase up to 14.26 times the concentration of aglycones in fermented autoclaved whole soybean flour. This was the maximum concentration found with fermentation conditions of 36 h at 35 °C, which increased the concentration of isoflavone aglycones to 77 % of total isoflavones.

Wu and Muir [82] evaluated the effect of *β*-glucosidase in isoflavones during enzymatic hydrolysis of defatted soy flour. The authors observed a 10-fold increase in the concentration of isoflavone aglycones when compared to the content found in non-fermented soy flour, indicating an alternative for increasing isoflavone aglycones in soy containing products.

Chen and Wei [83] studied the concentration of isoflavone aglycones in soybeans and their derivatives, including non-fermented (tofu, soy extract, and soy flour) and fermented products (natto, miso and sufu). The total content of isoflavones found in soybeans and their derivatives ranged from 109 to 508 mg / 100 g. The aglycone form was found at 2 to 19 % of the total isoflavones of soybean products and non-fermented soybeans, while the fermented soybean products presented from 4 to 94 % isoflavone aglycones, with higher concentrations in sufu and miso. The authors attributed the increase of aglycones to the action of *β*-glucosidase produced by microorganisms used in fermentation products. Some isoflavone isomers, such as genistein, are more effective in controlling cancer cells, and inhibiting the growth of prostate tumor cells in humans, as compared to its glycosylated form (genistin). Furthermore, it was reported that the antioxidant activity of genistein and daidzein were higher than those of their respective glycosylated isomers [84].

Both the absorption and retention of isoflavones by the human body increase according to its water solubility. Thus, genistein is absorbed more quickly than daidzein, which in turn is absorbed more quickly than glycitein. The absorption mechanism of genistein is already known; the isoflavone glycosides are initially absorbed by the intestinal mucosa with the continuity of the process in the liver. About 70 % of isoflavones go to bile 4 hours after ingestion and 25 % is excreted through urine [85].

According to Messina and co-workers [86], isoflavones had proven effective protection against carcinogenesis in animal tests (65 % positive results). Bradlow and Sepkovic [87] have also reported the protective action of isoflavones on prostate cancer, breast cancer and other hormone-dependent cancers in humans. This protective effect occurs through specific regulation of sex steroids and tumor growth factors, as identified by Lamartiniere and co-workers [88] who studied the effect of genistein intake in humans.

Studies with soy products without isoflavones showed loss of the benefits from using soy protein supplementation. However, an extract with high content of isoflavones by itself has demonstrated hypocholesterolemic activity. For this reason, soy protein for consumption must contain a minimal content of isoflavones [89].

Researchers suggest that the consumption of isoflavones must be approximately 50 mg per day to for beneficial effects to occur [90]. However, recent estimates based on soy-rich diets of Asian populations suggest that the average consumption of isoflavones by these populations varies from 11 to 45 mg isoflavones per day, while the average observed in Western countries varies from 0.5 to 2 mg isoflavones / day [91].

Fermented soy products have a higher content of peptides when compared with non-fermented soy products. During the fermentation process, various bioactive peptides were identified in soybeans as well as in soy protein hydrolysates. The profiles of peptides from soy protein showed an amino acid sequence with immunomodulatory, anticarcinogenic, antihypertensive, hypocholesterolemic, anti-obesity, antithrombotic and antioxidative properties. Soybeans provide 40 % protein, with higher concentrations of β-conglycinin and glycinin, representing approximately 65 % and 80 % of the total protein, respectively. Glycinin is the most important protein to yield bioactive peptides [10, 12, 71, 73, 92].

Wang and Mejia [93] reported that soybean protein-derived peptides have important role in physiological properties of the grain, especially to prevent chronic diseases. The biologically active peptides are found naturally in soybeans or may be obtained by protein hydrolysis by various methods, such as enzymatic digestion or fermentation.

Among soy peptides, lunasin may be highlighted. Dia and co-workers [94] isolated, purified and characterized lunasin from defatted soybean flour and evaluated its anti-inflammatory activity. The authors found three peptides with molecular weights of 5, 8 and 14 kDa that exhibited positive immunoreactivity towards lunasin antibody. This peptide also presented anti-inflammatory activity.

The bioactive peptides often act in the body as hormones and neurotransmitters, having an important physiological role. These peptides regulate metabolism (water, minerals and other nutrients) through the interaction with receptors and hormones, controlling the excretion glands, adjusting blood pressure and interfering with the body growth. They may also exert effects on sleep, intelligence, reasoning, sensitivity to pain, sexual behavior, and stress effects on the central nervous system [93, 95].

Currently, the potential of bioactive peptides has been identified in fermented soy products such as natto, tempeh, soy sauce, soybean paste and yoghurt. These peptides are: ACE inhibitors (angiostensin converting enzyme I), responsible for the hydrolysis of two important substrates involved in regulating blood pressure, angiotensin I and bradykinin that contain alanine, and phenylalanine and histidine, isolated from soybean fermented with *Bacillus natto* [96]. Other peptides with the same functions have also been found in fermented soybean paste (His-His-Leu) [97], soy sauce [98], natto and tempeh [10], and soy milk-based yogurt [99]. In addition, new bioactive peptide sequences may be synthesized during the fermentation process [10, 73]. Silva [81] reported that fermentation of whole autoclaved soybean flour using *Aspergillus oryzae* at 35 °C and 36 hour-incubation resulted the hydrolysis of soy proteins and formation of low molecular weight peptides (<20 kDa). This behavior was also observed by Hong, Lee and Kim [75], who evaluated the peptides from the fermentation of soybean flour using *Aspergillus oryzae* GB-107 incubated for 48 h. The authors observed a significant increase in small-peptide group (<20 kDa) and a reduction in larger peptides (> 60 kDa) when compared with non-fermented soybean flour, assigning the formation of lower molecular weight peptides to the action of proteolytic enzymes secreted by *Aspergillus oryzae* during fermentation. The fungus *Aspergillus oryzae* can produce enzymes such as aminopeptidase, serine endopeptidase, aspartate endopeptidase, among others [100].

Watanabe, Aoki and Fujimoto [101] studied the effect of dietary fermented soybean, such as tempeh, on calcium absorption in rats. The authors found that during fermentation with *Rhizopus microsporus* for 20 h, there was a reduction of phytate content and production of peptides in the range of 10 kDa, which allowed an increase in absorption of calcium from tempeh consumed by guinea pigs.

Beermann and co-workers [102] characterized the antioxidant properties of bioactive peptides from enzymatic hydrolysis of soy protein isolate with pancreatic trypsin and observed an increase in antioxidant activity when compared with the non-hydrolyzed protein isolate. Hydrolysis also increased the number of low molecular weight peptides (< 1 kDa), with a predominance of aromatic amino acid residues, which conferred functional properties to the product.

SOYBEAN GERMINATION PROCESS

The germination process provides increased nutritional value, by improving the digestibility and increasing the amount of protein efficiency ratio, reducing anti-nutritional factors in legumes, such as proteolytic inhibitors and lectins, causing hydrolysis of oligosaccharides (raffinose and stachyose) present in soybeans, which cause flatulence. Germination also provides higher levels of methionine, which is the first limiting amino acid in soy protein [103].

Although germination is an ample and complex phenomenon, it can be characterized when, under appropriate conditions, the embryo axis of the seed continues its development, which was interrupted by physiological maturity. A seed will germinate when certain conditions are favorable, such as adequate water supply, suitable temperature, composition of gases in the atmosphere, light (some seeds) and absence of germination inhibitors, and the first two conditions are the most crucial factors [104].

The direct use of soybeans in human foods has been limited by the presence of several anti-nutritional factors. The majority of processed soybean products have been derived from dry mature soybeans. However, the development of products from germinated soybean presents another option to increase the versatility and utilization of soybeans. Nevertheless, germination has been identified as an inexpensive and effective technology for improving the nutritional quality of soybeans [105]. Effects of germination on bioative compounds vary greatly with the plant species, seed varieties or cultivars, and germination conditions (temperature, light, moisture and germination time) [106, 107]. Lipids are the major source of compounds that are responsible for flavor in soybean protein products. Soybean seeds are a relatively rich source of lipoxigenases, which are an important factor in the generation of odor and off-flavor compounds from lipids that can deteriorate the palatability. Short periods of germination (72 h) can reduce substantially the odor and flavor scores of whole soybean flour because lipoxygenases activity is reduced during the germination process [108].

Paucar-Menacho and co-workers [109] studied the variation effects of germination time and temperature of Brazilian soybean cultivar BRS 133, using the Response Surface Methodology (RSM) with a 2^2 Central Composite Rotational Design. The independent variables studied were germination time (12, 21, 42, 63 and 72 h) and germination temperature (18, 20, 25, 30 and 32 °C). The germination conditions modified the content of bioative compunds in the flour obtained from germinated soybean seeds within the ranges studied.

The optimal germination conditions for soybean cultivar BRS 133 with high lunasin concentration, low lectin concentration and low lipoxygenase concentration, was exactly the central point, at 25 °C for 42 h. As compared to non-germinated soybean, it resulted in a significant increase in lunasin concentration from 12.29 to 21.20 mg / g (73.62 %), a significant decrease in lectin concentration from 16.96 to 7.62 mg / g (55.07 %) and a significant decrease in lipoxygenase activity from 13.3 to 4.0 % (69.92 %).

A significant increase in the concentration of isoflavone aglycones (daidzein and genistein) from 26.69 to 90.31 mg / g (238.36 %) and total saponins from 9.75 to 12.86 mg / g (31.89 %) was observed for 63 h germination time at 30 °C. In relation to genistein concentration in the non-germinated soybean, these germination conditions contributed to an increase from 14.40 to 44.97 mg / 100g of deffated sample (212.29 %).

Paucar-Menacho and co-workers [110] studied Brazilian soybean BRS 258 and concluded that both germination time and temperature have influenced the composition and concentration of bioactive compounds in germinated soybean flour. Low germination time of 12 h at 25 °C resulted in the highest lunasin concentration. Increasing germination time from 12 h to 72 h at 25 °C resulted in an increase of 31.9 % in soluble protein content, a decrease of 27.0 % in Bowman-Birk inhibitor concentration, and a decrease of 72.6 % in lectin concentration. The germination conditions of this study resulted in reduced Bowman-Birk inhibitor, lectin concentration and lipoxygenases activity. A significant increase in the concentration of isoflavone aglycone (daizein and genistein) and total saponin was observed in the combination of 63 h germination time at 30 °C.

Antioxidant and Anti-inflammatory Properties of Germinated and Hydrolysed Soy Flour

The bioactive peptides have low molecular weights or comprise short chains of amino acids produced by gut enzymatic digestion or during the process of soybean products, especially fermented, germinated or hydrolyzed soybeans [111]. Several bioactive peptides with different functions have been identified, such as those with antihypertensive, antithrombotic, anti-inflammatory, antioxidative and anti-obesity activities [12, 112, 113, 114].

Hydrolysis, germination and fermentation processes play an important role in formation of lower molecular weight and shorter chain compounds. The hydrolysed proteins are physiologically better than intact proteins, once the intestinal absorption appears to be more effective due to the increase in solubility and peptide content, making them attractive as a source of amino acids or peptides in human nutrition [115].

Oxidative stress and inflammation are two important factors involved in degenerative diseases [116]. The benefits of food antioxidants have been widely documented in the scientific literature [117]. The ability of peptides to behave as an antioxidant in oxidative reactions is attributed to the ability to trap free radicals. Peptides have a high ability to scavenge free radicals and form a protective membrane surrounding the fat, thus preventing the penetration of lipid oxidation initiators. The peptides from enzymatic hydrolysis have antioxidative properties that protect cells from damage caused by reactive oxygen species through the induction of genes involved in the inflammatory response. Oxidative stress occurs when reactive oxygen species overwhelm the cellular antioxidant defenses, or when the antioxidant defense system loses its ability to respond to environmental stress, leading to damage on vital cellular components such as oxidation or modification of proteins, DNA degradation, or oxidation of phospholipids in the cytoplasmic membrane. The chronicity of these changes due to the generation of excessive free radicals, in turn, may result in cell damage and therefore trigger various diseases including atherosclerosis, arthritis, diabetes and cancer [118]. Several authors have studied the antioxidant activity of peptides derived from soybeans [102, 114, 116, 119, 120]. Soybean flours germinated for 18 and 72 h and hydrolyzed with Alcalase for 1.2 and 3 h showed a high antioxidant capacity that ranged from 607.9 ± 32.1 to 740.6 ± 26.1 μmol Trolox equivalents (TE) / g flour. The antioxidant activity increased with the germination and hydrolysis process [121]. Darmawan and co-workers [114] reported that the enzymatic hydrolysis of soybeans by Alcalase showed antioxidant activity ranging from 57 to 81 μmol Trolox equivalents / g hydrolysate. The germination process can lead to the production of bioactive compounds with potent antioxidant properties. Paucar-Menacho and co-workers [109] found that germination might modify the concentrations of bioactive compounds such as lunasin, isoflavones and saponins of Brazilian soybean BRS 133.

Inflammation is an important part of normal immune response during infection and injury. However, aberrant or excessive inflammation may contribute to a number of acute and chronic diseases characterized by uncontrolled production of proinflammatory cytokines, eicosanoids derived from arachidonic acid, reactive oxygen species and adhesion molecules [122].

Several in vitro studies conducted in animals and humans have shown a potential anti-inflammatory effect associated with the consumption of soybeans and its derivatives. The beneficial effects are attributed to bioactive compounds such as isoflavones, saponins, lunasin and peptides from enzymatic hydrolysis of soybeans [94, 112, 116, 121, 123, 124, 125].

Inflammatory processes are characterized by abnormal production of cytokines such as TNFα (Tumor Necrosis Factor), and inflammatory mediators such as nitric oxide (NO), cyclooxygenase (COX-2) and prostaglandin E2 (PGE2) [126].

Substances such as COX-2 and PGE2 are important during inflammatory processes, since they mediate the production of other substances such as prostanoids, necessary to control vasodilation, vasoconstriction and platelet aggregation, which are critical during the immune response. An uncontrolled inflammatory process may lead to overproduction of these substances and hence to processes that can generate tissue damage and disease. A high production of Tumor Necrosis Factor, one of the most important proinflammatory mediators involved in systemic inflammation and that belongs to a group of cytokines that stimulate the acute phase reaction. TNFα, besides mediating the activation and recruitment of other immune system components to the site of infection and controlling different infections involving viruses, bacteria, and tumors, it may induce cellular apoptosis, tissue damage and inflammation that can lead to uncontrolled shock septic.

Currently, developing products for controlling inflammatory processes, with analgesic and anti-inflammatory capacity to inhibit the production of proinflammatory substances such as TNFα, NO, COX-2 and PGE2 is increasingly attracting attention of researchers, especially producing foods that limit the proinflammatory and oxidative responses to be a healthy food option to the population. Several studies have reported the reduction of inflammatory processes when soy protein is used [112, 115, 116, 121, 124, 125]. Five micrograms of soybean flour germinated for 18 and 72 h and hydrolysed by Alcalase for 1, 2 and 3 h showed inhibition of the secretion of NO, PGE2, TNF-α, iNOS and NF-kappaB induced by LPS in supernatant of macrophages cells, whereas proteins from non-hydrolyzed flours did not reduce the production of NO as compared to the hydrolysed flours.

Conclusion

The consumption of soybean has increased considerably in recent years due to its functional properties, with the presence of many bioactive compounds such as isoflavones, of which can provide health benefits to the consumer. New bioactive protein compounds are also being studied, such as lunasin, the Bowman-Birk inhibitor, lectin and non-protein bioactive compounds such as saponins.

The fermentation process promote a transformation of isoflavone glycoside isomer into aglycone. Germination process increase soluble protein concentration and antioxidant capacity of soybean flours. Germinated soybean flours followed by the hydrolysis promote a decrease on markers of inflammation such as NO, PGE2, TNF-a, iNOS; IKK-ainhibition.

Enzymatic hydrolysis and germination can enhance the biological activity of these compounds in soybean. Soy products and by-products with lower concentrations of antinutritional factors and with biologically active constituents are important for the promotion of health.

References

[1] Han, KK; Kati, LM; Haidar, MA; Girão, MJBC; Baracat, EC; Yim, DK; Carrão-Panizzi, MC. Efeito da isoflavona sobre os sintomas da síndrome de climatério. In *Simpósio brasileiro sobre os benefícios da soja para a saúde humana. Embrapa* , Londrina, 2001, 28-32.

[2] Trucon, C. Soja - Nutrição e Saúde; Editora Alaúde: São Paulo, 2009.

[3] Tecnologias de Produção de Soja, Região Central do Brasil 2004. http://www.cnpso.embrapa.br/producaosoja/SojanoBrasil.htm (accessed 02/02/2013), Empresa Brasileira de Pesquisa Agropecuária - Embrapa Soja.

[4] Grain: World Markets and Trade. http://usda01.library.cornell.edu/usda/current/grain-market/grain-market-12-11-2012.pdf (accessed 01/08/2013), United States Department of Agriculture. Foreign Agricultural Service.

[5] Liu, K. Soybeans: Chemistry, Technology and Utilization; Chapman & Hall: New York, 1997.

[6] Sales and Trends. http://www.soyfoods.org/soy-information/sales-and-trends (accessed 01/15/2013), Soyfood Association of North America.

[7] ASA. American Soybean Association. *Soy food*, 11, 2000.

[8] Grieshop, CM; Fahey, GC. Jr. *J. Agric. Food Chem.,* 2001, 59, 2669-2673.

[9] Grieshop, CM; Kadzere, CT; Clapper, GM; Flickinger, EA; Bauer, LL; Frazier, RL; Fahey, GC. Jr. *J. Agric. Food Chem.,* 2003, 51, 7684-7691.

[10] Gibbs, BF; Zougman, A; Masse, R; Mulligan, C. *Food Res. Int.*, 2004, 37, 123–131.

[11] Liu, K. Soybean as a Powwerhouse of Nutrients and Phytochemicals. In *Soybeans as Functional Foods and Ingredients*; Liu, K., AOCS Press: Champaign, 2004, 1-22.

[12] Gonzalez de Mejia, E; De Lumen, B. O. *Sexuality, Reproduction and Menopause*, 2006, 4, 91-95.

[13] Kerckhoffs, DAJM; Brouns, F; Hornstra, G; Mensink, RP. *J. Nutr.*, 2002, 132, 2494-2505.

[14] Singh, P; Kumar, R.; Sabapathy, SN; Bawa, AS. *Compr. Rev. Food Sci. Food Saf.*, 2008, 7, 14–28.

[15] Tikkanen, MJ; Adlercreutz, H. *Biochem. Pharmacol.*, 2000, 60, 1-5.

[16] Donkor, ON; Stojanovska, L; Ginn, P; Ashton, J; Vasiljevic, T. *Food Chem.*, 2012, 135, 950-959.

[17] Jenkins, DJA; Kendall, CWC; Vidgen, E; Vuksan, V; Jackson, CJ; Augustin, LSA; Lee, B; Garsetti, M; Agarwal, S; Rao, AV; Cagampang, GB; Fulgoni, V. *Metabolism*, 2000, 49, 1496-1500.

[18] Setchell, K. *Am. J. Clin. Nutr.*, 1998, 68S, 1333S-1346S.

[19] Anderson, JW; Johnstone, BM; Cook-Newell, ME. *N. Engl. J. Med.*, 1995, 333, 276-282.

[20] Messina, M; Messina, V. *J. Am. Diet. Assoc.*, 1991, 91, 836-840.

[21] Food Labeling: Health Claims; Soy Protein and Coronary Heart Disease, *Federal Register*, 64, 1999.

[22] López-Amarós, ML; Hernández, T; Estrella, I. *J. Food Compos. Anal.*, 2006, 19, 277-283.

[23] Peña-Ramos, EA; Xiong, YL. *J. Food Sci.*, 2002, 67, 2952-2956.

[24] Nakamori, T. *Food Style*, 21, 86-88.

[25] Abe, M; Suzuki, O; Tasaki, K; Tominaga, K; Wakasa, H. *Pathol Int.*, 1996, 46, 977–983.

[26] De Lumen, BO; Galvez, AF. (The Regents of the University of California). *Soybean protein nutraceuticals*. US Patent 6, 391, 848 B1, May 21, 2002.

[27] Jeong, HJ; Lam, Y; De Lumen, BO. *J. Agric. Food Chem.*, 2002, 50, 5903-5908.

[28] Jeong, HJ; Park, HJ; Lam, Y; De Lumen, BO. *J. Agric. Food Chem.*, 2003, 51, 7901-7906.

[29] De Lumen, BO. *Nutr. Rev.*, 2005, 63, 16-21.

[30] Soy Labs. Lunasin XPTM. Product overview. In *Soy Labs*, 2007, CD-ROM. Soy Labs LLC.

[31] Galvez, AF; Revilleza, MJR; De Lumen, BO. *Plant Physiol.*, 1997, 114, 1567-1569.

[32] Galvez, A; Chen, N; Macasieb, J; De Lumen, BO. *Cancer Res.*, 2001, 61, 7473-7478.

[33] Galvez, AF; De Lumen, BO. *Nat. Biotechnol.*, 1999, 17, 495-500.

[34] González De Mejía, E., Vasconez. M., De Lumen. B; Nelson. R. *J. Agric. Food Chem.*, 2004, 52, 5882-5887.

[35] Weder, JK; Haussner, K. *Z Lebensm Unters Forsch.,* 1991, 192, 535–540.

[36] Losso, JN. *Crit. Rev. Food Sci. Nutr.*, 2008, 48, 94–118.

[37] Wan, XS; Serota, DG; Ware, JH; Crowell, JA; Kennedy, AR. *Nutr. Cancer.*, 2002, 3, 167–173.

[38] Lippman, SM; Matrisian, LM. *Clin. Cancer Res.,* 2000, 6, 4599–4603.

[39] Kennedy, AR. *Am. J. Clin. Nutr.*, 1998, 68, 1406S-1412S.

[40] Meyskens, FL. Jr. *IARC Sci. Publ.*, 2001, 154, 49–55.

[41] Liener, IE. The nutritional significance of the plant lectins. In *Antinutrients and natural toxicants in foods*; Ory, R. L., Food & Nutrition Press: Westport, 1981, 143-157.

[42] Li, Z; Li, D; Qiao, S. *Arch. Tierernahr.*, 2003, 57, 369-380.

[43] Fasina, YO; Garlich, JD; Classen, HL; Ferket, PR; Havenstein, GB; Grimes, JL; Qureshi, MA; Christensent, VL. *Poult. Sci.*, 2004, 83, 559-571.

[44] Pusztai, A. *Plant Lectins*; Cambridge University Press: New York, 1991.

[45] Orf, JH; Hymowitz, T; Pull, SP; Pueppke, SG. *Crop Sci.*, 1979, 18, 899-900.

[46] Vasconez–Costa, M. *Effect of genotype, environment and processing on the level of lectin and lunasin in soybean.* Master Thesis. University of Illinois, Urbana-Champaign, 2004.

[47] González De Mejía, E; Prisecaru V. Lectins as bioactive plant proteins: A potential in cancer treatment. *Crit. Rev. Food Sci. Nutr.*, 2005, 45, 425-445.

[48] Valentiner, U; Fabian, S; Schumacher; U; Leathem, AJ. *Anticancer Res.*, 2003, 23, 1197–1206.

[49] Nielsen, IL; Williamson, G. *Nutr. Cancer*, 2007, 57, 1–10.

[50] Setchell, KD; Brown, NM; Desai, PB; Zimmer-Nechimias, L; Wolfe, B; Jakate, AS; Creutzinger, V; Heubi, JE. *J. Nutr.*, 2003, 133, 1027–1035.

[51] Omoni, AO; Aluko, RE. *Nutr. Rev.*, 2005, 63, 272–283.

[52] Clair, RS; Anthomy, M. *Handb. Exp. Pharmacol.*, 2005, 170, 301–323.

[53] Sugiyama, F; Wu, J; Fujioka, M; Ezaki, J; Takeda, K; Miyaura, C; Ishida, I; Yamada, K; Ishimi, Y. *J. Bone Miner. Metab.*, 2006, 24, 439–446.

[54] Wang, LJ; Yin, LJ; Li, D; Zou, L; Saito, M; Tatsumi, E; Li, LT. *Food Chem.*, 2007, 101, 1247–1253.

[55] Barbosa, ACL; Lajolo, FM; Genovese, MI. *Food Chem.*, 2006, 98, 757–766.

[56] Chun, J; Kim, JS; Kim, JH. 4. *Food Chem.*, 2008, 109, 278–284.

[57] Coward, L; Smith, M; Kirk, M; Barnes, S. *Am. J. Clin. Nutr.*, v. 68, 1486S-1491S, 1998.

[58] Yamabe, S; Kobayashi-Hattori, K; Kaneko, K; Endo, H; Takita, T. *Food Chem.*, 2007, 100, 369–374.

[59] Niamnuy, C; Nachaisin, M; Laohavanich, J; Devahastin, S. *Food Chem.*, 2011, 129, 899-906.

[60] Izumi, T; Osawa, S; Obata, A; Tobe, K; Saito, M. *J. Nutr.*, 2000, 130, 1695–1699.
[61] Kawakami, Y; Tsurugasaki, W; Nakamura, S; Osada, K. *J. Nutr. Biochem.*, 2005, 16, 205–212.
[62] Walsh, KR; Haak, SJ; Bohn, T; Tian, Q; Schwartz, SJ; Failla, ML. *Am. J. Clin. Nutr.*, 2007, 85, 1050–1056.
[63] Mandarino, JMG; Carrão-Panizzi, MC; Crancianinov, WS. *Teor de isoflavonas em cultivares de soja da Embrapa Soja*. Resumos do III Congresso de Soja do Mercosul - Mercosoja 2006. Rosário, Argentina, ACSOJA, 294-296, 2006.
[64] Berhow, MA; Kong, SB; Vermillion, KE; Duval SM. *J. Agric. Food Chem.*, 2006, 54, 2035-2044.
[65] Kudou, S; Tonomura, M; Tsukamato, C; Shimoyamada, M; Uchida, T; Okubo, K. *Biosci., Biotechnol., Biochem.*, 1993, 57, 546-550.
[66] Shiraiwa, M; Kudo, S; Shimoyamada, M; Harada, K; Okubo, K. *Agric. Biol. Chem.*, 1991, 55, 315-322.
[67] Heng, L; Vincken, JP; Hoppe, K; Koningsveld, GAV; Decroos, K; Gruppen, H; Boekel, MAJSV; Voragen, AGJ. *Food Chem.*, 2006, 99, 326-334.
[68] Kuo, LC; Cheng, WY; Wu, RY; Huang, CJ; Lee, KT. *Appl. Microbiol. Biotechnol.*, 2006, 73, 314-320.
[69] Villares, A; Rostagno, MA; García-Lafuente, A; Guillamón, E; Martínez, J. A. *Food Bioprocess Technol.*, 2011, 4, 27–38.
[70] Kim, H; Peterson, TG; Barnes, S. *Am. J. Clin. Nutr.*, 1998, 68, 1418S-1425S.
[71] Rayaprolu, JS; Hettiarachchy, NS; Chen, P; Kannan, A; Mauromostakos, A. *Food Res. Int.*, 2013, 50, 282-288.
[72] Silva, LH; Celeghini, RMS; Chang, YK. *Food Chem.*, 2011, 128, 640-644.
[73] Teng, D; Gao, M; Yang, Y; Liu, B; Tian, Z; Wang, J. *Biocatal. Agric. Biotechnol.*, 2012, 1, 32-38.
[74] Park, KY; Jung, KO; Rhee, SH; Choi, YH. *Mutat. Res.*, 2003, 523-524, 43-53.
[75] Hong, KJ; Lee, CH; Kim, SK. *J. Med. Food*, 2004, 7, 430-436.
[76] Kim, BN; Yang, JL; Song, YS. *Food Ind. Nutr.*, 1999, 4, 40–46.
[77] Valyasevi, R; Rolle, RS. *Int. J. Food Microbiol.*, 2002, 75, 231-239.
[78] Esaki, H; Watanabe, R; Onozaki, H; Kawakishi, S; Osawa, T. *Biosci., Biotechnol., Biochem.*, 1999, 63, 851-858.
[79] Caplice, E; Fitzgerald, GF. *Int. J. Food Microbiol.*, 1999, 50, 131–149.
[80] Aguiar, CL; Park, Y. K. *Boletim do Centro de Pesquisa de Processamento de Alimentos*, 2004, 22, 185-195.
[81] Silva, LH. *Efeito da fermentação utilizando Aspergillus oryzae sobre as características funcionais, tecnológicas e físico-químicas da farinha de soja integra e aplicação em pão de forma*. Thesis. Universidade Estadual de Campinas, Campinas, SP, 2009.
[82] Wu, J; Muir, AD. *Food Chem.*, 2010, 118, 328-332.
[83] Chen, TR; Wei, QK. *Nutr. Food Sci.*, 2008, 38, 540-547.
[84] Messina, J; Wood, CE. *Nutr. J.*, 2008, 7, 1-11.
[85] Hendrich, S; Murphy, PA. Isoflavones: source and metabolism. In *Handbook of nutraceuticals and functional foods*. Wildman, R. E. C. 2nd Edition, CRC Press: Boca Raton, 2001, 55-72.
[86] Messina, MJ; Persky, V; Setchell, KD; Barnes, S. *Nutr. Cancer*, 1994, 21, 113–121.
[87] Bradlow, HL; Sepkovic, DW. *Ann. N. Y. Acad. Sci.*, 2002, 963, 247–267.

[88] Lamartiniere, C. A; Cotroneo, M. S; Fritz, W. A; Wang, J; Mentor-Marcel, R; Elgavish, A. *J. Nutr.*, 2002, 132, 552S-558S.
[89] Sirtori, CR; Lovati, MR. *Curr. Atheroscler. Rep.*, 2001, 3, 47-53.
[90] Munro, IC; Hlywka, JJ; Stephen, AM; Doull, J; Flamm, G; Adlercreutz, H. *Nutr. Rev.*, 2003, 61, 1-33.
[91] Mortensen, A; Kulling, SE; Schwartz, H; Rowland, I; Ruefer, CE; Rimbach, G; Cassidy, A; Magee, P; Millar, J; Hall, WL; Birkved, FK; Sorensen, IK; Sontag, G. *Mol. Nutr. Food Res.*, 2009, 53, 266-309.
[92] Kwon, DY; Hong, SM; Ahn, S; Kim, MJ; Yang, HJ; Park, S. *Nutrition*, 2011, 27, 244-252.
[93] Wang, W; Mejia, EG. *Compr. Rev. Food Sci. Food Saf.*, 2005, 4, 63-78.
[94] Dia, V. P; Wang, W; Oh, V. L; Lumen, B. O; Meija, E. G. *Food Chem.*, 2009, 114, 108-115.
[95] Yimit, D; Hoxur, P; Amat, N; Uchikawa, K; Yamaguchi, N. *Nutrition*, 2012, 28, 154-159.
[96] Korhonen, H; Pihlanto, A. *Curr. Pharm. Des.*, 2003, 9, 1297-1308.
[97] Shin, Z; Yu R; Park, SA; Chung, DK; Ahn, CW; Nam, HS; Kim, KS; Lee, HJ. *J. Agric. Food Chem.*, 2001, 49, 3004-3009.
[98] Okamoto, A; Hanagata, H; Matsumoto, E; Kawamura, Y; Koizumi, Y; Yanadiga, F. *Biosci., Biotechnol., Biochem.*, 1995, 59, 1147–1149.
[99] Tomatsu, M; Shimakage, A; Shinbo, M; Yamada, S; Takahashi, S. *Food Chem.*, 2013, 136, 612-616.
[100] Liang, Y; Pan, L; Lin, Y. *Biosci., Biotechnol., Biochem.*, 2009, 73, 192-195.
[101] Watanabe, N; Aoki, H; Fujimoto, K. *J. Sci. Food Agric.*, 2008, 88, 2749-2752.
[102] Beermann, C; Euler, M; Herzberg, J; Stahl, B. *Eur. Food Res. Technol.*, 2009, 229, 637-644.
[103] Barcelos MF; Vilas Boas, EV; Lima, MA. *Ciênc. agrotec.*, 2002, 26, 817-825.
[104] Carvalho, NH; Nakagawa, J. *Semestes: ciência, tecnologia e produção*, 3rd ed, Fundacao Cargill: Campinas, 1988..
[105] Bau, H-M. ; Villaume, C; Nicolas, J-P; Méjean L. *J. Sci. Food Agric.*, 1997, 73, 1-9.
[106] Wuebker, E; Mullen, R; Hoelher, K. *Crop Sci.*, 2001, 41, 1857-1861.
[107] Gloria, B; Tavarez-Neto, J; Labanca, R. *J. Agric. Food Chem.*, 2005, 53, 7480-7485.
[108] Suberbie, F; Mendizrbal, D; Mendizabal, C. *J. Am. Oil Chem. Soc.*, 1981, 58, 192-195.
[109] Paucar-Menacho, LM; Berhow, M; Mandarino, JMG; Gonzalez de Mejia, E; Chang, YK. *Food Chem.*, 2010, 119, 636-642.
[110] Paucar-Menacho, LM; Berhow, M; Mandarino, JMG; Gonzalez de Mejia, E; Chang, YK. *Food Res. Int.*, 2010, 43, 1856-1865.
[111] Möller, NP; Scholz-Ahrens, KE; Roos, N; Schrezenmeir, J. *Eur. J. Nutr.*, 2008, 47, 171–182.
[112] Martinez-Villaluenga, C; Dia, VP; Berhow, M; Bringe, NA; Gonzalez De Mejia, E. *Mol. Nutr. Food Res.*, 2009, 53, 1007–1018.
[113] Wang, W; Bringe, NA; Berhow, MA; Gonzalez De Mejia, E. *J. Agric. Food Chem.*, 2008, 56, 4012–4020.
[114] Darmawan , R; Bringe, NA; Gonzalez De Mejia, E. *Plant Foods Hum Nutr.*, 2010, 65, 233-240.
[115] Kong, XZ; Guo, MM; Hua, Y. *Bioresour. Technol.*, 2008, 99, 8873–8879.

[116] Hernández-Ledesma, B; Hsieh, CC; De Lumen, BO. *Biochem. Biophys. Res. Commun.*, 2009, 390, 803–808.
[117] Psaltopoulou, T; Panagiotakos, DB; Pitsavos, C; Chrysochoou, C; Detopoulou, P; Skoumas, J; Stefanadis, C. *Nutr. Metab. Cardiovasc. Dis.*, 2010, 21, 561, 567.
[118] Sarmadi, B.H; Ismail, A. *Peptides*, 2010, 31, 1949–1956.
[119] Chen, H.M; Muramoto, K; Yamauchi, F; Nokihara, K. *J. Agric. Food Chem.*, 1996, 44, 2619–2623.
[120] Park, S; Lee, J; Baek, H; Lee, HG. *J. Food Biochem.*, 2010, 34, 120-132.
[121] Vernaza, MG; Dia, VP; Gonzalez de Mejia, E; Chang, YK. *Food Chem.*, 2012, 134, 2217–2225.
[122] Kumar, MV; Shimokawa, T; Nagy, TR; Lane, MD. *Proc. Natl. Acad. Sci.*, 2002, 99, 1921–1925.
[123] Kao, TH; Wu, WM; Hung, CF; Wu, WB; Chen, BH. *J. Agric. Food Chem.*, 2007, 55, 11068–11079.
[124] Hooshmand, S; Soung, DY; Lucas, EA; Madihally, SV. *J. Nutr. Biochem.*, 2007, 18, 609–614.
[125] Kang, JH; Sung, MK; Kawada, T; Yoo, H. *Cancer Lett.*, 2005, 230, 219–227.
[126] Hotamisligil, GS. *Nature*, 2008, 444, 860–867.

In: Seeds as Functional Foods and Nutraceuticals
Editors: R. Mora-Escobedo, J. De J. Berrios et al.
ISBN: 978-1-62808-489-4

Chapter 2

EFFECT OF GERMINATION ON BIOACTIVE COMPOUNDS OF SOYBEAN (*GLYCINE MAX*)

***Fabiola Araceli Guzmán-Ortiz*[1], *María del Carmen Robles-Ramírez*[1], *María Elena Sánchez-Pardo*[1], *José De J. Berríos*[2] *and Rosalva Mora-Escobedo*[1*]**

[1] Departamento de Ingeniería Bioquímica, Escuela Nacional de Ciencias Biológicas, I.P.N. Prol. Carpio y Plan de Ayala S/N, Col. Sto. Tomás, D.F. México
[2]USDA-ARS-WRRC, Processed Foods Research Unit, Albany, CA, US

ABSTRACT

Germination is the practice of soaking, draining, and keeping seeds until they produce sprouts. The increasing interest in functional and healthy food products has promoted the use of germinated soybean flour in the manufacture of foods for human consumption. It is well known that germination induces changes in nutritional and non-nutritional compounds. Therefore, the objective of this chapter was to provide a brief overview on the changes that take place along this process on bioactive compounds such as trypsin inhibitors, saponins, phytic acid, isoflavones, oligosaccharides, and lectins from soybean (*Glycine max*). These compounds have been determined in our laboratory and the results are included. Authors expect this chapter to provide interesting information about the effect of germination on the bioactive compounds of soybean.

GERMINATION

Germination is a feasible and low-cost process that allows the decrease of non-nutritional compounds as well as the increase of bioactive compounds in some legumes. During this process, a series of cellular events take place to facilitate the growth of the embryo.

* Corresponding author: rosalmora@hotmail.com.

Germination has three phases (**Figure 1**): the first one is hydration, where the tissues of the seed absorb the highest possible amount of water, and the respiration rate and volume of the seed increase (Simon, 1984). The second stage is known as germination, where the transformation of metabolites necessary for the development of the seedling takes place, the absorption of water is reduced, respiration and synthesis of new compounds increase, allowing cellular expansion and division, until the embryonic axis begins to express. During this stage, the storage nutrients (protein, lipids and carbohydrates) are mobilized in order to provide the necessary energy for protein synthesis and growth of the seedling. A relatively large fraction of these reserves is comprised by proteins, the globulins legumin and vicilin, which are known as glycinin and β-conglycinin in soy (Bon and Bowles, 1983; Robles, 2010). The third stage is known as growth, and it is related to the emergence of the radicle. In this stage, digestive enzymes are activated and the proteins are hydrolized to amino acids, which act as precursors for the synthesis of new proteins or other nitrogen-containing compounds (Wilson, 1986).

During germination, the main metabolic processes are respiration and mobilization of reserve compounds. Respiration increases in the first 12 h after imbibition starts and stabilizes between 12 and 24 h, since the seed coats that remain intact limit the access to oxygen. Enzymes are also activated and later a second increase in respiratory activity takes place, which is associated to the higher oxygen availability as a consequence of the rupture of the seed coat that has been produced by the emergence of the radicle. Another factor that contributes to the increase of the respiratory activity is the action of mitochondria and respiratory enzymes, which are synthesized in the cells of the developing embryonic axis. The number of mitochondria in the cotyledons also tends to increase and this phenomenon is associated to the mobilization of reserve compounds. In the last stage, once the nutrients in storage have been mobilized, cotyledons are disintegrated producing a decrease in respiratory activity (Robles, 2010).

On the other hand, mobilization of reserves takes place when the seed is completely hydrated and the synthesis of protein has been activated. Then hydrolytic enzymes are synthesized (among other proteins), which are able to hydrolyze part of the reserve compounds. In this manner, the embryonic axis uses the hydrolyzed contents for its respiration, elongation, and cell division, by absorbing them from the cotyledons. Finally, the radicle emerges and later originates the main root (Goyoaga, 2005). As germination mobilizes the nutrients required for the development of the seedling, it removes non-nutritional factors such as phytates, oligosaccharides, tannins and protease inhibitors (Robles-Ramírez et al., 2011).

The Soybean and Its Chemical Composition

Soy (*Glycine max*) belongs to the family of legumes, Angiosperm class, Papilionidae sub-family, *Glycine* genre, *max* species (Salunkhe, 1992). It is a yearly cultivar whose plant is generally 80 cm tall and its vegetative cycle ranges between three to seven months. The stem is rigid and erect; the main root reaches up to one meter deep. The soybean is produced in pods of 4 to 6 cm in length, each containing 3 to 4 grains. Soy develops optimally in warm and tropical regions, it is sowed according to the irrigation system: for rainfed cultivation, between June 20th and July 30th; when working with residual moisture, the optimum is October 20th to November 25th, and when an irrigation system is available, the recommendation is between December 20th and January 20th (Sanz et al., 2007).

From the nutritional point of view, soybeans are a rich source of protein, found in the range of 35 to 44% (Sanz et al., 2007). The nutritive value of soy protein depends on a series of factors such as the amino acid profile, digestibility, and content of essential amino acids. The WHO (World Health Organization) established that soy protein has all the essential amino acids in sufficient amounts to cover the requirements of protein ingestion of each age and biological situation in human beings (Howes et al., 2006; Rinder et al., 2006; FAO 2005; Torres and Tovar, 2009). According to the classification of proteins by solubility (Osborne, 1924), soybean proteins are rich in a heterogenous mix of globulins (60-75% of the total protein content) and albumins. On the other hand, soy proteins may also be classified according to their sedimentation coefficient in the ultracentrifuge, as measured in Svedberg units. In this manner, the following fractions are obtained: 2S (trypsin inhibitors, cytochrome C, α-conglycinin, AL1 y AL3,) 7S (hemaglutinins, lipoxygenase, α-amylase, β-conglycinin, γ-conglycinin, P34 vacuolar protein), 11S fraction (glycinin) and 15S (glycinin polymer) (Wang and González 2005).

Proteins are modified as an effect of germination. Lee and Karunanithy (1990) observed an increase in the content of total crude protein of more than 21% as compared to the non-germinated seeds. In our laboratory, *Glycine max* soybeans germinated for 2, 3 and 6 days were analyzed; it was observed that protein content tends to increase along germination time (**Table 1**). However, no significant difference was found ($p<0.05$) for the samples, this may be explained by the fact that protein is not yet mobilized in the first stages of germination, because the first energy source are carbohydrates and changes in the contents of nutrients were not observed until after the sixth day. Increase in protein content may be due to the synthesis of proteolytic enzymes or to the change in composition of other constituents during germination. González-Espinosa et al., (2011) found similar behavior in their soybean studies with crystalline variety.

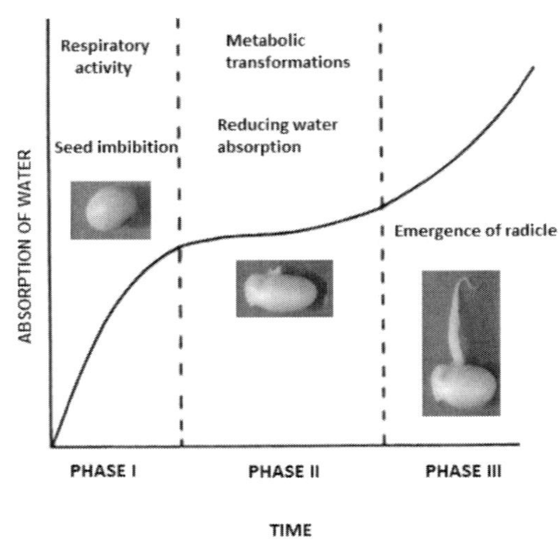

Figure 1. Role of time in the metabolic activity during germination (modified from Nonogaki et al., 2010).

Table 1. Chemical composition of germinated soybean from 0, 2 , 3 and 6 days (%)

Germination time (days)	Protein	Lipids	Ashes	Carbohydrates
0	39.93 ± 0.22^{a}	23.00 ± 0.45^{b}	4.76 ± 0.03^{b}	25.6
2	39.70 ± 0.16^{a}	21.43 ± 0.06^{c}	4.85 ± 0.52^{bc}	29.65
3	40.11 ± 0.13^{a}	23.99 ± 0.66^{ab}	4.54 ± 0.10^{c}	27.38
6	41.21 ± 1.63^{a}	26.19 ± 2.37^{a}	6.56 ± 0.23^{a}	20.45

Results are expressed in dry basis (DB) as the mean value of triplicates ± standard deviation, samples with the same letter do not show significant differences as indicated by the LSD test ($p \leq 0.05$).

Other important components of soybean are lipids. Their total content in soybeans is approximately 20%, and it is mainly constituted by triacylglycerides, with approx. 12% of saturated fatty acids, 20-22% oleic acid, 54-60% linoleic acid and 4-7.5% linolenic acid. Phospholipids, sterols and tocopherols are also found (Sanz et al., 2007).

It has been reported that the lipid content decreases gradually as germination goes on. Chandrasiri et al., (1987) reported that lipid content decreases up to the 5th day of germination. We observed that lipid content decreased in about 8.6% after two days of germination. However, on days 3 and 6 it increases progressively (**Table 1**). Between these days no significant difference was observed ($p<0.05$) among germinated samples, but it was significant when compared to the non-germinated sample. Differences are also observed when comparing the reports by other authors. González-Espinosa, (2011) observed a progressive increase of lipids up to the third day of germination, and Jiménez et al., (1985) also reported an increase of lipids up to the third day and then a decrease on the fifth. The degradation of reserve nutrients (lipids and carbohydrates) during germination is meant to provide the energy required for protein synthesis in the growing seedling (Bau et al., 1997). Triglycerides are degraded by the action of lipases from the fatty bodies that act as storage, which may explain why during short periods of germination the level of lipids decreases while the protein content increases. We observed that ash is also affected, since the content of this component increases along the germination process.

Regarding carbohydrates in soybean, the content is of approximately 34% and may be divided into water and ethanol-insoluble polysaccharides (arabinogalactans, arabinans, xylans, galactomannans and cellulose), which represent approximately 50% of total carbohydrates. Water-soluble oligosaccharides are also present, such as verbascose and stachyose (3.8%), as well as raffinose (1.1%), which cause the flatulence associated to the consumption of legumes. There is also sucrose (4.5%) and other monosaccharides in lesser amounts, mainly glucose and arabinose (De Luna, 2007). Soybean contains 9% of dietary fiber that mainly consists of lignin, cellulose and hemicellulose (arabinogalactans) (Steggerda, 1996; Rinder et al., 2006; Guillon and Champ, 2002; Campos et al., 2010).

Soy has become a highly-consumed legume due to its favorable effects on human health. However, the presence of anti-nutritional compounds limits its possible uses in foods. Germination is an alternative process that may improve the biological and nutritional quality of legumes, allowing a wider use in both human and animal nutrition.

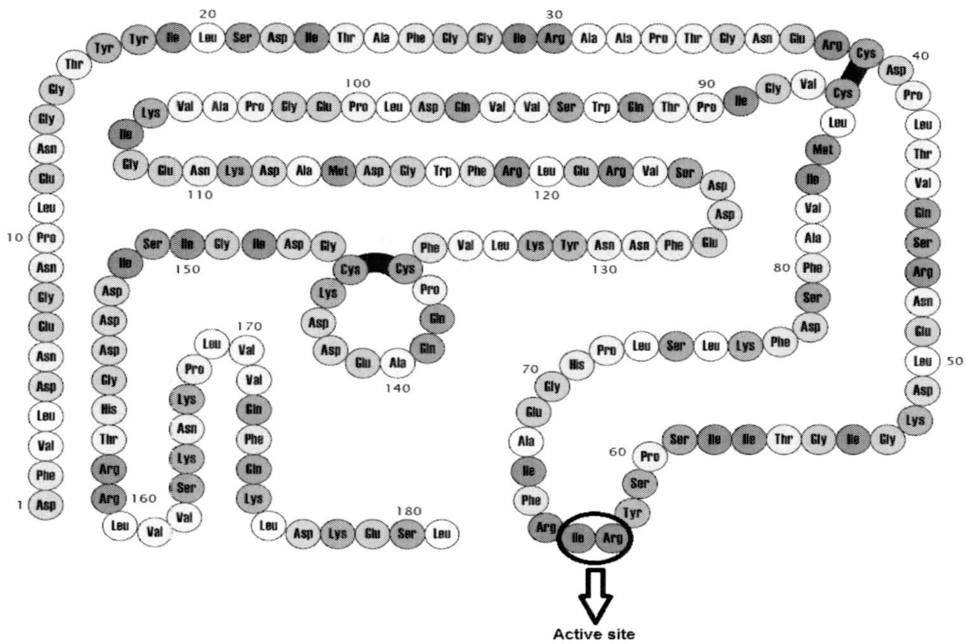

Figure 2. Structure of the Kunitz inhibitor.

TRYPSIN INHIBITORS

Trypsin inhibitors are proteins of low molecular weight, which inhibit the enzyme systems of predators (microorganisms or insects), or have a regulatory role in the processes of proteolysis of the cells. They have been identified as the Kunitz and Bowman-Birk inhibitors, located in fraction 2S (**Figures 2 and 3**). The Kunitz inhibitor is formed by 181 amino acids, has a molecular weight of 21kDa, two disulfide bridges and interacts stoichiometrically with trypsin, having a single active site located in arginine63-isoleucine64 (Guerrero et al., 2009).

The Bowman-Birk inhibitor is constituted by a polypeptide chain of 71 amino acids; it has a molecular weight of 8 kDa with seven disulfide bonds, and has a specific inhibiting ability for chymotrypsin. It is called double-headed inhibitor because it has separate binding sites for trypsin and chymotrypsin. The active sites are lysine16-serin17 and leucin44-serin45 for trypsin and chymotrypsin, respectively (Barac et al., 2005; Guerrero et al., 2009). Trypsin inhibitors have the capacity to intervene in the digestion of proteins through the inhibition of the proteolytic activity by joining to the active site of the enzyme and forming complexes that block its activity, causing hypertrophy/hyperplasia of the pancreas and increase in the secretion of pancreatic enzymes. They are also able to slow the growth process in animals (Dittrich et al., 1998).

The content of trypsin inhibitors in soybean varies according to the specific variety, climate and farming conditions. Pedrosa et al., (2012) reported an amount of 86.1 trypsin inhibitory units (TIU)/mg on a dry basis for untreated soybeans. Trugo et al., (2000) reported 46 TIU/mg, Haddad et al., (2007) found 41.6 TIU/mg. However, these compounds may be affected by germination due to the presence of proteases K1, B1 and B2, which favor proteolysis of both the Kunitz and the Bowman-Birk inhibitors. Proteases are stored in inactive form in the dry seed; as germination proceeds, they become active and they reach an optimum point after four days for K1 and B1 at an optimum pH 4. B2 protease has been reported to appear after nine days of germination. These enzymes favor proteolysis of both inhibitors, therefore causing their degradation (Papastoitsis and Wilson 1991).

Mostafa et al., (1987) and Trugo et al., (2000) reported that after two days of germination, the quantity of inhibitors was reduced in 25 to 32%. This variation in results may be due to the farming conditions of the different soy varieties. Trugo et al. (2000) observed that after two days of germination of *Phaseolus vulgaris* and *Glycine max* there was a reduction of 20 and 25% of inhibitor activity, respectively. In the analysis performed at our laboratory (**Figure 4**), we observed that these compounds decrease along germination (2 and 6 days), and the lowest concentration was achieved at the 6th day, with a significant difference ($p<0.05$) among the samples and the highest reduction being of 27.4%.

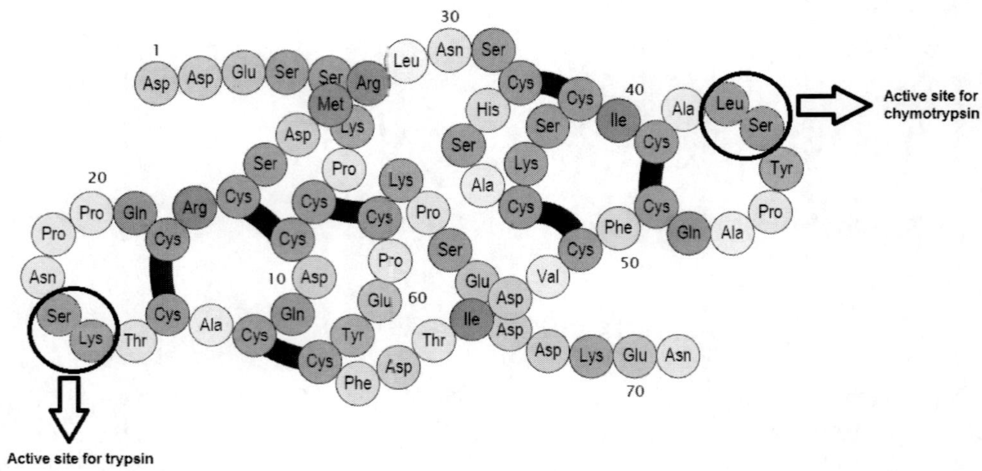

Figure 3. Structure of the Bowman-Birk inhibitor.

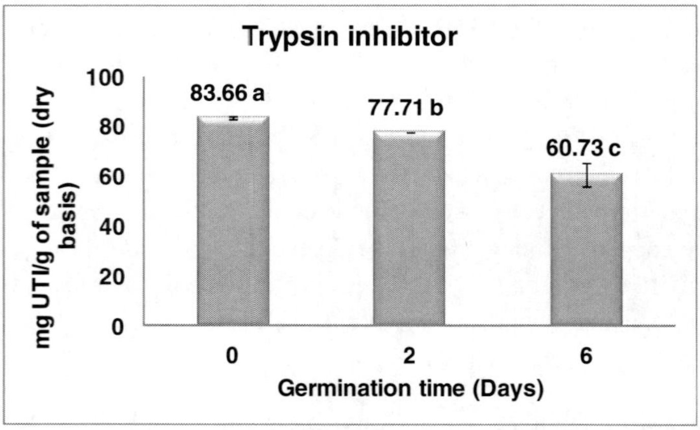

[a/]Results are expressed as the mean of three determinations ± standard deviation. Samples with the same letter do not show significant differences according to the LSD test ($p \leq 0.05$).

Figure 4. Effect of germination on the content of trypsin inhibitors in soybean germinated for 0, 2 and 6 days.

SAPONINS

Soy saponins are glucosides formed by the union of an aglycone sapogenin with one or more sugar molecules. When united to sugars, they transform into glycosylated structures. Sapogenin may be a steroid (C_{27}) or a triterpene (C_{30}) (Torres and Tovar, 2009), while sugars may be galactose, glucose, pentose or methyl pentose (Schwarz, 1993). Aglycones are divided into groups A, B, C and E. Saponins in group A consist of bidesmosidic glycosides of soyasapogenol A, when they have two saccharide units. Group B of soy saponins consist of monodesmosidic glycosides of soyasapogenol B or soyasaponin E, in this case they have joined up to three saccharide units (tridesmosidic). Each of the groups has 2 to 5 saccharide units and may be linear or ramified (**Figure 5**) (Haralampidis, et al., 2002).

Legumes such as soy are rich in saponins, which have a variety of biological activities (Shiraiwa et al., 1991; Hubert et al., 2005). Soy contains only saponins from groups A and B, and B exist in plant tissues in an intact form as conjugates of 2,3-dihydro-2,5-dihydroxy-6-methyl-4*H*-pyran-4-one (DDMP) at the 22-hydroxyl position (Kuduo et al.,1993).

Consumption of saponins through the diet helps in the prevention of chronic diseases such as cancer or cardiovascular disorders, as well as the reduction of cholesterol levels (Barrón et al., 2009). The mechanism is the union of saponins to bile salts and cholesterol in the gastrointestinal tract forming complexes; they also create holes that increase the permeability of membranes, thus reducing the reabsorption and increasing the excretion in the gastrointestinal tract (Amarowicz et al., 1994). Another action of saponins is the anti-inflammatory effect of the prostate, since they inhibit the synthesis of prostraglandins and have an anti-edematose effect due to triterpenes. This may inhibit glycosil transferase, an enzyme related to depolymerization of proteoglycans in the perivascular connective tissue during the inflammatory process (Dvorkin et al., 2002).

In soy, the highest concentration of saponins has been found in the hypocotyl. A variety in the content of soy saponins on dry basis has been reported, from 9 to 53 mg/100 g (Anderson and Wolf, 1995) to 744 mg/100 g (Paucar et al., 2010b).

Germination has an important effect on the concentration of saponins, increasing their concentration. This is due to the synthesis and activation of different enzymes that favor a higher production of secondary metabolites, while causing the degradation of the DDMP conjugate that allows the detection of saponins (Kurosawa et al., 2002; Gu et al., 2002). Paucar et al., (2010a) reported a content of saponins of 6.7 a 23.5mg/g in germinated soy seeds, the highest concentration being after 63 h at 30ºC. Shimoyamada and Okubo (1991) found a 15-fold increase in soy saponins concentration under light conditions and an increase of 8-fold in dark conditions, after 8 days of germination. Jyothi et al., (2007) reported that soy saponins concentration increased from 2.8% to 8.9% after 4 days of germination. The variation in the results on the increase of saponins observed by different authors may be due to the specific variety of the seeds and germination conditions of each specific study.

Our research was carried out in soy germinated for 0, 2 and 6 days and shows that the concentration of saponins tends to increase along germination time. A significant difference ($p<0.05$) was observed after day 2, however this value was not significantly different ($p<0.05$) when compared to day 6, even though this last sample showed a higher concentration of saponins (**Figure 6**).

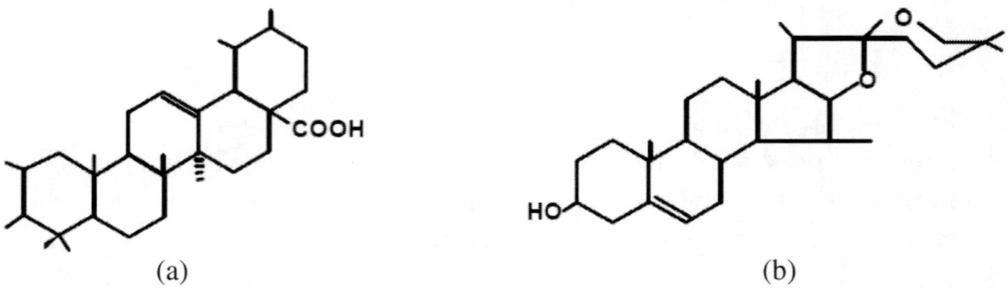

Figure 5. Saponin structure; (a) Triterpenoid sapogenin (b) Steroidal sapogenin (Modified from Sparg et al., 2004).

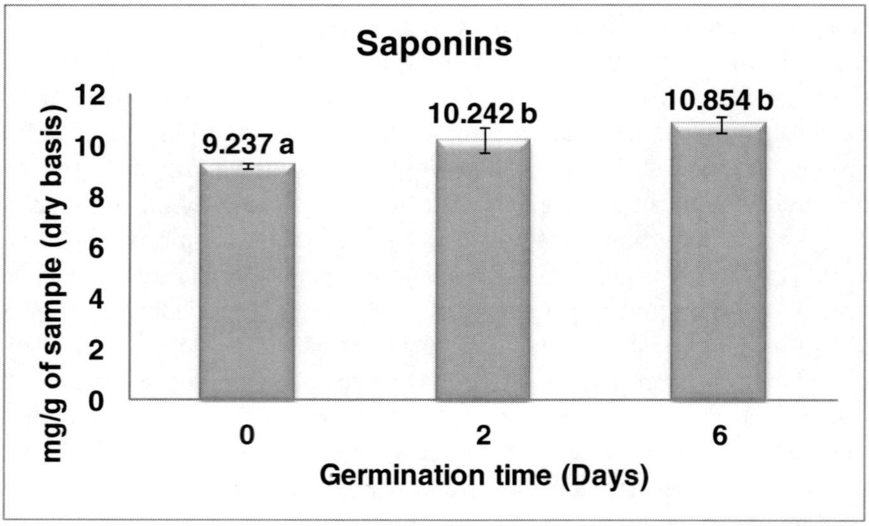

[a/]Results are expressed as the mean of three determinations ± standard deviation. Samples with the same letter do not show significant differences according to the LSD test ($p \leq 0.05$).

Figure 6. Effect of germination on the content of soy saponins after 0, 2, and 6 days.

PHYTIC ACID

Phytic acid or myo-inositol 1,2,3,4,5,6-hexaquisphosphate (Ins P6) was discovered in 1872 by Pfeffer in vegetable seeds (Erdman, 1979) (**Figure 7**). It acts as the primary storage form of phosphorus (P) in seeds of cereals and legumes, is deposited during seed development as a mixed phytate salt of iron (Fe), zinc (Zn), calcium (Ca), magnesium (Mg), and potassium (K). These salts, known as phytates, regulate various cellular functions such as DNA repair, chromatin remodeling, endocytosis, nuclear messenger RNA export, and potentially hormone signaling, all important for plant and seed development (Dil et al., 2010, Campos et al., 2010).

Phytic acid shows anti-nutritional activity due to its capacity to form insoluble complexes with minerals and proteins, turning them into inassimilable compounds. It decreases the availability of zinc, manganese, copper, molibdene, calcium, magnesium, iron and proteins, which may affect plasmatic cholesterol. When joined to a protein it induces a decrease in solubility and functionality (Torres and Tovar, 2009). Many of these complexes are insoluble, and when considering solubility as an essential factor for nutrients to be absorbed, the presence of phytic acid causes a reduction in nutrient bioavailability and as a consequence, decreases the nutritional value of foods.

Nevertheless, phytic acid may have positive health effects, since it may decrease cancer risk through different mechanisms: when joined to Fe^{2+} decreases the formation of free radicals during lipid oxidation, which is catalyzed by this ion. It also joins Zn^{2+}, an ion necessary for DNA synthesis that directly reduces cellular proliferation. Furthermore, as it reduces cell proliferation, it increases differentiation of malignant cells with possible reversion to the normal phenotype and is involved in host defense mechanism and tumor abrogation (Shamsuddin, 2002). Phytic acid retards digestion of starch, helping it to reach the colon where it is fermented by bacteria, producing short-chain fatty acids that have a protective action towards cancer (Graf et al., 1990; Thompson, 1993).

Soy contains approximately 4 mg/g of phytic acid (Bau et al., 1994). In our laboratory a content of 2.9 mg/g on a dry basis was found. The quantity of phytic acid may vary according to the farming conditions. It is mainly located in the cotyledon, most probably in the form of soluble potassium phytate (Bau et al., 1994). Phytates are affected by germination just as trypsin inhibitors and saponins. A reduction of these compounds in several legumes has been reported, and this behavior has been attributed to the action of phytase (myo-ionositol-hexaquisphosphato 3-phosphohydrolase), which increases 800-2000% during germination (Reddy et al., 1982). This enzyme has a high affinity for phytic acid and catalyzes the hydrolysis of phytic acid in myo-inositol, myo-inositol phosphate and inorganic phosphates; its optimal pH of activity is 4.5-4.8 and 55°C (Bau et al., 1997; Martínez et al., 2002). With the activation of the pre-existing phytases and the synthesis of new phytases during germination, myo-inositol hexaquisphosphate present in the seed may be hydrolyzed. From this compound, the inorganic phosphate required for the growth of the plant is obtained, and the myo-inositol is utilized *in situ* for the synthesis of the cell wall (Goyoaga, 2005). We observed that after 6 days of germination phytates decreased up to 33% (**Figure 8**). Bau et al., (1994) reported a decrease of 17% on the content of phytic acid after 5 days of germination. Other researchers have reported a reduction of 21 and 20% for lentils and chickpeas after 24 h of germination, respectively (Reihaneh et al., 2007).

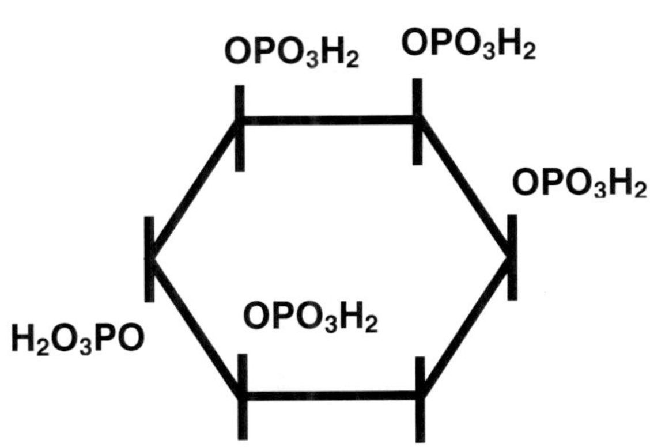

Figure 7. Chemical structure of phytic acid (myo-inositol 1,2,3,4,5,6-hexaquisphosphate).

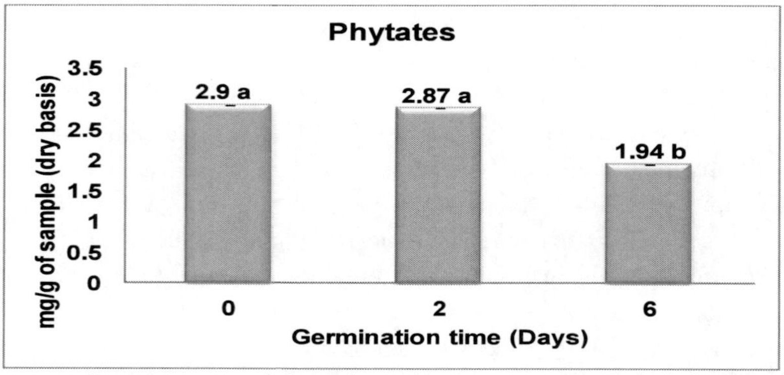

[a/]Results are expressed as the mean of three determinations ± standard deviation. Samples with the same letter do not show significant differences according to the LSD test ($p \leq 0.05$).

Figure 8. Effect of germination on the content of phytic acid in germinated soy after 0, 2, and 6 days.

ISOFLAVONES

Isoflavones are another bioactive group present in soy. They have a flavonoid structure, are non-nutritive and chemically similar to estrogens, therefore they are considered phyto-estrogens (Barnes, 2004). These compounds are derived from hetero-cyclic phenols, being synthesized biologically from the union of an aromatic precursor (hydroxycinamil coenzyme-A-ester) with an aliphatic (malonyl coenzyme-A) (Ludueña et al., 2007).

The basic structure of aglycone isoflavones is 3-phenylchroman, which is hydroxylated in positions 4' and 7' (**Figure 9**). The specific type of aglycone depends on the substituents in carbons 5 and 6 (daidzein, genistin and glycitein). If aglycones are joined to a sugar they are known as glucosides (daidzin, genistin and glicitin). All glucosides may be esterified in the position 6''-O- with an acetyl or malonyl group, thus forming other compounds (acetyldaidzin, acetylgenistin, acetylglycitin, malonyldaidzin, malonylgenistin, and malonylglycitin) (Peñalvo et al., 2004)

Aglycones

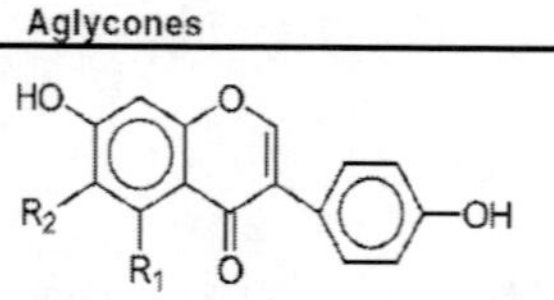

R_1	R_2	Compound
H	H	Daidzein
OH	H	Genistein
H	OCH_3	Glycitein

Glucosides

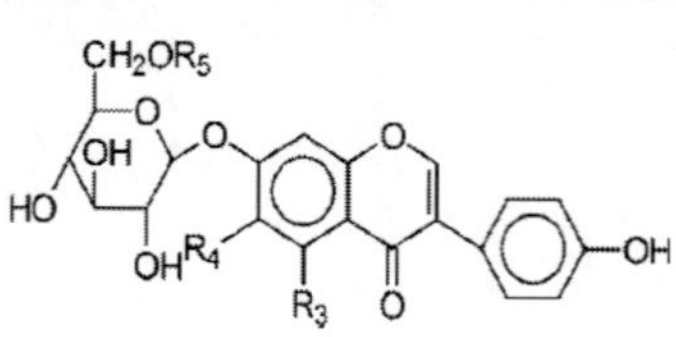

R_3	R_4	R_5	Compound
H	H	H	Daidzin
OH	H	H	Genistin
H	OCH_3	H	Glycitin
H	H	$COCH_3$	Acetyldaigzin
OH	H	$COCH_3$	Acetylgenistin
H	OCH_3	$COCH_3$	Acetyglycitin
H	H	$COCH_2COOH$	Malonyldaidzin
OH	H	$COCH_2COOH$	Malonylgenistin
H	OCH_3	$COCH_2COOH$	Malonylglycitin

Figure 9. Chemical structure of soy isoflavones (Shimoni, 2004).

Isoflavones are natural compounds that may act as estrogens in the body. They are ingested in the form of glucosides, then hydrolyzed by intestinal glucosidases and bacterial enzymes into daidzein, genistein and glicitein (Anderson and Wolf, 1995; Murkies et al., 1998; Setchell et al., 1999; Setchell, 2000). Daidzein may also form equol and o-desmethlylangolensin by the action of bacteria (Salazar, 2006).

Isoflavones are capable of joining estrogenic receptors. An estrogenic receptor (ER) joins to molecules of different nature, even of non-steroid structure, as long as this molecule has certain characteristics, such as the presence of an aromatic ring, hydroxyl groups and hydrophobic character (Hu and Aizawa, 2003). The resulting ligand-receptor complex is functionally equivalent to the one formed by 17β-estradiol, capable of stimulating the transcription of cellular DNA, which produces a specific cellular response. The intensity of this response is lower in the case of isoflavones as compared to estradiol and varies according to the specific receptor type (α or β), even though the affinity of isoflavones for ERβ is higher than for ERα. This difference in affinities towards both iso-forms of the estrogenic receptor is probably due to the different sequence of amino acids. Even though isoflavones, specifically genistein, join the ERβ with almost the same efficiency as estradiol, the required concentration to induce the transcriptional activity is 10^4 times higher for genistein than estradiol. A similar phenomenon takes place with daidzein, equol or glicitein (Kuiper, et al., 1998). When isoflavones join the β receptor, estrogenic activity increases producing benefits to the cardiovascular and osseous systems.

Isoflavones also show anti-carcinogenic activity *in vivo*. Laboratory animals that were fed a soy-fortified diet show a lower proliferation of tumor mammal cells after stimulation with direct and indirect inducing agents (Murkies et al., 1998; Coward et al., 1993). It is possible that the antioxidant effect of isoflavones increases cardiovascular protection of the arterial wall by inhibiting the formation of free radicals, oxygen peroxide and peroxide anions. In this sense, equol inhibits the expression of a NADPH oxidase by causing a decrease in the production of the peroxynitrite radical, which in turn decreases the oxidation of low-density lipoproteins and limits the formation of atheromas. It has been described that genistein also increases the activity of several antioxidant enzymes producing an effect similar to equal in oxygen- and nitrogen-containing species (Navarro, 2005)

Soy is the main source of isoflavones, the content varying between 1200 and 4200 μg/g (Anderson and Wolf, 1995). Paucar et al., (2010a y b) analyzed soy varieties BRS133 and BRS258, finding a total content of isoflavones of 3900 μg/g and 223.7 μg/g, respectively. During germination, these compounds are modified under certain conditions, some investigations report that the concentration increases (Lin and Lai, 2006). Other reports mention that isoflavone aglycones increase after the first day of germination due to the release of aglycones produced by the activity of glucosidases; the biosynthesis of isoflavones is also increased through the malonate and phenylpropanoid pathways (Paucar et al., 2010b).

Robles (2010) reported that protein isolates obtained from soybeans germinated for 6 days increase their content of genistein from 26 μg/g to 86 μg/g and of daidzein from 49 μg/g to 293 μg/g. However, it was observed that the content of genistein decreased from 350 μg/g to 113 μg/g, which may be due to the conversion of this compound onto another type of isoflavone (Zhu et al., 2005). González-Espinosa (2011) reported an increase of up to 203% and 262% in genistein and daidzein, respectively, in protein isolates of soybeans germinated for 6 days. Paucar et al., (2010b) observed an increase of isoflavones from 222.37 mg/100g to 294.8 mg/100g in soybeans germinated for 11 days, the maximum value being found after two days of germination. However, these authors did not find glicitein or any acetylated forms of isoflavones.

OLIGOSACCHARIDES

Soybean is a rich source of oligosaccharides, which are non-reducing sugars of low molecular weight, water soluble and formed by residues of –D-galatopyranoside joined to sucrose by the galactose unit. This is the basic unit from which raffinose, stachyose and verbascose are formed as units of galactose that are added to form different oligosaccharides (**Figure 10**) (Qiushuang et al., 2007).

Oligosaccharides are reserve compounds that are synthesized and stored in cotyledons during the development of the seed. They are not digested by monogastric animals or humans, since enzyme α-1,6-galactosidase is not found in their intestinal mucosa (De Lumen, 1992). Therefore, these compounds are not digested and pass to the colon where they are fermented by intestinal bacteria and considered responsible for flatulence, due to the production of gasses and short-chain fatty acids such as butyric, propionic and acetic. The production of short-chain fatty acids has some health benefits such as the reduction of colon cancer risk, as it has been observed that butyric acid induces apoptosis of tumor cells *in vitro*. Butyrate is considered to be the main nutrient of epithelial cells that cover the colon, and stimulates its growth (De Dios et al., 2009).

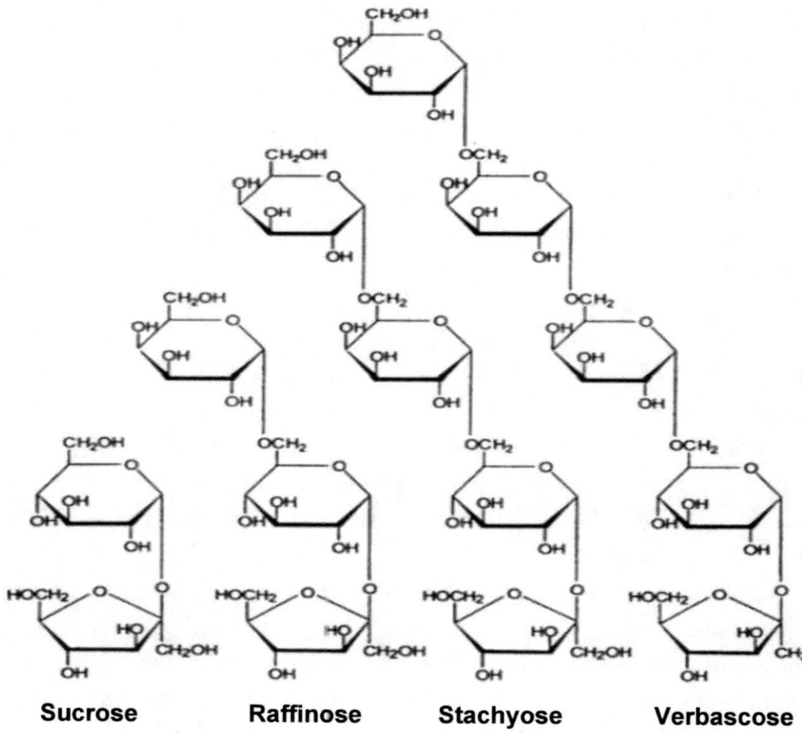

Figure 10. Chemical structure of sucrose and oligosaccharides of the raffinose family.

Soybean contains 22.1mg/g - 43.1mg/g of sucrose, 6.7mg/g - 9.7mg/g of raffinose and 27.5mg/g - 41.3mg/g of stachyose (Qiushuang et al., 2007; Martín et al., 2008; Pedrosa et al., 2012). Germination causes losses in the amounts of oligosaccharides due to the increase of the activity of α-galactosidase, which hydrolyzes stachyose first, followed by raffinose in the α-1-6-galactoside bonds (Martín et al., 2008). Up to 75% of sucrose is degraded by germination after 4 days, staying constant up to day 6; stachyose and raffinose were also progressively degraded (Bau et al., 1994). Pazur et al., (1962) reported that all α-galactosides disappear after 6 days of germination. Martín et al., (2008) observed that the concentration of sucrose increased from 22.1mg/g to 47.3mg/g after 12 h of germination and then decreased back to 26.1mg/g after 24 h. In raffinose, the same authors found a decrease from 6.7mg/g to 2.7 mg/g; verbascose was not detected at 12 h but it was found at 24 h at a concentration of 2.1mg/g. The increase in sucrose after 12 h of germination may be due to the conversion of oil and soluble oligosaccharides, such as raffinose, onto sucrose to be used in the rapid expansion and growth of the seedling. Afterwards, sucrose is metabolized into simple sugars through the activity of invertase, increasing the concentrations of glucose and fructose (Kuo et al., 1990).

LECTINS

Lectins, also known as hemagglutinins, are non-immune proteins that irreversibly bind to a carbohydrate unit of glycoproteins, glycolipids or polysaccharides. They are synthesized during the development of the seed together with storage proteins and they are named after the plant that produces them, their biological activity is variable (Nasia et al., 2009). Lectins from legumes are generally constituted by 2 or 4 identical subunits of 25 to 30 kDa, each subunit contains a union site for the metallic ions Ca^{2+}, Mn^{2+} and Mg^{2+}. They are formed by approximately 250 amino acids, distributed in 12 β anti-parallel sheets and connected by loops, which produces a flat, dome-shaped structure. The four loops located in the upper part of the monomer form the carbohydrate-recognition site (Hernández et al., 2005).

Lectins pose certain health risks since they combine with glycoproteins of the red blood cell membranes, causing their coagulation or agglutination. Due to this effect, they are known as phytohaemagglutinins. Their action *in vivo* is due to their high specificity to recognize carbohydrates, and their main effect is related to the adherence to carbohydrates on the surface of the small intestine, thus affecting the absorption and transfer of nutrients (De Dios et al., 2009). It has been reported that lectin toxicity is due to the epithelial lesions in the intestine that favor the proliferation of bacteria and decrease nutrient availability.

However, lectins may have a similar effect to the one produced by prebiotics, as they produce fermentable material in the colon. Some authors mention that lectins have biological activity by interacting with specific molecules in different blood groups, mitogenic stimulation of lymphocytes and inhibition of phagocytosis, promotion of cellular adhesion, inhibition of mycotic growth and a similar effect to insulin by acting as insulin-mimetic agent. It has also been reported that lectins inhibit tumor growth due to a lower nutrient absorption (De Dios et al., 2009). Concentration of lectins in soybean may vary, since it depends on the variety under study, among other factors. Table 2 shows the concentration of lectins reported by different authors.

Table 2. Concentration of lectins in different soy products

Type of soy product	Concentration of lectins (mg/g)	Reference
Variety BRS133	4.21	Paucar et al., (2010a)
Variety 258	5.68	Paucar et al., (2010b)
Variety Jinong7	2.81	Gu et al., (2010)
Variety Jilin45	6.25	Gu et al., (2010)
Soy germinated for 3 days at 25°C	2.16	Paucar et al., (2010b)
Soy germinated for 63 h at 30°C	1.40	Paucar et al., (2010a)

There are few investigations related to the behavior of lectins during germination. Chen et al., (1977) reported that germination decreases the blood agglutination activity of lectins. However, some authors mention that the concentration of these proteins tends to increase by germination, due to a disorganized increase in protein bodies (Rougé and Pere, 1982; Díaz et al., 1990). Paucar et al., (2010b), observed that after 3 days of germination at 25ºC, lectins decreased approximately 74%. In a different study, Paucar et al., (2010a) found that lectins in the BRS133 soy variety also decreased at 30ºC and a germination time of 63 h (Table 2).

Conclusion

Soybean has a role in preventing a wide range of diseases such as obesity, coronary heart diseases, cancer, and osteoporosis. These, together with its known nutritional value, are current consumer demands. The potential of germinated soybean is clear, although it is important to bear in mind the germination time. We observed a significant reduction of trypsin inhibitors and phytic acid due to this process; hence germination makes sprouts nutritionally more attractive than their non-sprouted seeds. It is possible that the increase of bioactive phytochemicals such as saponins, isoflavones, and other phenolic compounds improve the antiobesity, coronary heart diseases and anticancer properties of soybean.

References

Amarowicz, R., Shimoyamada, M. & Okubo, K. (1994). Hypocholesterolemic effects of saponins. *Roczniki Panstwowego Zakladu Higieny*, *45*(1–2), 125–130.

Anderson, R. L. & Wolfm, W. J. (1995). Compositional changes in trypsin inhibitors, phytic acid, saponins and isoflavones related to soybean processing. *Journal Nutrition*, *125*, 581S- 588S.

Barnes, S. (2004). Soy Isoflavones– Phytoestrogens and What Else? *Journal Nutrition*, *134*, 1225S-1228S.

Barac, M. B., Stanojevic, S. P. & Pesic, M. B. (2005). Biologically active components of soybean and soy protein products-A review. *Acta Periodica Technologica*, *36*, 155-168.

Barrón-Yánez, M. R., Villanueva-Verduzco, C., García-Mateos, M. R. & Colinas-León, M. T. (2009). Valor nutricio y contenido de saponinas en germinados de huauzontle

(*Chenopodium nuttalliae Saff.*), calabacita (*Cucurbita pepo L.*), canola (*Brassica napus L.*) y amaranto (*Amaranthus leucocarpus* S. Watson syn. *hypochondriacus L.*). *Revista Chapingo, Serie horticultura, 15*(3), 237-242.

Bau, H. M., Villaume, C., Chandrasiri, V., Nicolas, J. P. & Mejean, L. (1994). Effect de la germination sur la composition et la valeur nutritive des graines de soja chez le rat. *Science des Aliments, 14*, 688-695.

Bau, H. M., Villaume, C., Jean-Pierre, N. & Mé jean, L. (1997). Effect of germination on chemical composition, biochemical constituents and antinutritional factors of soya bean (*Glycine max*) seeds. *Journal Science Food Agricultural, 73*, 1-9.

Bon, H. M. & Bowles, D. J. (1983). Characterization of soybean endopeptidasa activity using exogenous and endogenous substrates, *Plant Physiology, 72*, 345-350.

Campos Vega, R., Loarca Piña, G. & Oamah, B. D. (2010). Minor components of pulses and their potential impact on human health. *Food Research International, 43*, 461-482.

Chandrasiri, V., Bau, H. M., Villaume, C., Giannangeli, F., Lorient, F. & Me• jean, L. (1987). Effet de la germination de la graine de soja sur la composition et la valeur nutritionnelle de sa farine. *Sciences des Aliments, 7* (hors se• rie), 139-150.

Chen L. H., Thacker, R. R. & Pan, S. H. (1977). Effect of germination on hemagglutinating activity of pea and bean seeds. *Journal of Food Science, 42*, 1666–1670.

Coward L., Barnes, N., Setchell, K., Barnes, S. (1993). Genistein, daidzein, and theri B-glycoside conjugates: antitumor isoflavones in soybean foods from american and asian diets. *Journal Agricultural Food Chemistry, 41*, 1961-1967.

De Dios Elizalde, A., Porrilla, Y. & Chaparro, D. C. (2009). Factores anitinutricionales en semillas. *Facultad de Ciencias Agropecuarias. 7*(1), 45-54.

De Lumen, O. B. (1992). Molecular strategies to improve protein quality and reduced flatulence in legumes: a review. *Food Structure, 11*, 33–46.

De Luna Jiménez., A. (2006). Valor nutritivo de la proteína de soya. *Investigación y Ciencia. 14*(36), 29-34.

Díaz, C. L., Hosselet, M., Logman, G. J. J., Van Driessche, E., Lutenberg, B. J. J. & Kijne, J. W. (1990). Distribution of glucose/mannose-specific isolectins in pea (Pisum sativum L.) seedlings. *Planta, 181*, 451-461.

Dil Thvarajah, Pushparajah Thvarajah, Chai Thiam See & Albert Vandenberg. (2010). Phytic acid and Fe and Zn concentration in lentil (Lens culinaris L.) seeds is influenced by temperature during seed filling period. *Food Chemistry, 122*, 254-259.

Dittrich, K., Leitzmann, C. (1998). Los alimentos bioactivos. Editorial Integral. *Guía de los alimentos que curan y protegen de las enfermedades*. Barcelona España.

Dvorkin, L. & Song, K. Y. (2002). Herbs for beign prostatic hyperplasia. The *Annals of Pharmacotherapy, 36*(9), 1443-52.

Erdman, L. W. (1979). Oilseed phytate: nutritional implications. *Journal American Oil Chemistry Society, 56*, 736-741.

FAO. (Organización de las Naciones Unidas para la Agricultura y la Alimentación). "Cereales en la alimentación humana". (On line). Accessed 2 August, 2005. Available in: http://www.fao.org/documents/show_cdr.asp?url_file=/DOCREP/006/W0073S /w0073s0 um

González-Espinosa, L. A., Robles-Ramírez, M. C. & Mora-Escobedo R. (2011). Soybean and obesity. In: Tzi Bun Ng editor. *Soybean: Biochemistry, Chemistry and Physiology*. Rijeka Croatia, 555-576.

Goyoaga Jorba C. (2005). Estudio de factores no nutritivos en "Vicia Faba I": influencia de la germinación sobre su valor nutritivo. Tesis Doctoral. Departamento de Nutrición y Bromatología II. Universidad Complutense de Madrid. Madrid, España.

Graf, E. & Easton, J. W. (1990). Antioxidant function of phytic acid, free radical. *Biology and Medicine*, *8*, 61–69.

Gu, L., Tao, G., Gu, W. & Prior, R. (2002). Determination of soyasaponins in soy with LC–MS following structural unification by partial alkaline degradation. *Journal of Agricultural and Food Chemistry*, *50*(24), 6951–6959.

Guillon, F. & Champ M. M. J. (2002). Carbohydrate fractions of legumes: uses in human nutrition and potential for health. *British Journal of Nutrition*, *88*(3), S293-S306.

Guerrero-Beltrán, J. A., Estrada-Giron, Y., Swanson Barry, G. & Barbosa-Cánovas, G. V. (2009). Pressure and temperature combination for inactivation of soymilk trypsin inhibitors. *Food Chemistry*, *116*, 676–679.

Haddad, J. & Karim A. (2007). A study of the impact of instantaneous controlled pressure drop on the trypsin inhibitors of soybean. *Journal of Food Engineering*, *79*, 353-357

Haralampidis, K., Trojanowska, M. & Osbourn, A. E. (2002). Biosynthesis of triterpenoid saponins in plants. *Advances in Biochemical Engineering/ Biotechnology*, *75*, 31-49.

Hernández Cruz, P., Pérez Campos, E., Martínez Martínez, L., Ortiz, B. & Martínez, G. (2005). Las lectinas vegetales como modelo de estudio de las interacciones proteína-carbohidratos. *Revista de Educacion Bioquímica*. *24*(001), 21-27.

Hubert, J., Berger, M. & Daydé, J. (2005). Use of simplified HPLC–UV analysis for soyasaponins B determination: study of saponin and isoflavone variability in soybean cultivars and soy-based health food products. *Journal of Agricultural and Food Chemistry*, *53*, 3923–3930.

Howes, L. G., Howes, J. B. & Knight, D. C. (2006). Isoflavone therapy for menopausal flushes: A systematic reviewand meta-analysis. *Maturitas*, *55*(3), 203-11

Hu, J. Y. & Aizawa, T. (2003). Quantitative structure-activity relationships for estrogen receptor binding affinity of phenolic chemicals. *Water Research*, *37*, 1213-22.

Jiménez, M. J., Elias, L. G., Bressani, R., Navarrete, D. A., Gómez, R. & Molina, M. R. (1985). Biochemical and nutritional studies of germinated soybean seeds. *Archivos Latinoamericanos de Nutricion*, *35*(3), 480-90.

Jyothi, T. C., Sindhu Kanya, T. C. & Appu Rao, A. G. (2007). Influence of germination on saponins in soybean and recovery of soy sapogenol I. *Journal of Food Biochemistry*, *31*, 1–13.

Kuduo, S., Tonomura, M:, Tsukamatc, C., Uchida, T., Sakabe, T., Tamura, N., et al. (1993). Isolation and structural elucidation of DDMP-conjugated soyasaponins as genuine saponins from soybean seeds. *Bioscience, Biotechnology, and Biochemistry*, *57*(4), 546–550.

Kuiper, G. G. J. M., Nilsson, S. & Gustaffson, J. A. (1998). Characteristics and funtion of the novel estrogen receptor β. *Endocrinology*, *142*, 89-112.

Kuo, T. M., Doehlert, D. C. & Crawford C. G. (1990). Sugar metabolism in germinating soybean seeds. *Plant Physiology*, *93*, 1514-1520.

Kurosawa, Y., Takahara, H. & Shiraiwa, M. (2002). UDP-glucuronic acid: Soyasapogenol glucuronosyltransferase involved in saponin biosynthesis in germinating soybean seeds. *Planta*, *215*(4), 620–629.

Lin, P. Y. & Lai, H. M. (2006). Bioactive compounds in legumes and their germinated products. *Journal of Agricultural and Food Chemistry*, *54*, 3807–3814.

Lee, C. K. & Karunanithy, R. (1990). Effects of germination on the chemical composition of Glycine and Phaseolus beans. *Journal Science Food Agricultural*, *51*, 437-445.

Ludueña, B., Mastandrea, C., Chichizola, C. & Franconi, M. C. (2007). Isoflavonas en soja, contenido de daidzeína y genisteína y su importancia biológica. Asociación Bioquímica Argentina. *Bioquímica y Patología Clínica*, *71*(1), 54-66.

Martín-Cabrejas, M. A., Felicia-Díaz, M., Aguilera, Y., Benítez,V., Mollá, E. & Esteban Rosa M. (2008). Influence of germination on the soluble carbohydrates and dietary fibre fractions in non-conventional legumes. *Food Chemistry*, *107*, 1045–1052.

Martínez-Domínguez B., Ibañez-Gómez M. V. & Rincón-León F. (2002). Acido fítico: aspectos nutricionales e implicaciones analíticas. *Archivos Latinoamericanos de Nutrición*, *52*(3), 219-231.

Mostafa, M. M., Rahma, E. H. & Rady, A. H. (1987). Chemical and nutritional changes in soybean during germination. *Food Chemistry*, *23*, 257-275.

Murkies, A., Wilcox, G. & Davis, S. (1998). Phytoestrogens. *Journal of Clinical Endocrinology and Metabolism*, *83*(2), 297-303.

Nasia, A., Picariellob, G. & Ferrantia, P. (2009). Proteomic approaches to study structure, functions and toxicity of legume seeds lectins. Perspectives for the assessment of food quality and safety. *Journal of Proteomics*, *72*, 527-538.

Nonogaki, H., Bassel, G. W. & Bewley, J. D. (2010). Germination—Still a mystery, *Plant Science*, *179*, 574–581.

Osborne, T. B. (1924). Osborne, The vegetable proteins (2nd Ed), Longmans, Green, New York.

Papastoitsis, G. & Wilson, K. A. (1991). Initiation of the degradation of the soybean Kunitz and Bowman-Birk trypsin inhibitors by a cystein protease. *Plant Physiology*, *96*, 1086-1092.

Paucar-Menacho, L. M, Berhow, M. A., Gontijo-Mandarino, J. M., González de Mejia, E. & Chang Y. K. (2010)(a). Optimization of germination time and temperature on the concentration of bioactive compounds in Brazilian soybean cultivar BRS 133 using response surface methodology. *Food Chemisty*, *119*, 636-642.

Paucar-Menacho, L. M., Berhow, M. A., Gontijo-Mandarino, J. M., Chang, Y. K. & González de Mejia, E. (2010)(b). Effect of time and temperature on bioactive compounds in germinated Brazilian soybean cultivar BRS 258. *Food Research International*, *43*, 1856-1865.

Pazur, J. H., Skadaksharaswamy, M. & Meidell, G. E. (1962). The metabolism of oligosaccharides in germinating soybean Glycine max. *Archives of Biochemisty and Biophysics*, *99*, 78-85.

Pedrosa, M. M., Cuadrado, C., Burbano, C., Allaf, K., Haddad,J., Gelencser, E., Takacs, K., Guillamon, E. & Muzquiz, M. (2012). Effect of instant controlled pressure drop on the oligosaccharides, inositol phosphates, trypsin inhibitors and lectins contents of different legumes. *Food Chemistry*, *131*, 862-868.

Peñalvo, J. L. Tarha, N. & Adlercreutz, H. (2004). A simplified HPLC method for total isoflavones in soy products. *Food Chemistry*, *87*, 297-305.

Qiushuang, W., Leqin, K., Dongmei, Y., Bili, B., Jianmei, J. & Tiejin, Y. (2007). Change in oligosaccharides during processing of soybean sheet. *Asia Pacific Journal of Clinical Nutrition*, *16* (Suppl 1), 89-94.

Reddy, N. R., Sathe, S. K. & Salunkhe, D. K. (1982). Phytates in legume and cereals. *Advances Food Research*, *28*, 1-92.

Reihaneh, A. G. & Jamuna, P. (2007). The impact of germination and dehulling on nutrients, antinutrients, *in vitro* iron and calcium bioavailability and in vitro starch and protein digestibility of some legume seeds. *LWT-Food Science and Technology*, *40*, 1292-1299.

Rinder, E., et al., (2006). Soja propiedades nutricionales y su impacto en la salud. Grupo Q S. A. *Sociedad Argentina de Nutrición*, 1-100.

Robles Ramírez María del Carmen (2010). Efecto de la germinación sobre las propiedades anticancerosas de la proteína de soya. Tesis Doctoral. Instituto Politécnico Nacional. México.

Robles-Ramírez M. C., Ramón-Gallegos E. & Mora-Escobedo R. (2011). Soybean germination and cancer disease. In: Jason E. Maxwell Editor. *Soybeans: Cultivation, Uses and Nutrition*. New York, NY. *Nova Science Publishers* Inc, 223-249.

Rougé, P. & Pere, D. (1982). Occurrence of lectin during the life cycle of Lathyrus species. In: BOG- HANSEN, T. C. (Ed.) *Lectins: Biology*, *Biochemistry*, *Clinical Biochemistry*. Berlin, New York, Walter de Gruyter & Co., *2*, 137-150.

Salazar Morales A. (2006). Isoflavonas de soya contra osteoporosis, cáncer de mama y síntomas de la menopausia. *Infarmate*. *1*(6), 1-5.

Salunkhe, D. K., Chavan, J. K., Adsule, R. N. & Kadam, S. S. (1992). Soybean. In: Salunke et al., (Eds). World Oil Seeds, *Chemistry*, *Technology and Utilization*. New York, NY: Van Nostrand Reinhold, 9-58.

Sanz Pérez, B., Herrera Castilloa, E., Yuste Grijalaba, F. J., Haya Palazuelos, J., Kr Imperfort, L., Hernández Rodríguez, M., Riobó Serván, P. & Jiménez Hernández, R. (2007). *La salud y la soya*. Editorial Edimsa. ISBN-10:84-7714-251-3. Madrid.

Schwarz, M. W. Saponins. In: B. Elevers, S. Hawkins, W. Eussery, & G. Schulz (Eds.). (1993). *UllmannÕs encyclopedia of industrial chemistry*. New York: VCH Publishers. A23, 485– 498.

Setchell, K. & Cassidy, A. (1999). Dietary Isoflavones: Biological effects and relevance to human health. *Journal of Nutrition*, *129*, 758-767.

Setchell, K. (2000). Absorption and metabolism of soy isoflavones from food to dietary supplements and adults to infants. *Journal of Nutrition*, *130*, 654S-655S.

Shamsuddin Abulkalam, M. (2002). Anti-cancer function of phytic acid. *International Journal of Food Science & Technology*, *37*(7), 796-782.

Shimoni, E. (2004). Stability and shelf life of bioactive compounds during food processing and storage: Soy Isoflavones. *Journal of Food Science*, *69*, 161-166.

Shimoyamada, M., Okubo, K. (1991). Variation in saponin contents in germinating seeds and effect of light irradiation. *Agricultural and Biological Chemistry*, *55*, 577-579.

Shiraiwa, M., Harada, K. & Okubo, K., (1991). Composition and structure of ''group B saponin'' in soybean seed. *Agriculture Biological Chemistry*, *55*, 911-917.

Simon, E. U. (1984). Early events in germination. In. Morray, D. R. Editor. *Seed Physiology* Vol. *III*, Germination and reserve mobilization. Academic Press, 77-115. New York

Sparg S. G., Light M. E. & van Staden, J. (2004). Biological activities and distribution of plant saponins. *Journal of Ethnopharmacology*, *94*, 219-243.

Steggerda, F. R., Richards E. A. & Rackis J. J. (1996). Effects of various soy bean products of flatulence in the adult man. *Sociology and Experimental Biology and Medicine*, *121*, *1235*-1239.

Thompson, L. U. (1993). Potential health benefits and problems associated with antinutrients in foods. *Food Research International*, *26*, 131-49.

Torres-Torres, N. & Tovar-Palacio A. R. (2009). La historia del uso de la soya en México, su valor nutricional y su efecto en la salud. *Salud pública de México*. *51*(3), 246-257.

Trugo, L. C., Donangelo, C. M., Trugo, N. M. F. & Bach Knudsen, K. E. (2000). Effect of heat treatment on nutritional quality of germinated legume seeds. *Journal of Agricultural Food Chemistry*, *48*, 2082-2086.

Wang, W. & González de Mejía, E. (2005). A new frontier in soy bioactive peptides that may prevent age-related chronic diseases. *Comprehensive Reviews in Food Science and Food Safety*, *4*, 63-78.

Wilson, K. A., Rightmire B. R., Chen J. C. & Tan-Wilson A. L. (1986). Differential proteolysis of glycine and β-conglycinin polypeptides during soybean germination and seedling growth. *Plant Physiology*, *82*, 7176.

Zhu, D., Hettiarachchy, N., Horax, R. & Chen, P. (2005). Isoflavone concentrations in germinated soybean seeds. *Plant Foods for Human Nutrition*, *60*, 147–151.

In: Seeds as Functional Foods and Nutraceuticals
Editors: R. Mora-Escobedo, J. De J. Berrios et al.
ISBN: 978-1-62808-489-4

Chapter 3

THE USE OF SOYBEAN PROTEIN AS A FUNCTIONAL FOOD TO CONTROL LIPID METABOLISM IN SOME CHRONIC DISEASES

Nimbe Torres*[*], *Andrea Jiménez-García and Armando R. Tovar
Departmento de Fisiología de la Nutrición, Instituto Nacional de Ciencias Médicas y Nutrición Salvador Zubirán, México, México

ABSTRACT

The soybean is a high-protein legume with a protein digestibility chemical amino acid score (PDCAAS) of 1, indicating that is a high-quality protein. For this reason, during the early 70's, the consumption of soy protein in combination with other vegetable proteins were used to fight against malnutrition in developing countries. In the last two decades, advances in molecular biology have permitted the understanding of the molecular mechanism of action of this legume. It has been demonstrated that the amino acid composition as well as the presence of bioactive compounds such as isoflavones play an important role in the regulation of lipid metabolism mediated by the transcription factor SREBP-1, which in turn regulate insulin secretion and insulin sensitivity. Recent studies have demonstrated that long term consumption of soy protein plays a significant role in the management of some chronic diseases related with obesity such as lipotoxicity, diabetes and fatty liver disease by regulating the expression of several transcription factors involved in fatty acid synthesis, fatty acid oxidation, cholesterol synthesis and leptin. Now, the nutrigenomics have permitted the development of translational nutrition to use the discoveries in nutrition and their application to benefit both society and the food industry. Recently, we have designed a different type of dietary portfolio (combination of 2 or more functional foods) including soy protein for specific diseases and to reduce biochemical abnormalities frequently observed in subjects with some chronic diseases.

* Corresponding author E mail: nimbester@gmail.com, Departmento de Fisiología de la Nutrición. Instituto Nacional de Ciencias Médicas y Nutrición Salvador Zubirán. Vasco de Quiroga No 15. Col Sección XVI. México, D.F 14000, México. Tel/fax (52) (55) 56553038.

INTRODUCTION

Soybean is a legume with a protein digestibility chemical amino acid score (PDCAAS) of 1, indicating that is a high-quality protein. For this reason, during the early 70's, the consumption of soy protein in combination with other vegetable proteins were used to fight against malnutrition in developing countries. In the last three decades, advances in molecular biology have permitted the understanding of the molecular mechanism of action of this legume. It has been demonstrated that the amino acid composition as well as the presence of bioactive compounds, such as isoflavones, play an important role in the regulation of insulin secretion and insulin sensitivity that in turn control lipid metabolism mediated by the transcription factor SREBP-1 (sterol receptor element binding protein). Thus, recent studies have demonstrated that long term consumption of soy protein plays a significant role in the management of some chronic diseases related with obesity such as lipotoxicity, diabetes and fatty liver disease by regulating the expression of several transcription factors involved in fatty acid synthesis, fatty acid oxidation, cholesterol synthesis and leptin. The knowledge of the mechanism of action of soy protein using nutrigenomics has allowed the development of translational nutrition to apply research findings in soybeans for the benefit of both society and the food industry. Based on this research, we have designed different types of dietary portfolios (combination of 2 or more functional foods) including soy protein for specific diseases and reduced biochemical abnormalities frequently observed in subjects with some chronic diseases.

SOYBEAN SEEDS: ORIGIN AND TAXONOMY

Soybeans are oleaginous legumes that originated in Southeast Asia and were first domesticated by Chinese farmers around 1100 BC [40]. They are the key food element in the Chinese diet, and it is considered an ancient sacred grain. Soybeans did not reach Europe until 1739 and the first evidence of their use in the United States was recorded in 1765 in Savannah, Georgia [8]. By the early 1800s, U.S farmers first grew soybeans and raised a variety for soy sauce. During the Civil War, soldiers used soybeans to brew "coffee" when real coffee became scarce [46]. In the late 1800s, soybean was used by farmers as forage for cattle since it is a source of high protein food for the livestock. It also enriches the soil with nitrogen, making it a good element for crop rotations; and it grew well on acid soils and during droughts [45]. By the 1930's, a remarkable growth of interest occurred in using soybean oils and proteins to manufacture a remarkable array of industrial products ranging from paints and soaps to plastics and glues.

Soybeans have been processed into comparatively simple food products for many centuries in the Orient. Processing includes water extraction (soy milk) with coagulation by calcium salt (tofu), roasting (kinako), and fermentation (miso, tempeh, and soy sauce). The bulk of the world soybean crop is processed into edible oil and soybean meal [61].

The taxonomic classification of soybean is the following: Kingdom: Plantae; Order: Fabales; Family: Fabaceae; Subfamily: Faboideae; Genus: Glycine; Species: *G. max*; Common name: soybean [37] (**Figure 1**).

Figure 1. Soybean seeds.

In Central America, soybean formed an import role in the early 60`s because of the severe malnutrition problems due to the protein crisis in these countries. Quality protein became an important approach to fight malnutrition in areas where milk and other food products of animal origin are costly or limited. Soybean protein and other vegetable protein mixtures were suitable for human feeding with high nutritional value.

By 1970, Mexico was one of the main countries in Latin America that started to develop different soy products and using soy protein to enrich and substitute animal protein in numerous products, lowering production costs. One project included enriching tortilla with 8% soy protein to improve protein efficiency. However, this project was irrelevant due to alterations in elasticity and flavor of tortilla. Nowadays, new technology has accomplished this project and soy tortillas are available in the market. Following this, soya beverages were introduced, but higher costs were a limitation for its usage. By 1975, the Food Technology Program of Social Interest was created by the National Food Program (PRONAL, for its acronym in Spanish) to accomplish a better food intake in malnourished populations by designing high nutritional value products with low cost and easy conservation and distribution like soy products [49].

The Global Soybean Seed Market

Soybean is one of the most important crops in the world. Soybeans can produce at least twice as much protein per acre than any other major vegetable or grain crop, even more than corn or milk, and 15 times more protein per acre than land set aside for meat production [38].

One hundred and ninety five million tons of soybeans are produced in the world every year [44], with the United States of America being the largest soybean producer. By 2011, the United States accounted for 33% of the world's soybean production, generating 83.2 million metric tons, followed by Brazil with a 29% generating 72 million metric tons. The seed accounts for around 56% of the world's oilseed production [46]. Though the US is still the largest soy exporting country, other countries like Brazil, Argentina, China and India have all become major producers as the world´s demand for soy increases. Already, South America produces more soybeans than North America (combined U.S. and Canada production) [63], due to the large regions of fertile land and low labor costs. Brazil by itself has numerous available arable land and water resources, which makes this nation a potential leader in soybean production [44].

Soybean protein costs vary depending on the extraction and purification process of the seeds. Sixty-four grams of defatted soy flour protein cost one US dollar, while soy protein isolate gives only 16 grams of protein per dollar [38]. Isolated soy protein cost can be even five to seven times higher than that of defatted soy flour.

About 85% of the world´s soybean crop is processed into meal and vegetable oil. When soybean is processed, a 60-pound bushel yields for about 11 pounds of crude soybean oil and 47 pounds of soybean meal [40]. The seed's oil is primarily used for food industry and human consumption. Almost all the meal is used in animal feed. Approximately two percent of the soybean meal is processed into flours and proteins for food use. Human consumption of soy products includes milk, soy flour, soy protein, tofu, and meat alternatives. Every component of the seed is used (**Figure 2**).

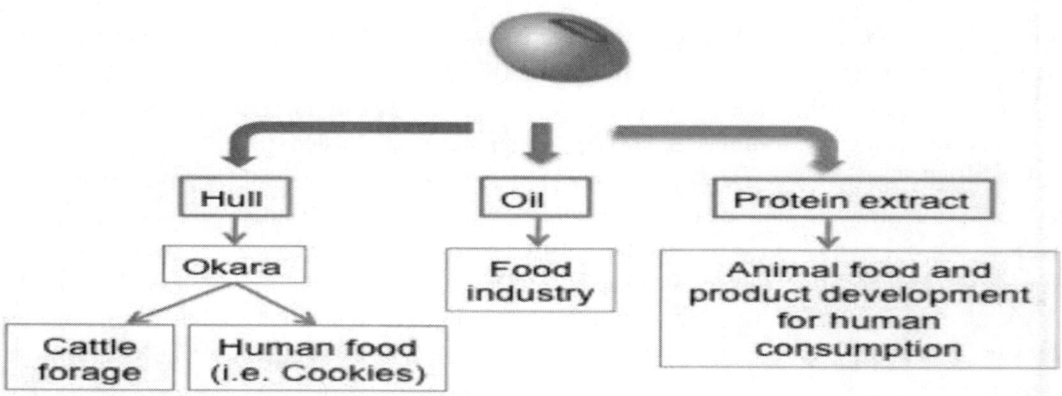

Figure 2. Use of soybean components.

Some soy ingredients have become important in the food manufacturing industry. Lecithin, for example, is a fatty substance used as a food supplement for its emulsifying capacity. Partially hydrogenated soybean oil has been a main ingredient for processed food like snacks, baked goods and salad dressings. Soy oil may also be used for biodiesel production and processed into hydraulic oil, grease, solvent, ink, and plastic products [44].

The demand of soybean production has increased significantly in the last 40 years. Soybeans are and will continue to be the main source of animal fed. A steady demand for soy food, especially healthy food, will ensure an increased production in the market.

FORMER SCOPE AND FUTURE TRENDS OF SOYBEAN

Legumes have historically acted as a critical source of food insurance against times of famine. Specifically, soybean has been widely used as a source of protein and healthy food to overcome problems of malnutrition in the world [11] (**Figure 3**). Due to the high quality protein content, soybean protein foods have been utilized in school feeding programs around the world. Nowadays, soybean isolates can be combined with many other functional foods to create specific dietary portfolios. Dietary portfolios are the combination of two or more functional foods designed to normalized biochemical abnormalities of specific diseases.

In addition of the use of soybean products as nutritional healthy food in malnutrition, during the last 3 decades there has been a substantial increase in the knowledge of the biochemical and molecular mechanisms of action of soy protein in lipid metabolism. Evidence in experimental animal models demonstrated that consumption of soy protein decrease total serum cholesterol and LDL (low density lipoprotein) cholesterol concentration [59]. Several human studies similarly demonstrate that at least 25g of soy protein intake decrease serum lipid levels (**Table 1**).

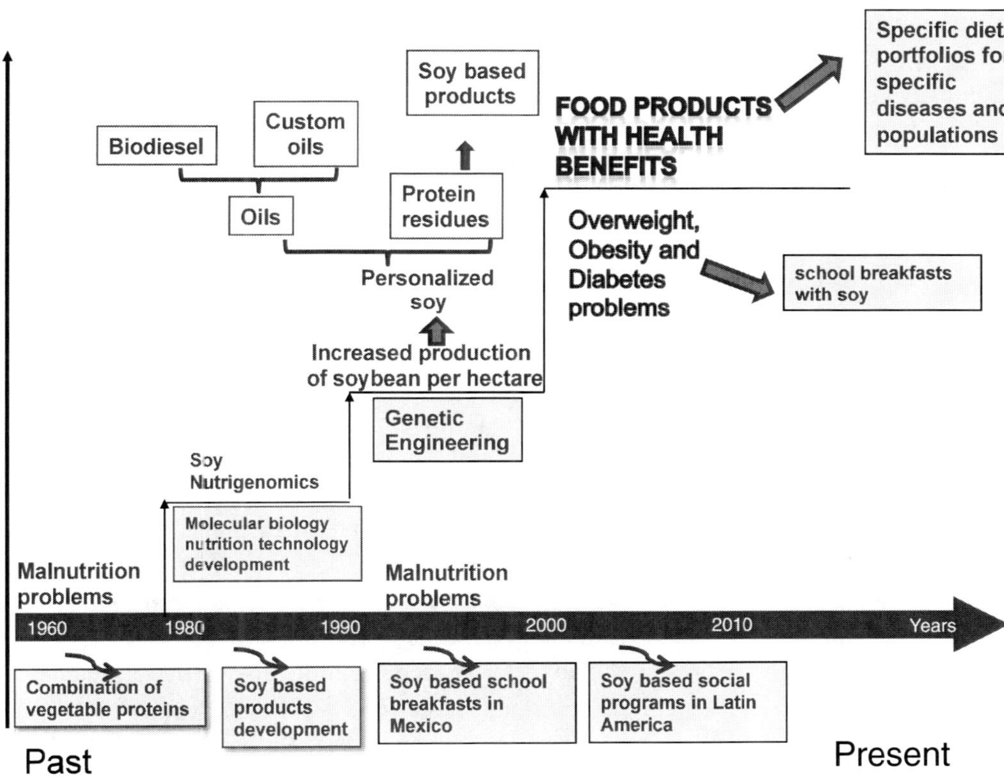

Figure 3. Different approaches in the evolution of the use of soy.

Table 1. Meta-analysis of soybean hypolipidemic effects

	Anderson [3]	Jenkins *et al* [25]	Zhan and Ho [65]	Anderson [4]
Total cholesterol	9.3%	22.0%	3.8%	NR
LDL	12.9%	28.6%	5.3%	5.5%
Triglycerides	10.5%	7.7%	7.3%	10.7%
HDL	2.4 %	6.7%	3.0%	3.2%

LDL: cholesterol (bad cholesterol), low density lipoprotein; HDL: cholesterol (good cholesterol), high density lipoprotein; NR, not reported.

In 1999, the Food and Drug Administration (FDA) authorized the use, on food labels and in food labeling, of health claims stating that the consumption of soy protein reduces the risk of coronary heart disease [15]. The daily dietary intake level of soy protein that has been associated with reduced risk of coronary heart disease is 25 grams or more per day of soy protein [15]

In the last 20 years, the molecular mechanism of action of soy protein on serum lipids have been studied and established, as discussed later in this chapter. It is important to point out that the effect of soy protein on lipid metabolism is due to the combination of the amino acid profile of soy protein and to the presence of isoflavones [18].

The new trends in nutrition involve the use of a combination of functional foods and soy protein to enhance the beneficial health effects. When foods independently known to lower cholesterol, such as almonds, are combined in a healthy way of eating, the beneficial effects are additive. Patients with elevated LDL cholesterol levels, treated with a diet containing almonds and other nuts, plant sterols (also found in nuts), soy protein, and soluble fiber (in high amounts in beans, oats, pears) reduced blood levels of all LDL fractions including small dense LDL (the type that most increases the risk for cardiovascular disease) with near maximal reductions seen after only 2 weeks [26].

COMPOSITION: PHYSICAL AND CHEMICAL PROPERTIES

The composition of soybean depends on several factors, including genotype, geographic location, growing season weather condition, soil condition, maturity, and agronomic practices. Soybeans, as most plants, are grown from seed and cultivated into moist ground. About four to seven days after planting, soybeans sprout and small plants begin to grow. Small flowers blossom and at the same time the soybean pods begin to grow. By the time the soybeans begin to mature, the leaves will turn brown and fall off. This is the time when the mature pods are exposed and soybeans are ready to be harvested.

The use of soybean products has shown a steady increase in the use of these products as ingredients in a wide variety of fabricated or engineered foods. The widespread adoption of soybean proteins in food industry has stimulated research on their physical and chemical properties in many laboratories [62].

1. Physical properties

Soybeans are oval shaped and their sizes are variety-dependent. They are odorless seeds and their common color is yellow, although it may vary up to black or different brown tones. The seed is composed of three major parts: seed coat or hull; cotyledon; and germ or hypocotyls. The cotyledon comprises 90.3% of the seed, while the hull and hypocotyl make up for 10% [20].

This vegetable product has a low water content of only 8%. This way they have a peculiar storage life and are well suited for long periods of transportation and storage. Shape of grain and characteristic dimensional properties such as length, width, thickness, the arithmetic average diameter and the geometric average diameter, increased linearly depending on the increase of moisture content [27]. If the critical water content of the soybeans is exceeded, fat cleaving enzymes are activated. Light and heat may also accelerate this process. Fat cleavage may be limited by low temperatures.

2. Chemical composition of soybean

Soybean is a good source of macro and micronutrients including protein, lipids, carbohydrates, dietary fiber, vitamins and minerals (**Table 2**). The main component of soy is protein, due to its high concentration compared to other legumes. It contains significant amounts of all the essential amino acids that must be provided to the human body.

Table 2. Nutrients in raw soybean seed (100g dry wt)

Nutrient	Soybean
Complex carbohydrates (g)	21
Simple carbohydrate (g)	9
Stachyose (mg)	3300
Raffinose (mg)	1600
Protein (g)	36
Total fat (g)	19
Insoluble fiber (g)	10
Soluble fiber (g)	7
Calcium (mg)	276
Magnesium (mg)	280
Potassium (mg)	1797
Iron (mg)	16
Zinc (mg)	4.8

Source: [5,58]

Table 3. Nutrients in cooked soybean (100g)

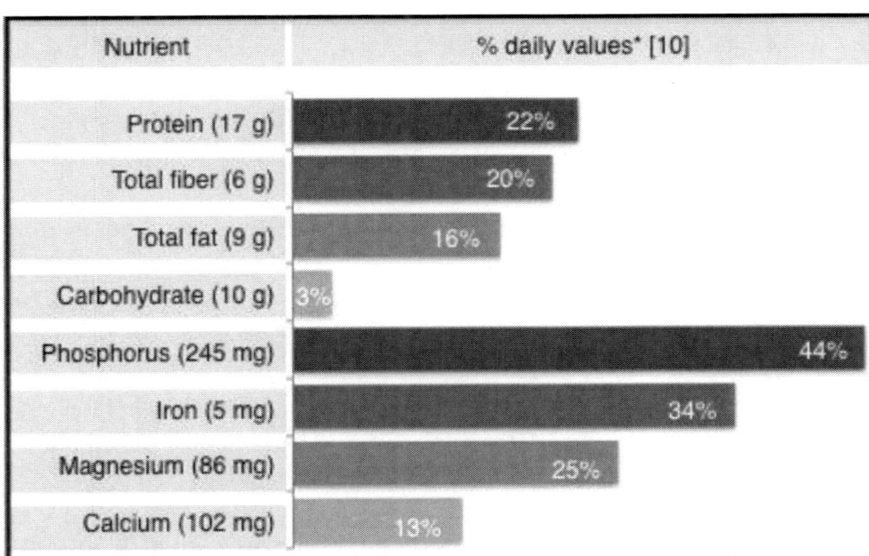

* Based on a 2000 kilocalorie diet.

Source [5, 58]

Soybeans can be utilized as the whole seed or used by fractions depending on the component of interest. For human consumption, whole soybean seed must be cooked with wet heat to destroy the trypsin inhibitors, which are toxic to humans. Once cooked, soybeans are still a great source of protein and other nutrients (**Table 3**).

NUTRIENTS

Soy Protein

The dietary requirement for protein will be the minimum intake which satisfies metabolic demands and which maintains appropriate body composition and growth rates. According to this, soy protein is a vegetable protein with an amino acid profile that meets the requirement for each amino acid in humans and rats for growth and maintenance [54].

Table 4. Indispensable Amino acid profile in soybean (mg/g protein)

	Isolated soy protein	FAO/WHO pattern Children (2-5 years)
His	26	19
Ile	49	28
Leu	82	66
Met+Cys	26	58
Phe+Tyr	90	63
Thr	38	34
Trp	13	11
Val	50	35

Source: [49]

Soybean has been qualified as a source of high quality protein. Soy protein is the only widely available plant protein that has been shown to be a complete protein based on numerous nitrogen balance studies using isolated soy protein (ISP) or soy protein concentrate (SPC) [64]. These results indicate that for healthy adults, the isolated soy protein is of high nutritional quality, comparable to that of animal protein sources, and that the methionine content is not limiting for adult protein maintenance [64].

The importance of soy protein has been undervalued because the limiting amino acid in soy protein is methionine. However, despite the fact that the rat requirement for methionine is ~50% higher than that for humans, the weight gain of growing animals is similar to that of rats fed with casein. Furthermore, methionine is a precursor of homocysteine (Hcy), which is an independent cardiovascular risk factor. Rats fed with a soy protein diet show a lower serum Hcy concentration than do those fed with a casein diet. Similarly in humans, soy intake is inversely associated with serum Hcy levels [36], indicating that substitution of animal protein for soy protein can result in lower Hcy levels, potentially reducing coronary heart disease risk [50].

The protein digestibility-corrected amino acid score (PDCAAS) is the most widely recognized and approved method for evaluating protein quality of food proteins. This measurement was internationally approved in 1991 by the Food and Agriculture Organization of the World Health Organization (FAO/WHO). This method is based on comparing the amino acid profile of a food protein to a reference value for children of 2 to 5 years old, to determine an amino acid score. The amino acid score is corrected through multiplying by the true digestibility of the protein, resulting in a PDCAAS value. A PDCAAS value of 1.00 or 100% would indicate that a protein provides adequate amounts of all of the essential amino acids when fed in nutritionally appropriate amounts [35]. Soy protein is the only vegetable protein source equivalent to the animal protein with a PDCAAS=1 [49] (**Table 5**). The nutritional advantage of soy is the fact that it is a great source of protein. They especially play a beneficial role in vegetarian's diets and in malnutrition.

Table 5. PDCAAS values selected foods

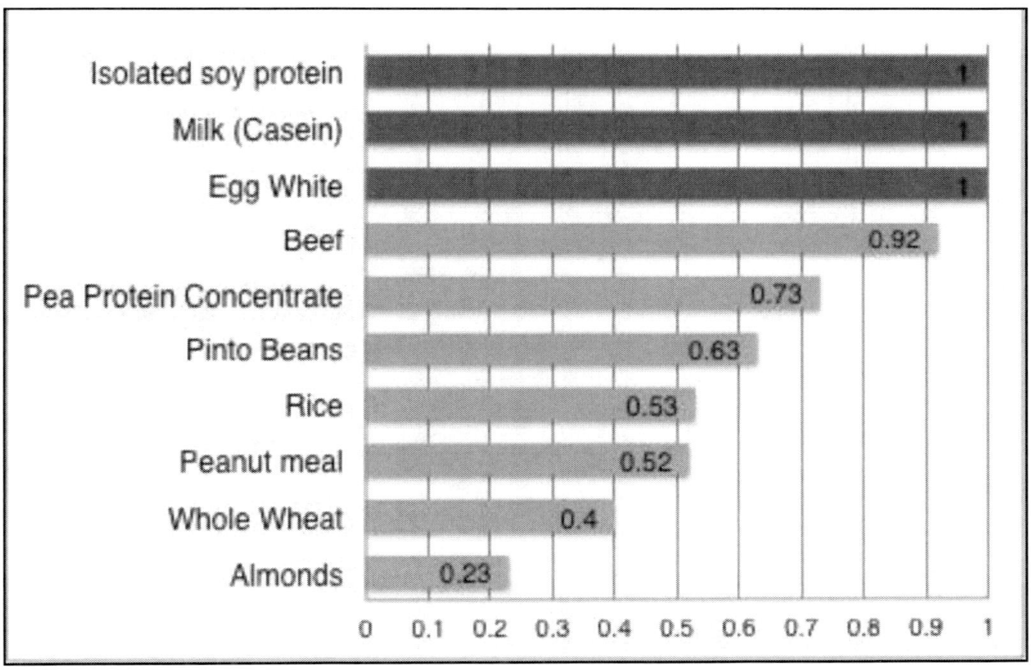

Source [23]

The strength of high quality proteins to balance the amino acid pattern of a human diet is extremely relevant, proving that 6.2 g of soy protein are the equivalent of 1.2 g of casein (milk protein) [42]. The amount of soy required to cover the needs of sulfur amino acids is 0.4 to 0.5 g of protein/kg/day. The consumption of 0.6 g of soy protein/kg/day provides a nitrogen equilibrium equivalent to the one generated with 0.4 g/kg/day of egg white [49]. Methionine complementation has proved to increase the bioavailability of soy protein by allowing a minor amount of soy needed to cover the amino acid requirement.

The relative low sulfur amino acid content in soybean may actually provide an advantage in terms of calcium retention. Human studies showed that the consumption of soy protein is associated with a markedly lower urinary calcium excretion compared with the consumption of similar amounts of whey protein or a mixture of animal proteins [34].

Soy Fat

The second most important component in soy is oil. Due to the high oil content, soybean is also considerate as an oleaginous. Typically soybean seeds contain about 20% oil on a dry weight basis [31]. Soybean oil is the most predominant vegetable oil produced in the world. It is the world's second most consumed vegetable oil, just 5% below palm oil [46].

Oil obtainment is achieved by solvent extraction or mechanical pressing. It consists primarily of neutral lipids, which include tri-, di- and monoacylglycerols, free fatty acids, and polar lipids such as phospholipids [20]. Soybean oil has a high content of α-linoleic acid and a lower level of α-linolenic acid (**Table 6**). The ratio of α-linoleic to α-linolenic acid in soybeans is ≈7.5:1 (α-linolenic acid makes up ≈7–8% of the total fat), whereas typical western diet has a ratio of 10 to 30:1 [34]. These polyunsaturated fatty acids are both essential fatty acids for humans and therefore of dietary importance. Nonetheless, they are also the cause of oxidative instability of this oil.

Table 6. Lipid composition in edible soybean oil (1 tbsp:13.6g) [29,58]

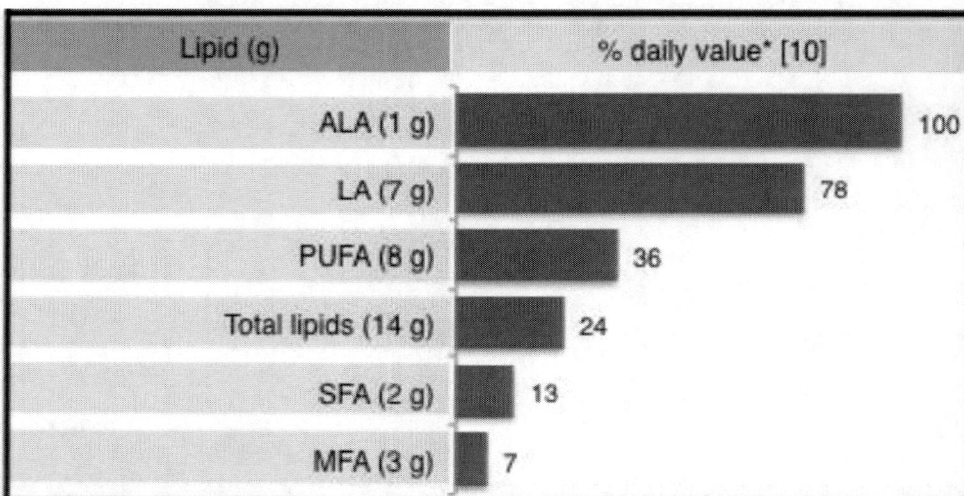

* Based on a 2000 kilocalorie diet. ALA, α-linolenic 18:3n-3; LA, α-linoleic 18:2n-6; PUFA, polyunsaturated fatty acids; SFA, saturated fatty acids; MFA, monounsaturated fatty acids.

Edible soybean oil is also rich in phytoesterols. They are plant sterols structurally similar to cholesterol. Soybean oil contains the most common phytoesterols like stigmasterol, campesterol and beta-sitosterol. Several studies suggest that phytoesterols may offer protection in common cancers, but these statements require additional study [7].Processing techniques such as hydrogenation or genetic transformation have been used to modify the fatty acid composition to improve functional properties or decrease its oxidative characteristic. Increased oleic acid in soybean is important because of health benefits and increased oxidative stability. It has been reported that consumption of oleic acid could reduce the risk of cardiovascular disease [33].

CARBOHYDRATES

Carbohydrates are present in soybeans mainly in the form of oligosaccharides. They are large molecules consisting of three or more sugar units. Raffinose is present in a 0.8-1.0% and stachyose in a 4-4.5%. These carbohydrates are not very well digested in the body. They cannot be hydrolyzed by the α–galactosidase in the intestinal mucosae, managing to pass indigested to the colon. Fermentation occurs in the colon and generates short chain fatty acids and gas (methane, hydrogen and carbon dioxide). It is possible to remove substantial amounts of oligosaccharides and to markedly reduce flatulence by changing the water in which beans are boiled one or more times. However, there may be some beneficial effects associated with oligosaccharide consumption. The oligosaccharides, because of their growth-promoting effect on bifidobacteria, have been hypothesized to promote the health of the colon, increase longevity, and decrease colon cancer risk [34].

NONNUTRITIVE COMPONENTS

Trypsin Inhibitors

Soybean trypsin inhibitors are a group of small proteins that account for a minor portion of the proteins present in the seeds. Trypsin inhibitors may reduce the efficiency of protein digestion by inhibiting or neutralizing protein digesting enzymes. However, these inhibitors are mostly destroyed when soybeans are heat processed to make soy foods. The most common trypsin inhibitors found in soy are the Bowman-Birk Inhibitor (BBI) and the Kunits Trypsin Inhibitor (KTI). Questions have been raised on whether inhibitors are safe to consume due to many studies where soybean-derived inhibitors developed rat pancreatic lesions leading to neoplasia or tumor formation [16]. Pancreatic enlargement resulting from consumption of inhibitors is also reported to occur in chickens and growing guinea pigs, but not in dogs, pigs, calves and monkeys. Monkeys studies fed diets containing low levels of trypsin inhibitors for up to 4 years show no adverse effects on the pancreas [21]. Although the possible significance of trypsin inhibitors for human health is yet to be resolved, any concern could preferentially be avoided by eliminating the inhibitors from soybean foods.

Phytate

Phytates are known as inhibitory factors in soy protein isolates because they have a strong binding to many minerals, especially iron. They have been associated with reducing non-heme food iron absorption. Depending on the type of food, commercial soy foods may contain around 76-528 mg of phytate per serving [2]. Decreasing the content on phytic acid on food may improve iron absorption [24].

Phytochemicals: Isoflavones

Soybeans, as well as other seeds, have a number of non-nutritive bioactive compounds found naturally in food that are known as phytochemicals. These phytochemicals provide a healthy effect of soy. The most important ones are saponin and isoflavones, which are alcohol-soluble components. Saponins are glycoside phytochemicals consisting of a triterpenoid or steroid bound to a sugar molecule, while isoflavones are also glycosides that produce the metabolite equol.

The three major isoflavones in soybeans are genistein, daidzein and glycltein (Figure 4). They are classified as phytoestrogens because they are plant derived compounds with estrogenic activity. They bound to estrogen receptors α and β and their estrogenic activity is 1/1000 less than natural estrogens [49]. The isoflavone content in soy foods varies widely depending on the specific soy product and the processing techniques used to produce it [32]. Isoflavones are maintained in soy protein preparations that are not extracted with alcohol.

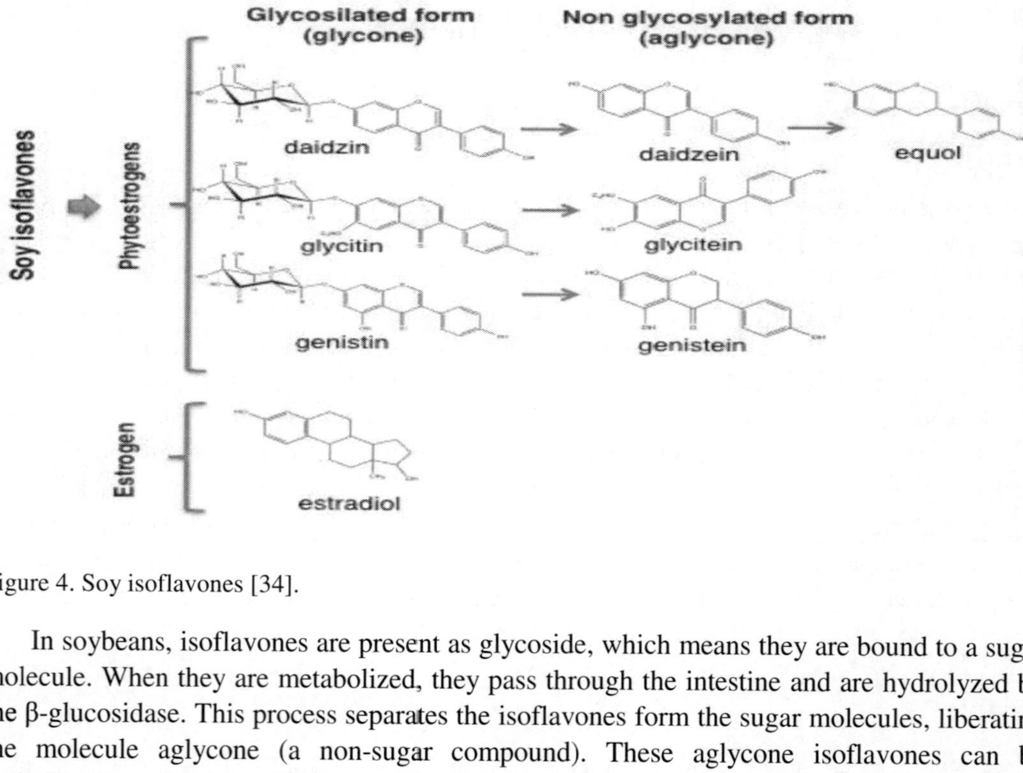

Figure 4. Soy isoflavones [34].

In soybeans, isoflavones are present as glycoside, which means they are bound to a sugar molecule. When they are metabolized, they pass through the intestine and are hydrolyzed by the β-glucosidase. This process separates the isoflavones form the sugar molecules, liberating the molecule aglycone (a non-sugar compound). These aglycone isoflavones can be assimilated or can be further metabolized into equol or O-desmethylangolensin. The conversion of these components by the microflora is independent in every individual. Differences in intestinal microflora make equol production to occur in only around 1 of every 3 individuals consuming soyfoods. In nonfermented soyfoods, the isoflavones appear mostly as the conjugate, whereas in fermented soy products such as miso, the aglycones dominate. Soybeans and soy products contain ≈1–3 mg isoflavones/g protein [34] (**Table 4**). Isoflavones have received considerable attention in recent years. They are being studied for their potential role in the improvement of insulin sensitivity and preventing fatty acid oxidation.

Table 4. Isoflavones content in daily food (mg/100 g of edible portion)

	Daidzein	Genistein	Glycitein	Total isoflavones
Beans, black, mature seeds, raw	0	0	0	0
Instant beverage, soy, powder, not reconstituted	40.07	62.18	10.90	109.51
Soy cheese, American	5.75	8.7	3.5	17.95
Soy drink	2.75	5.1		7.85
Tofu yogurt	5.7	9.4	1.2	16.3
Soyburger, unprepared				
Soy flour, defatted	64.55	87.31	15.08	150.94
Soy meal, defatted, raw	80.77	114.71	16.12	209.58
Soy protein concentrate, by alcohol extraction	5.78	5.26	1.57	11.49
Soy protein concentrate, aqueous washed	38.25	52.81	4.94	94.65
Soy protein isolate	30.81	57.28	8.54	91.05

Source: [57]

HEALTH BENEFITS OF SOY

Substantial data from epidemiological surveys and nutritional interventions in humans and animals indicate that soy protein has beneficial effects in the regulation of insulin levels, obesity, dyslipidemia and some forms of chronic renal disease. Replacement of animal protein in diet with soybean protein has been reported to lower the level of plasma lipids in human subjects. Four meta-analyses revealed that soy protein intake reduced total serum cholesterol, low-density lipoprotein (LDL) cholesterol and triglycerides by 3.8-9.3%, 5.3-12.9% and 7.3-10.7%, respectively, and increased HDL cholesterol by 2.4-3.2% (**Table 1**). The mechanisms of reducing serum lipids have not been fully established. Full effects of soy protein on lipid metabolism have been observed when soy protein contains isoflavones [50]. Several mechanisms of action include the enhancement of bile acid excretion, changes in the insulin:glucagon ratio, thyroid hormone concentrations, and an amino acid composition that causes changes in cholesterol metabolism [6].

Mechanism of Action of Soy Protein

The hypocholesterolemic effect of soy protein compared to animal protein has been demonstrated in humans and a variety of animal models. The molecular mechanisms by which soy protein decreases serum lipids had begun to be elucidated. Hepatic lipid metabolism is controlled in part by the transcription factors termed sterol regulatory element binding proteins (SREBPs). These transcription factors belong to the basic helix-loop-helix-leucine zipper family. There are 3 isoforms of SREBPs that are membrane-bound to the endoplasmic reticulum and nuclear envelope, i.e., SREBP-1α, SREBP-1c, and SREBP-2. The NH_2-terminal domain of the mature forms of these transcription factors is released after 2 proteolytic cleavages in the Golgi apparatus by 2 specific proteases, SP1 and SP2. The NH_2-terminal of SREBPs travels into the nucleus and binds to the cis-acting element termed the sterol regulatory element in the promoter or enhancer regions of genes involved in cholesterol or fatty acid synthesis. In the liver, SREBP-1c and SREBP-2 are the predominant forms. SREBP-2 preferentially stimulates transcription of the genes of cholesterol synthesis, including hydroxymethyl-glutaryl-CoA (HMG-CoA) reductase and HMG-CoA synthase, whereas SREBP-1c stimulates transcription of the genes of fatty acid biosynthesis, such as acetyl CoA carboxylase and fatty acid synthase (FAS). SREBP-2 is activated in response to a low cholesterol content of the cell, and SREBP-1c by insulin. In both instances, activation is mediated through the SREBP cleavage-activating protein [22].

It has been demonstrated that serum insulin concentration increases 3.5-fold over the basal levels after the consumption of a casein diet, whereas ingestion of a soy protein diet increases serum insulin by only 50% [53]. Insulin selectively stimulates SREBP-1 expression in liver [43]. Our study group demonstrated that short- and long-term ingestion of a soy protein diet maintained normal serum insulin concentrations compared with rats fed a casein diet. This in turn was accompanied by a low expression of SREBP-1 mRNA in liver and by decreasing expression of the lipogenic enzymes, FAS and malic enzyme [6] (**Figure 5**).

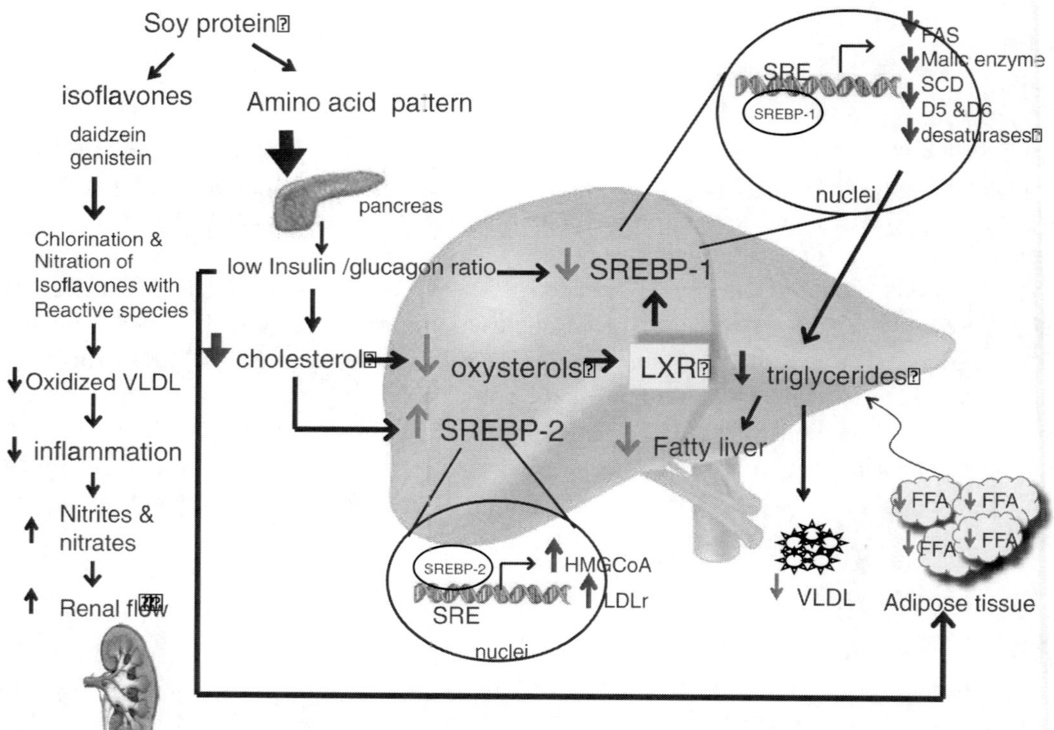

Figure 5. Mechanism of action of soy protein [50]. Soy protein consumption decreases the insulin/glucagon ratio and reduces the expression of SREBP-1, which in turn decreases the expression of lipogenic genes. Isoflavones stimulate SREBP-2, increasing serum cholesterol clearance. Low hepatic cholesterol reduces oxysterol concentrations, preventing the stimulation of LXR-α, which additionally reduces SREBP-1 expression. As a consequence of soy protein consumption, there is a low concentration of VLDL particles. In addition, low serum cholesterol levels observed with soy protein diets prevent kidney damage, decreasing inflammatory response and increasing NO production. VLDL, very low density lipoprotein; Lys, lysine; Arg, arginine; SREBP, sterol receptor element binding protein; HMGCoA, hydroxymethylglutaryl-CoA; LXR, liver receptor X; FFA, free fatty acid; FAS, fatty acid synthetase; SCD, stearoyl-CoA desaturase.

NEPHROTIC SYNDROME

The hallmark of nephrotic syndrome (NS) is the increased glomerular permeability to protein, which leads to proteinuria. Homeostatic mechanisms are unable to cope with such urinary protein losses, and the resulting hypoalbuminemia frequently leads to edema. The glomerular damage characteristic of NS can also result in urinary loss of lipids and plasma proteins including some of the apolipoproteins, resulting in plasma lipid abnormalities. Hyperlipidemia promotes messangial damage, glomerular sclerosis and an increment in the permeability of the glomerular capillary wall. Taken together, these lipid abnormalities are dangerous and may accelerate the atherogenic process. Restriction of protein intake ameliorates proteinuria in nephrotic patients and rats. Low protein diets improve not only proteinuria but also cholesterolemia and blood urea nitrogen. However, despite their beneficial effects, low protein diets retard growth in young individuals and reduce tissue protein in adults [52]. It has been proposed that consumption of diets containing vegetable protein may retard kidney damage by lowering the glomerular filtration rate and urinary albumin excretion [28]. Soy protein is a vegetable protein that sustains adequate growth rate in rats and infants. It also has a high arginine/lysine ratio, which is associated with lower insulin secretion compared with protein of animal origin. Anderson and colleagues [3], in their meta-analysis, showed that the consumption of soy protein significantly decreased serum concentrations of total cholesterol, LDL cholesterol and triglycerides (TG), with a larger decrease in subjects with moderate or severe hypercholesterolemia.

Consumption of soy protein improves abnormal lipid profile and reduces kidney damage in individuals with renal disease [54]. Compared to casein, 20% soy protein diet in rats with nephrotic syndrome lowers total urinary protein excretion (proteinuria) and improves creatinine clearance. The evidence of the reduction in kidney damage reduces lipid abnormalities characteristic in nephrotic rat, primarily by lowering serum VLDL-TG and LDL cholesterol. This study demonstrates how the consumption of a soy protein diet in nephrotic rats prevents glomerular injury, consequence of a lesser development of glomerular sclerosis and reduced production of the proinflammatory cytokines (interleukine)-1α, TNF (tumor necrosis factor)-α and TGF (tumor growth factor)-β. One of the major benefits of this intervention is that it provides a non-protein restricted diet that may be beneficial in the long-term treatment of chronic renal diseases. However, further assessment is necessary before its adoption in humans [54].

REGULATION OF INSULIN SECRETION BY SOY PROTEIN AND ISOFLAVONES

It was suggested that the consumption of vegetable proteins decreases insulin concentration and increases glucagon concentration. However, subsequent investigation demonstrated that serum concentrations of these hormones are dependent by the type and amount of protein. After the consumption of an adequate amount of protein in a diet meeting the requirement (18-20%) for rodents, the serum insulin concentration will depend on the type of protein. Animal feds with 18% soy protein show significantly low serum insulin concentration than rats fed casein diet. When animals are fed with a high protein diet (50%), only animals fed casein diet showed high serum insulin concentration. These results indicate that animal protein produce high serum insulin concentration [53] (Table 7). Regarding glucagon, animals fed 18% protein show low glucagon concentration, whereas animals fed high protein have elevated postprandial serum glucagon concentration regardless the type of protein, indicating that glucagon secretion will increase after high dietary protein consumption [53].

Table 7. Serum insulin and glucagon concentrations in rats fed casein or soy protein

	Serum insulin (ng/ml)				Serum glucagon (pg/ml)			
	Soy protein(%)		*Casein (%)*		*Soy protein(%)*		*Casein (%)*	
	18	50	18	50	18%	50%	18%	50%
Fasting	0.5	0.4	0.5	1	50	60	35	35
Postprandial	0.6	1.0	2.5	2.2	55	210	100	230

Source: [53]

It is believed that eating high protein diets produce high concentrations of serum insulin, however recent evidence indicates that after consumption of a high protein diet, an increases results in both insulin and glucagon concentrations. However, it was not clear whether the effect of consumption of soy protein on serum insulin concentration was associated with changes in the mechanism of insulin secretion or peripheral insulin sensitivity. Hyperglycemic clamp showed that long-term soy protein consumption stimulates to a lesser extent insulin secretion by the pancreas, even in the presence of saturated fat, when compared with animals fed the casein diet. This effect was associated with a decrease in pancreatic islet area, insulin, and PPARγ (peroxisome proliferator activator receptor gamma) mRNA abundance, which in turn prevents the induction of GLUT-2 (glucose transporter) and, therefore, insulin secretion. These effects were also observed in isolated pancreatic islets incubated with the amino acid concentration resembling those found after soy protein consumption as well as with isoflavones. Our data also reveled that infusion of isoflavones rapidly reduced insulin secretion in vivo. Furthermore, animals fed soy protein showed an improvement in insulin sensitivity measured by euglycemic clamp when compared with those fed the casein diet, indicating that these animals require less insulin to maintain glucose in the normal range. These results suggest that insulin secretion is regulated by the type and concentration of amino acids sensed by the pancreas as well as the presence of isoflavones [39].

Soy Protein and Lipotoxicity

Obesity is characterized by an increase in adipose tissue mass to compensate for an excessive triglyceride accumulation. However the adipocytes become dysfunctional and release excessive amounts of free fatty acids, leptin and tumor necrosis factor alpha (TNF)-α to non-adipocyte tissues to counteract the free fatty acid overload. Leptin resistance appears and non adipose tissues such as liver, skeletal muscle and pancreas accumulate excessive amounts of non-oxidized lipids. These lipids are shuttled to the formation of ceramids, a process termed lipotoxicity [51].

We have demonstrated that rats fed a soy protein diet can control serum and hepatic lipid concentration by modulating serum insulin concentration. However, it was not known whether soy protein had the same effect on hepatic steatosis in animal models that develop hyperinsulinemia. The obese Zucker diabetic rat (ZDF obese fa/fa) develops hyperglycemia, hyperinsulinemia, hyperlipidemia and hepatic steatosis. Our results showed that the consumption of a soy protein diet by ZDF obese fa/fa rats ameliorates fatty liver and decrease hepatic cholesterol and triglyceride levels to a greater extent than a casein diet does, despite the fact that the rats had high serum insulin concentration [55]. The consumption of soy protein reduced hepatic triglyceride levels by 50%, the same extent reported in obese mice with inactivation of SREBP-1c. In addition, dietary soy protein reduced serum cholesterol and triglycerides accompanied with LDL cholesterol and VLDL triglycerides. The accumulation of lipids in the liver could be regulated by different transcription factors involved in the expression of genes for fatty acid biosynthesis. SREBP-1c has been established as the principal regulator of hepatic fatty acid biosynthesis [55].

As a result of low hepatic SREBP-1 mRNA concentration in rats fed soy protein, there was a concomitant reduction of FAS mRNA abundance, leading to a decrease in endogenous fatty acid biosynthesis. Additionally, low SREBP-1 expression also reduced the abundance of SCD-1 (stearoyl-CoA desaturase) and Δ5 and Δ6 desaturase mRNAs. SCD-1 is one of the enzymes whose expression is increased in ob/ob mice and corrected by leptin administration [12]. SCD-1 mRNA levels were highly increased in this ZDF fa/fa rats fed the casein diet, whereas rats fed the soy protein diet showed an ~80% decrease, indicating that this enzyme can be regulated by SREBP-1 in spite of hyperleptinemia. The reduction in SCD-1 expression by soy protein could be associated with subsequent reductions in hepatic lipids and VLDL synthesis. In the liver, fatty acids are esterified with glycerol to form triglycerides or with cholesterol to form cholesteryl esters that can: *1)* accumulate, leading to increased hepatic lipid content; *2)* be packaged into VLDL for transport to other tissues; or *3)* be oxidized. Therefore, if SCD-1 is low, there will be low monosaturated fatty acids, the enzymatic products of SCD-1 required for phospholipid and cholesteryl ester synthesis, leading to low VLDL production [55].

OBESITY

Chronic consumption of high fat or carbohydrate diet is associated with the development of obesity, however it is not well established weather dietary protein place a role in the development of abnormalities of lipid metabolism that occur during obesity [47]. It is clearly established that long-term consumption of a high-fat diet accelerates the development of obesity. However, the type of some nutrients, such as dietary protein (**Figure 6**), can play an important role in the progression of obesity. We have demonstrated that rats fed a high fat-soy protein diet for 180 days gained significantly less weight than those fed a high fat-casein diet, and the difference was mainly due to a lower body fat content rather than changes in body protein content or cumulative energy intake [47]. The difference in weight gain between these groups was not associated with the quality of the protein consumed, because rats fed with casein or soy protein grew at the same rate during the first 50 d of the study, when the protein requirement is higher than during the maintenance period. The possible reason why different types of proteins in a high-fat diet may contribute more or less to the development of obesity could be in part due to their specific amino acid patterns on the release of hormones such as insulin [39] and possibly glucagon [53].

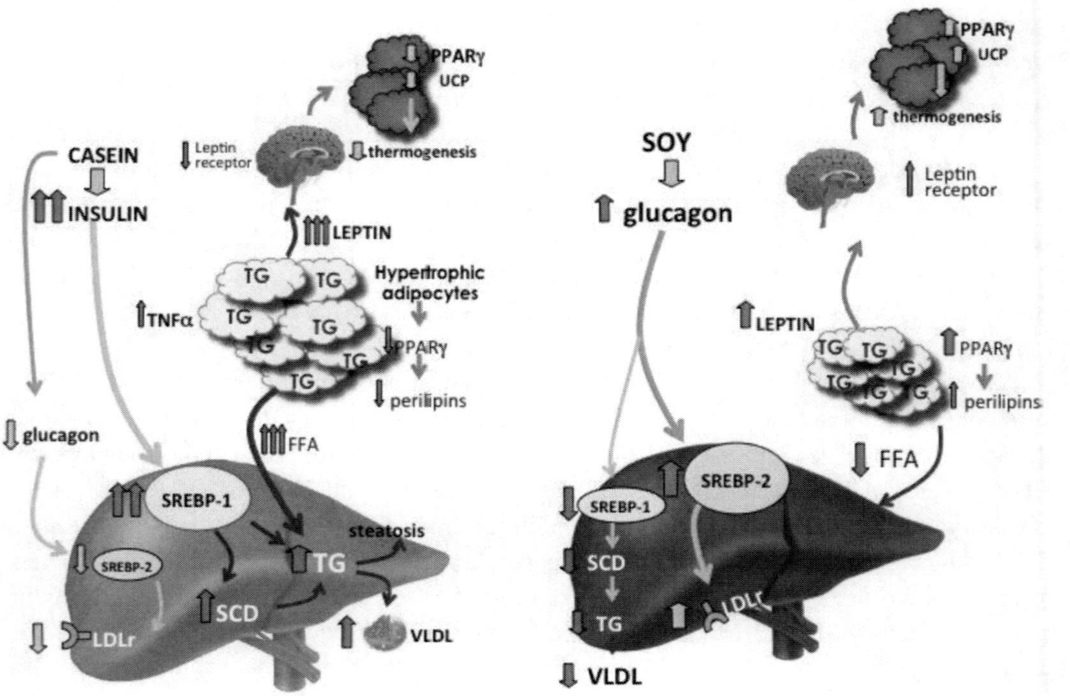

Figure 6. The type of dietary protein differentially modulates the expression of lipogenic genes in several organs and tissues. Soy protein reduces the insulin/glucagon ratio leading in a decrease of SREBP-1 and its target gene SCD-1 and increase SREBP-2 and LDLr in liver preventing the appearance of hepatic steatosis. In addition, soy protein up-regulates PPARγ in white and brown adipose tissue maintaining their functionality of the adipose tissue. In addition, soy protein increases the expression of leptin receptor improving leptin sensitivity leading in a less weight gain. Finally there is an increase in UCP in brown adipose tissue increasing fatty acid oxidation. PPARγ, peroxisome proliferator activated receptor; UCP, uncoupling protein; TNFα, tumor necrosis factor alpha; FFA, free fatty acid; SREBP, sterol receptor binding protein; TG, triglycerides; LDLr, low density lipoprotein receptor; SCD-1, stearoyl-CoA desaturase; VLDL, very low density lipoprotein.

During the development of obesity, there are gradual changes in serum insulin and leptin concentration that, in a given time, will result in hyperinsulinemia or hyperleptinemia. Furthermore, in vivo and in vitro studies with pancreatic islets have shown that soy protein or its isoflavones stimulate insulin secretion in a lower extent than casein [39]. At the end of this study, insulin concentration increased in rats fed high-fat diets, whereas serum leptin concentration increased depending on fat content and the type of protein consumed. These results indicate that the consumption of a specific type of dietary protein can modulate the insulin: glucagon ratio, which in turn can control hepatic lipogenesis mediated by SREBP-1. As a consequence, there was decreased accumulation of in the liver and reduced assembly and release of VLDL-TG particles in rats consuming soy protein and high fat-soy protein diets. It has been shown that insulin upregulates SREBP-1 activity and expression, whereas glucagon has the opposite effect. Our results indicated that the insulin:glucagon ratio could be modified by consumption of soy protein despite the high-fat content in the diet preventing lipid deposition in the liver and the development of hepatic steatosis [47].

In addition, leptin resistance may contribute to the development of fatty liver and its progression to cirrhosis and fibrosis [14]. We observed that rats fed high fat-casein diet had fatty liver and collagen deposition, associated with increased body weight, and hyperleptinemia that induces leptin resistance. Rats fed soy protein even in the presence of a high-fat diet had serum leptin concentrations in the normal range, resulting in low hepatic lipid accumulation due to the leptin action to stimulate FA oxidation [47].

Serum glucagon concentration was also correlated with SREBP-2 (sterol receptor element binding protein) mRNA abundance, a transcription factor that regulates the expression of many genes involved in cholesterol synthesis and uptake in liver [22]. When cellular sterol levels fall, SREBP-2 is transcriptionally active and activates cholesterogenic target genes to increase synthesis and uptake of cholesterol in the liver. Our results showed that soy protein consumption decreased liver cholesterol and increased SREBP-2 [47].

The establishment of obesity is accompanied by concomitant alterations not only in the liver but also in adipose tissue, where adipocytes become dysfunctional. Functional adipocytes release several adipocytokines to protect nonadipose tissues from lipid overload. Long-term consumption of high-fat or high-carbohydrate diets induces adipocyte hypertrophy and eventually dysfunction, characterized by an excessive release of free fatty acids (FFA) to circulation [13]. Our results showed that soy protein also may play a role in the adipocyte-decreasing serum FFA associated with small adipocytes.

The release of fatty acids from adipocytes depends of the rate of TG lipolysis mediated by the hormone-sensitive lipase. The activity of this enzyme on the lipids stored in adipocytes is restricted by the abundance of perilipins surrounding the lipid droplet membrane. Interestingly, groups fed soy protein and high-fat- soy protein showed higher perilipin expression than those fed casein and high-fat casein, limiting the release of fatty acids [47]. At the end of the study, only rats fed the high-fat casein diet showed hyperleptinemia and it was associated with lower amounts and mRNA abundance of leptin receptor in the arcuate nucleus, indicating less sensitivity to the leptin action in the hypothalamus as a result of leptin resistance. Because the hypothalamus controls BAT (brown adipose tissue) thermogenic activity, the decrease in leptin action in rats fed casein and high fat-casein diets was associated with reduced PPARγ (peroxisome proliferator activated receptor) and UCP-1 (uncoupling protein) mRNA in BAT. A major characteristic of brown adipocytes is the presence of multilocular lipid droplets enriched with mitochondria. However, rats fed high fat-casein resulted in numerous lipid-filled unilocular adipocytes. These results suggest that BAT from rats fed high fat-casein diet oxidizes fatty acid to a lesser extent than BAT from rats fed high fat-soy protein and as a consequence, fatty acid storage and lipid droplet area increased, resembling white adipose tissue morphology. These results may explain in part the higher body weight gain in rats fed high fat-casein. Stated evidence demonstrated how long-term consumption of soy protein is associated with beneficial effects on liver and adipose tissue despite the consumption of a high-fat diet [47].

There has been little evidence that dietary protein type may influence the gene expression profile and its consequences on the biochemistry and physiology of the adipose tissue, particularly during the consumption of a high-fat diet. Our results show that the type of dietary protein has a significant influence on adipose tissue phenotype. Soy protein diets reduced adipocyte size due to a reduction in the triglyceride/DNA ratio, indicating an increase in fat cell number and thus preventing hypertrophy. In contrast, casein diets produced the opposite effect. We believe that these changes are in part the result of hormonal modifications. Soy protein, by maintaining normal serum insulin and leptin concentrations, leads to an improved sensitive response to these hormones in target organs and tissues, including adipocytes. Several studies have shown that serum leptin not only regulates hypothalamic and peripheral tissues but also regulates adipocyte metabolism by a paracrine mechanism. Microarray analysis revealed an important finding that soy protein consumption modified the expression of 90 genes involved in metabolic functions and inflammatory response in adipose tissue and that one of the central modified genes (or nodes) was leptin [17]. Leptin expression was down-regulated in the adipose tissue of rats fed soy or soy high fat diets as compared to those fed casein or casein high fat diet. Addition of fat to diet caused an up-regulation of leptin in adipose tissue and decreased FAS and acetyl-CoA carboxylase (ACC) expression. Furthermore, down- regulation of leptin expression by soy protein was accompanied by a decrease in the expression of the hormone-sensitive lipase (HSL), an important enzyme in the regulation of adipocyte lipolysis with no significant change in the expression of uncoupling protein 1, which is involved in the thermogenic process. It has been recently described that central action of adipokines such as leptin has a direct communication with white adipose tissue to regulate lipolysis [17].

ANTIOXIDANT ACTIVITY

Soybean contains many ingredients that are considered to have health benefits. Besides soy protein, some phytochemical compounds have an important role in health. Isoflavones have antioxidant properties. The antioxidant activity has been assessed using three methods, namely human low-density lipoprotein (LDL) oxidation, the ferric reducing-antioxidant power (FRAP) and the anti-DPPH free radical assays. It was found that soybean was rich in genistein and daidzein but they were present mainly in form of glycosides. Both the FRAP and the anti-DPPH assays demonstrated that these glycosides, as well as genistin, daidzein, glycitin, malonyl glycitin and malonyl genistein possessed similar antioxidant activities to their corresponding aglycones, genistein and daidzein. In contrast, the antioxidant potency of these glycosides was much weaker than their corresponding aglycones, genistein and daidzein, in the LDL oxidation assay. The present results demonstrate that genistein and daidzein were less effective, as antioxidants in the three assays, than two well-defined antioxidants, green tea epicatechin and α-tocopherol. It is concluded that soybean isoflavones and their glycosides possess antioxidant activity but they are ineffective antioxidants compared with tea epicatechins and α-tocopherol [30].

In addition to its hypocholesterolemic properties, soy protein may reduce kidney damage by a second mechanism involving soy isoflavones. The main isoflavones present in the soy protein, genistein and daidzein, may reduce glomerular damage during nephrosis by protecting LDL particles from oxidation [60], although their antioxidant capacity is limited. Also, isoflavones can react with reactive oxygen species. These species are produced by neutrophils as part of the inflammatory response in kidney during NS. The main reactive oxygen species are hydrogen peroxide and superoxide anion, which react with chloride and nitric oxide, leading to the formation of hypochlorous acid and peroxynitrite. Both compounds react with tyrosine residues on proteins, producing chlorinated and nitrated proteins with abnormal function. Because isoflavones have a phenolic ring, they can react competitively with the hypochlorous acid and peroxynitrate, reducing the chlorination and nitration of proteins [9], preventing glomeruli damage. Therefore, soy protein may reduce kidney damage during NS by reducing circulating concentrations of cholesterol and by preventing the production of abnormal proteins as a consequence of the inflammatory response [54].

Isoflavones as Regulators of Cholesterol Metabolism

Several biological effects of soy protein have been attributed to a group of phytochemicals associated with this protein that are known as isoflavones. Recent evidence has demonstrated that isoflavones are involved in the mechanism by which soy protein reduces serum cholesterol.

Cholesterol metabolism is controlled by several transcription factors that regulate the expression of genes involved in elimination of cholesterol via the formation of bile acids and also of genes involved in reverse transport cholesterol which remove the excess of intracellular cholesterol into the circulation or into the intestinal lumen. The key transcription factor is the liver X receptor (LXR). There are two isoforms of LXR designated as LXR-α and LXR-β. These transcription factors are nuclear receptors that are naturally activated by oxysterols that are formed from cholesterol. Thus, when cholesterol increases in the cell, LXR isoforms are activated to eliminate the excess of cholesterol, since it is known that elevation of intracellular cholesterol can damage the cells [66]. In addition, LXR-α also regulates the expression of genes involved in fatty acid and triglyceride biosynthesis. Isoflavones, particularly genistein is able to regulate LXR via activation of adenosine monophosphate kinase (AMPK), which in turn phosphorylate LXR. [18].

Interestingly, it has been shown that AMPK can phosphorylate both isoforms of LXR, but phosphorylation of LXR-α down-regulate its biological activity resulting in a decrease expression of its target genes, including genes of fatty acid biosynthesis, whereas phosphorylation of LXR-β up-regulate its biological activity increasing the expression of its target genes such as those for bile acid synthesis and reverse cholesterol transport. Thus, these findings can now explain in part how isoflavones, particularly genistein, can decrease lipogenesis in organs such as the liver, but at the same time can increase the elimination of cholesterol increasing fecal cholesterol and bile acids excretion leading in low serum cholesterol concentration [18].

Soy Polysaccharides and Glycemic Index

Soy polysaccharides are a non-gel-forming fiber source composed of various types of fibers. They improve the glucose tolerance curve in adult healthy subjects and improve the return in plasma glucose levels toward baseline in obese noninsulin dependent diabetic patients [56]. Dietary fiber has many beneficial effects in glucose tolerance because it delays gastric emptying and alters the rate of nutrient absorption.

Soy products with low carbohydrate concentrations are a good food option for patients diabetes. This is because they are considered with low glycemic index (GI). GI is a numeric way to indicate how fast or slow the carbohydrates reach the blood stream. Soy products, like soy milk (6 g protein/200-250 ml), are considered food with low GI, gaining a score of 55 or less. Viscosity is augmented by its fiber and oil content, consequently reducing the absorption rate and preventing glucose from rapidly elevating in the blood stream. This helps the pancreas to avoid an exhaustive insulin secretion and helps keep normal glucose concentration [48].

Phytic Acid

Phytic acid is another component of soy with known beneficial effects. It has a mineral binding capability in the gut that decreases bioavailability of Fe, Ca, Zn and Cu, which can indirectly alter plasmatic cholesterol. Chelation of Fe may suppress oxidative damage to lipids and proteins in blood, decreasing initiation of atherosclerotic lesion formation [41]. This molecule can also reduce the risk of colon cancer due to its antioxidant effect [49].

The Use of Soy Protein in the Development of Dietary Portfolio for a Specific Population

As mentioned above, legumes have historically acted as a critical source of food insurance against times of famine. However, nowadays, due to the high protein quality of soybeans, they can be combined with many other functional foods to create specific dietary portfolios with health benefits in the control of chronic degenerative diseases such as obesity, diabetes or metabolic syndrome. A dietary portfolio is the combination of 2 or more functional foods to normalize the biochemical abnormalities of diseases related with the nutrition. Recently, a dietary portfolio was developed with soy protein, oat, dehydrated nopal and chia seed for metabolic syndrome in the Mexican population [19]. Diagnosis of metabolic syndrome requires that three of the following criteria be satisfied: elevated waist circumference (population and country-specific definitions), elevated triglyceride concentrations (>1.69 mmol/L, low high density lipoprotein cholesterol (<1.04 mmol/L in men and 1.30 mmol/L in women), elevated fasting glucose (>5.6 mM), and systolic and diastolic blood pressure >130 mm Hg and 85 mm Hg, respectively. The consumption of this dietary portfolio showed decreases in serum triglyceride, C-reactive protein, and insulin AUC (area under the curve). The most interesting finding in this study was that individuals with a specific polymorphism in the ABCA1 gene (R230C) were better responders to this dietary portfolio. This polymorphism has been associated with low levels of serum HDL, obesity and Type 2 Diabetes in the Mexican population [1]. The additional beneficial effects of this dietary portfolio in patients with the R230C polymorphism in the ABCA gene were greater reduction in body weight and a greater increase in serum adiponectin concentration [19]. The study shows the importance of soybean in a specific dietary portfolio for individual needs.

Conclusion

Soybean is a legume with high-quality protein. For this reason, during the early 70's, the consumption of soy protein in combination with other vegetable proteins were used to fight against malnutrition in developing countries. In the last three decades, advances in molecular biology have permitted to understand the molecular mechanism of action of this legume. Soybean has been one of the most studied legumes in the world. Because of that, there are numerous amounts of information available regarding soybean research *in vitro*, in animals, and in humans. The main components of soybean are oil, fiber, soy protein, and isoflavones. All components of soy are used in the food technology industry, for human use, and for biodiesel. Several meta-analyses have shown that soy protein has beneficial effects on human health by decreasing serum lipids. We have focused to study the possible molecular mechanism of action of soy protein in liver, adipose tissue, kidney, and pancreas in different models such as obesity, renal disease, lipotoxicity and diabetes. Our initial findings demonstrated that rats fed soy protein diet that meets the protein requirement in rodents increased in a lesser extent insulin concentration than animals fed casein diet. This, in turn, also induced in a lesser extent the gene expression of the transcription factor SREBP-1, which is involved in fatty acid synthesis also called lipogenesis. Thereafter, we found that not only the amino acid pattern of soy protein, but also the isoflavones tightly bounded to the soy protein, were also responsible for the decrease in insulin secretion by the pancreas. The reduction in insulin secretion was associated with a reduction in the accumulation of lipids in liver, muscle, heart and adipose tissue. These results in part explained the hypolipidemic effect found in some animal and human studies. Recently, we have demonstrated that also soy isoflavones regulate the expression of genes involved in cholesterol metabolism and transport through the transcription factor LXR reducing serum cholesterol concentration. Long-term studies with growing animals have demonstrated that soy protein is able to sustain growth at the same rate that casein diet, indicating that it is a high quality protein. In addition, in the adult life, animals fed soy protein diet gained less body weight due to a less body fat deposition even in the presence of a high saturated fat diet. An interesting finding was that these rats had lower serum leptin concentration than rats fed a casein high-fat diet, despite similar energy intake. Rats fed the soy high fat diet had significantly smaller adipocytes than those fed the casein high-fat diet. Microarray analysis revealed that soy protein consumption modified the expression of 90 genes involved in lipid metabolism, inflammatory response and renin-angiotensin system in adipose tissue. The network analysis showed that soy protein regulated the expression of leptin and it was identified as a central regulator of the expression of lipid metabolism genes in adipose tissue. Hence, soy protein maintains the size and functionality of adipose tissue through biochemical adaptations, adipokine secretion, and global changes in gene expression.

Taking advantage of the functional properties of soybean protein, we developed specific dietary portfolios (a combination of functional foods) with the purpose of controlling biochemical abnormalities associated with specific chronic degenerative diseases. Soy protein, in combination with other functional foods, synergizes the effect of a single component to provide a beneficial outcome by reducing total serum cholesterol and triglycerides in hyperlipidemic subjects or reducing glucose intolerance, serum insulin, C-reactive protein, and triglycerides concentrations in subjects with metabolic syndrome. These dietary portfolios could be more personalized if the genetic variants of specific populations are taken into account. Recent work has shown the susceptibility of Hispanics to metabolic diseases related to their Native American heritage. A frequent cholesterol transporter ABCA1 gene variant (R230C), apparently exclusive to Native American individuals, was associated with low HDL-C levels, obesity, and type 2 diabetes in Mexican Mestizos. However these subjects were better responders to dietary treatments with soy protein and soluble fiber than subjects with the R230R genotype.

This work shows how the nutrigenomics help to study the mechanism of action of a nutrient through the regulation of gene expression, protein and metabolites formation in different organs and disease models after the consumption of soy protein. Also, the knowledge of the mechanism of action of a nutrient can help to design and develop new products with the purpose to have new dietary strategies to control the development of chronic degenerative disease. Finally, with the tools provided by the nutrigenetics, it will possible to design a personalized nutrition.

REFERENCES

[1] Acuña-Alonzo V, Flores-Dorantes T, Kruit JK, Villarreal-Molina T, Arellano-Campos O, Hünemeier T, Moreno-Estrada A, Ortiz-López MG, Villamil-Ramírez H, León-Mimila P, Villalobos-Comparan M, Jacobo-Albavera L, Ramírez-Jiménez S, Sikora M, Zhang LH, Pape TD, Granados-Silvestre Mde A, Montufar-Robles I, Tito-Alvarez AM, Zurita-Salinas C, Bustos-Arriaga J, Cedillo-Barrón L, Gómez-Trejo C, Barquera-Lozano R, Vieira-Filho JP, Granados J, Romero-Hidalgo S, Huertas-Vázquez A, González-Martín A, Gorostiza A, Bonatto SL, Rodríguez-Cruz M, Wang L, Tusié-Luna T, Aguilar-Salinas CA, Lisker R, Moises RS, Menjivar M, Salzano FM, Knowler WC, Bortolini MC, Hayden MR, Baier LJ, Canizales-Quinteros S. A functional ABCA1 gene variant is associated with low HDL-cholesterol levels and shows evidence of positive selection in Native Americans. *Hum Mol Genet*, 2010. 19(14): p. 2877-85.

[2] Al-Wahsh I.A., Horner H.T., Palmer R.G., Reddy M.B., Massey L.K. Oxalate and phytate of soy foods. *J Agric Fod Chem*. 2005. 53(14): p. 5670-4.

[3] Anderson, J.W., B.M. Johnstone, and M.E. Cook-Newell, Meta-analysis of the effects of soy protein intake on serum lipids. *N Engl J Med*, 1995. 333(5): p. 276-82.

[4] Anderson, J.W., B.M. Smith, and C.S. Washnock, Cardiovascular and renal benefits of dry bean and soybean intake. *Am J Clin Nutr*, 1999. 70(3 Suppl): p. 464S-474S.

[5] Anderson, J.W. and H.M. Bush, Soy protein effects on serum lipoproteins: a quality assessment and meta-analysis of randomized, controlled studies. *J Am Coll Nutr*, 2011. 30(2): p. 79-91.

[6] Ascencio, C., Torres, N., Isoard-Acosta, F., Gómez-Pérez, F.J., Hernández-Pando, R., Tovar, A.R. Soy protein affects serum insulin and hepatic SREBP-1 mRNA and reduces fatty liver in rats. *J Nutr*, 2004. 134(3): p. 522-9.

[7] Awad A.B., Fink C.S. Phytosterols as anticancer dietary components: evidence and mechanism of action. *J Nutr*. 2000. 130(9): p. 2127-30.

[8] Barnes, S., The biochemistry, chemistry and physiology of the isoflavones in soybeans and their food products. *Lymphat Res Biol*, 2010. 8(1): p. 89-98.

[9] Boersma, B.J., Patel, R.P., Kirk, M., Jackson, P.L., Muccio, D., Darley-Usmar, V.M., Barnes, S. Chlorination and nitration of soy isoflavones. *Arch Biochem Biophys*. 1999. 368(2): P. 265-75.

[10] Bourges, H., Casanueva, E., Rosado, J.L. (2009). Recomendaciones de ingestión de nutrimentos para la población mexicana. México: INNCMSZ.

[11] Bressani, R., et al., Protein quality of a soybean protein textured food in experimental animals and children. *J Nutr*, 1967. 93(3): p. 349-60.

[12] Cohen, P., and Friedman J.M. Leptin and the control of metabolism: role for stearoyl-CoA desaturase-1 (SCD-1). *J Nutr*. 134: p. 1455-2463.

[13] DeFronzo, R.A. Dysfunctional fat cells, lipotoxicity and type 2 diabetes. *Int J Clin Pract*. 2004. 143: p. 9-21.

[14] Ding, X., Saxena, N.L., Lin, S., Xu, A., Srinivasan, S., Anania, F.A. The roles of leptin and adiponectin: a novel paradigm in adipocytokine regulation of liver fibrosis and stellate cell biology. *Am J Pathol*. 2005. 166: p. 1655-69.

[15] Food and Drug Administration. Health claims meeting significant scientific agreement (SSA). Available from: *http://www.fda.gov/Food/LabelingNutrition/LabelClaims/HealthClaimsMeetingSignificant ScientificAgreementSSA/default.htm*

[16] Friedman, M., Brandon, D.L. Nutritional and health benefits of soy proteins. *J Agric Food Chem*. 2001. 9(3): p. 1069-86.

[17] Frigolet, M.E., Torres, N., Uribe-Figueroa, L., Rangel, C., Jimenez-Sanchez, G., Tovar, A.R. White adipose tissue genome wide-expression profiling and adipocyte metabolic functions after soy protein consumption in rats. *J Nutr Biochem*. 2011. 22(2): p. 118-29.

[18] González-Granillo, M., Steffensen, K.R., Granados, O., Torres, N., Korach-André, M., Ortíz, V., Aguilar-Salinas, C., Jakobsson, T., Díaz-Villaseñor, A., Loza-Valdes, A., Hernandez-Pando, R., Gustafsson, J.Å., Tovar, A.R. Soy protein isoflavones differentially regulate liver X receptor isoforms to modulate lipid metabolism and cholesterol transport in the liver and intestine in mice. *Diabetologia*. 2012. 55(9): p. 2469-78.

[19] Guevara-Cruz, M., et al., Increase in HDL-C concentration by a dietary portfolio with soy protein and soluble fiber is associated with the presence of the ABCA1R230C variant in hyperlipidemic Mexican subjects. Mol Genet Metab, 2010. 101(2-3): p. 268-72.

[20] Gunstone, F.D., ed. Vegetable Oils in Food Technology: Composition, Properties and Uses. 1 ed., ed. R.J. Hamilton. 2002, Blackwell Publishing.

[21] Harwood J.P., Ausman L.M., King N.W., Sehgal P.K., Nicolosi R.J., Liener I.E., Donatucci D., Tarcza J. Effect of long-term feeding of soy-based diets on the pancreas of Cebus monkeys. *Adv Exp Med Biol*. 1986. 199: p. 223-37.

[22] Horton, J.D., Goldstein, J.L., Brown, M.S. SREBPs: activators of the complete program of cholesterol and fatty acid synthesis in the liver. *J Clin Invest*. 2002. 109: p. 1125-31.

[23] Hughes G.J., Ryan D.J., Mukherjea R., Schasteen C.S. Protein digestibility-corrected amino acid scores (PDCAAS) for soy protein isolates and concentrate: criteria for evaluation. *J Agric Food Chem*. 2011. 59(23): p. 12707-12.

[24] Hurrel R.F., Juillerat M.A., Reddy M.B., Lynch S.R., Dassenko S.A., Cook J.D. Soy protien, phytate, and iron absorption in humans. *Am J Clin Nutr*. 1992. 56(3): p. 573-8. 10.

[25] Jenkins, D.J., et al., Effects of a dietary portfolio of cholesterol-lowering foods vs lovastatin on serum lipids and C-reactive protein. *JAMA*, 2003. 290(4): p. 502-10.

[26] Jenkins, D.J., et al., Assessment of the longer-term effects of a dietary portfolio of cholesterol-lowering foods in hypercholesterolemia. *Am J Clin Nutr*, 2006. 83(3): p. 582-91.

[27] Kibar, H., Ozturk, T., Physical and mechanical properties of soybean. *Int Agrophysics*, 2008. 22: p. 239-244.

[28] Kontessis, P., Jones, S., Dodds, R., Trevisan, R., Nosadini, R., Fioretto, P., Borsato, M. Sacerdoti, D., Viberi, G. Renal, metabolic and hormonal responses to ingestion of animal and vegetable proteins. *Kidney Int*. 1990. 38: p. 136-144.

[29] Kremmyda L.S., Tvrzicka E., Stankova B., Zak A. Fatty acids as biocompounds: their role in human metabolism, health and disease: a review. part 2: fatty acid physiological roles and applications in human health and disease. *Biomed Pap Med Fac Univ Palacky Olomouc Czech Repub*. 2011. 155(3): p. 195-218.

[30] Lee, C.H., Yang, L., Xu, J.Z., Yeung, S.Y.V., Huang, Y., Chen, Z.Y. Relative antioxidant activity of soybean isoflavones and their glycosides. *Food Chemistry*. 2005. 90(4): p. 735-741.

[31] Lee, J.D., K.D. Bilyeu, and J.G. Shannon, Genetics and breeding for modified fatty acid profile in soybean seed oil. *J. Crop Sci. Biotech*., 2007. 10: p. 201-210.

[32] Lichtenstein, A.H., Soy protein, isoflavones and cardiovascular disease risk. *J Nutr*, 1998. 128(10): p. 1589-92.

[33] Lopez-Huertas, E. Health effects of oleic acid and long chain omega-3 fatty acids (EPA and DHA) enriched milks. A review of intervention studies. Pharmacol Res. 2010. 61(3): p. 200-7.

[34] Messina M.J. Legumes and soybeans: overview of their nutritional profiles and health effects. *Am J Clin Nutr*. 1999. 70(3 Suppl): p. 439S-450S.

[35] Millward DJ, Layman DK, Tomé D, Schaafsma G. Protein quality assessment: impact of expanding understanding of protein and amino acid needs for optimal health. *Am J Clin Nutr*, 2008. 87(5): p. 1576S-1581S.

[36] Nagata C, Shimizu H, Takami R, Hayashi M, Takeda N, Yasuda K. Soy product intake is inversely associated with serum homocysteine level in premenopausal Japanese women. *J Nutr*, 2003. 133(3): p. 797-800.

[37] National Center for Biothechnology Information. Taxonomy Browser. Available from: *www.ncbi.nlm.nih.gov*.

[38] National Soybean Research Laboratory. Soy Benefits. Available from: *http://www.nsrl.uiuc.edu/soy_benefits.html*

[39] Noriega-López, L., Tovar, A.R., Gonzalez-Granillo, M., Hernández-Pando, R., Escalante, B., Santillán-Doherty, P., Torres, N. Pancreatic insulin secretion in rats fed a soy protein high fat diet depends on the interaction between the amino acid pattern and isoflavones. *J Biol Chem*. 2007. 282(28): p.20657-66.

[40] North Carolina Soybean Producers Association, Inc. The history of soybean. Available from: *http://www.ncsoy.org/ABOUT-SOYBEANS/History-of-Soybeans.aspx*.

[41] Potter, S.M., Overview of proposed mechanisms for the hypocholesterolemic effect of soy. *J Nutr*, 1995. 125(3 Suppl): p. 606S-611S.

[42] Schaafsma, G., The protein digestibility-corrected amino acid score. *J Nutr*, 2000. 130(7): p. 1865S-7S.

[43] Shimomura, I., Bashmakov, Y., Ikemoto, S., Horton, J.D., Brown, M.S., Goldstein, J.L. Insulin selectively increases SREBP-1c mRNA in the liver of rats with streptozotocin induced diabetes. *Mol Cell Biol*. 1999. 19: p. 3760-3768.

[44] Soyatech. News and Intelligence for the Soybean and Oilseed Industries. Available from: *www.soyatech.com*.

[45] Soyinfo Center. Available from: *http://www.soyinfocenter.com/HSS/history.php*.

[46] Soystats. *A reference guide to important soybean facts and figures*.; Available from: .

[47] Torre-Villalvazo I., Tovar, A.R., Ramos-Barragán, V.E., Cerbón-Cervantes, M.A., Torres, N. Soy protein ameliorates metabolic abnormalities in liver and adipose tissue of rats fed a high fat diet. *J Nutr*. 2008. 138(3): p. 462-8.

[48] Torres y Torres N., Palacios-González B., Noriega-López L., Tovar-Palacio A.R., [Glycemic, insulinemic index, glycemic load of soy beverage with low and high content of carbohydrates]. *Rev Invest Clin*. 2006. 58(5): p. 487-97

[49] Torres y Torres, N. and A.R. Tovar-Palacio, [The importance of soy in Mexico, its nutritional value and effect on health]. *Salud Publica Mex*, 2009. 51(3): p. 246-54.

[50] Torres, N., I. Torre-Villalvazo, and A.R. Tovar, Regulation of lipid metabolism by soy protein and its implication in diseases mediated by lipid disorders. *J Nutr Biochem*, 2006. 17(6): p. 365-73.

[51] Torres, N., Torre-Villalvazo, I., Tovar, A.R., Future directions in reducing hepatic lipotoxicity. *Future Medicine*. 2006. 1(3). p. 331-341.

[52] Tovar, A.R., Santos, A., Halhali, A., Bourges, H., Torres, N. Hepatic histidase gene expression responds to protein rehabilitation in under-nourished growing rats. *J Nutr*. 1998. 128: p. 1631-1635.

[53] Tovar, A.R., C. Ascencio, and N. Torres, Soy protein, casein, and zein regulate histidase gene expression by modulating serum glucagon. *Am J Physiol Endocrinol Metab*, 2002. 283(5): p. E1016-22.

[54] Tovar AR, Murguía F, Cruz C, Hernández-Pando R, Aguilar-Salinas CA, Pedraza-Chaverri J, Correa-Rotter R, Torres N. A soy protein diet alters hepatic lipid metabolism gene expression and reduces serum lipids and renal fibrogenic cytokines in rats with chronic nephrotic syndrome. *J Nutr*, 2002. 132(9): p. 2562-9.

[55] Tovar AR, Torre-Villalvazo I, Ochoa M, Elías AL, Ortíz V, Aguilar-Salinas CA, Torres N. Soy protein reduces hepatic lipotoxicity in hyperinsulinemic obese Zucker fa/fa rats. *J Lipid Res*. 2005. 46(9): p. 1823-32.

[56] Tsai AC, Vinik AI, Lasichak A, Lo GS. Effects of soy polysaccharide on postprandial plasma glucose, insulin, glucagon, pancreatic polypeptide, somatostatin, and triglyceride in obese diabetic patients. *Am J Clin Nutr*, 1987. 45(3): p. 596-601.

[57] U.S. Department of Agriculture. Agricultural Research Service. USDA Database for the Isoflavone Content of Selected Foods. Available form: *http://www.ars.usda.gov/nutrientdata/isoflav)*

[58] United States Department of Agriculture. Nutrient Data Laboratory: National Nutrient Database for Standard Reference. Available from: *http://ndb.nal.usda.gov/*

[59] van Raaij JM, Katan MB, Hautvast JG, Hermus RJ. Effects of casein versus soy protein diets on serum cholesterol and lipoproteins in young healthy volunteers. *Am J Clin Nutr*, 1981. 34(7): p. 1261-71.

[60] Wiseman H., O'Reilly J.D., Adlercreutz H., Mallet A.I., Bowey E.A., Rowland I.R., Sanders T.A. Isoflavone phytoestrogens consumed in soy decrease F(2)-isoprostane

concentrations and increase resistance of low-density lipoprotein to oxidation in humans. *Am J Clin Nutr*. 2000. 72(2): p. 395-400.

[61] Wolf, W.J., Legumes: Seed Composition and Structure, Processing Into Protein Products and Protein Properties. United States Department of Agriculture: p. 291-314.

[62] Wolf, W.J., Physical and chemical properties of soybean proteins. *J Am Oil Chem Soc*, 1977. 54(2): p. 112A-117A.

[63] World Soy Foundation. Soy is Part of the Solution. Available from: *http://www.worldsoyfoundation.org/whysoy/solution.html*

[64] Young V.R., Puig M, Queiroz E, Scrimshaw NS, Rand WM. Evaluation of the protein quality of an isolated soy protein in young men: relative nitrogen requirements and effect of methionine supplementation. *Am J Clin Nutr*. 1984. 39(1): p. 16-24.

[65] Zhan, S. and S.C. Ho, Meta-analysis of the effects of soy protein containing isoflavones on the lipid profile. *Am J Clin Nutr*, 2005. 81(2): p. 397-408.

[66] Zhao C., Dahlman-Wright K., Liver X receptor in cholesterol metabolism. *J Endocrinol*. 2010. 204(3): p. 233-40.

In: Seeds as Functional Foods and Nutraceuticals
Editors: R. Mora-Escobedo, J. De J. Berrios et al.
ISBN: 978-1-62808-489-4

Chapter 4

LEGUME PULSES AS A SOURCE OF FUNCTIONAL DIETARY FIBER INGREDIENTS

Ma. Montaña Cámara Hurtado[1], Ma. Cortés Sánchez-Mata[1], Patricia Morales Gómez[1] and José De J. Berrios[2]
[1]Dpto. Nutrición y Bromatología II. Facultad de Farmacia. Universidad Complutense de Madrid (UCM). Madrid, Spain
[2]USDA-ARS-WRRC, Processed Foods Research Unit, Albany, CA, US

ABSTRACT

Dietary fiber is currently defined as carbohydrate polymers with 10 or more monomeric units, which are not hydrolyzed by the endogenous enzymes in the small intestine of humans and belong to the following categories: edible carbohydrate polymers naturally occurring in the food consumed; carbohydrate polymers, which have been obtained from food raw material by physical, enzymatic or chemical means and which have been shown to have a physiological benefit to health, as demonstrated by generally accepted scientific evidence to competent authorities; synthetic carbohydrate polymers that have been shown to have a physiological benefit to health, as demonstrated by generally accepted scientific evidence to competent authorities. Plant components that fall within this definition include non-starch polysaccharides such as celluloses, some hemi-celluloses, gums and pectins, as well as lignin, resistant dextrins and resistant starches. Non-digestible oligosaccharides are also included in the fiber fraction of foods. Fiber intake can be obtained from foods naturally rich in dietary fiber, such as pulse seeds, or through the consumption of foods fortified with fiber as a functional ingredient. Although, still there is no official consensus on what constitutes a functional food, in 2011 the Functional Food Center has adopted a new definition of functional food "Functional Food is a Natural or processed food that contains known biologically-active compounds which when in defined quantitative and qualitative amounts provides a clinically proven and documented health benefit, and thus, an important source in the prevention, management and treatment of chronic diseases of the modern age". In this sense, it is well known that insufficient consumption of fiber in Western societies is closely linked to certain health problems, such as the so-called developed society's diseases. This chapter includes the latest information on the use of pulses as source of dietary fiber and its components as novel or functional ingredients, and the evaluation and application of their technological and biological functionality.

1. Introduction

Fiber Definition and Constituents

The term "dietary fiber" was introduced for the first time by Hipsley in 1953 [1], as non-digestible constituents of the cell walls, the definition of this concept has since then substantially evolved. In the 70's, different researchers established the nature of the substances taking part in dietary fiber, such as polysaccharides and lignin, and began the process of developing a worldwide consensus on fiber definition and analytical methodology for dietary fiber. Since 1985, AOAC Method 985.29 has been globally adopted as a standard method for determination of total dietary fiber in foods. The dietary fiber definition has been dependent on the analytical method used. Some enzymatic methods include the residue obtained from hydrolysis of food components using different enzymes and conditions. Several modifications to analytical methods of dietary fiber led to include the entire components determined as dietary fiber, such as resistant starch. In the early 90's, the consensus on the physiological dietary fiber definition was reaffirmed, as well as the inclusive components, with the agreement of AOAC International and FAO/WHO. Since 2000, the idea of functionality in the human body has been often included in the definition of fiber making reference to health benefits of fiber in the body, as laxation and modulation of blood cholesterol and glucose, as well as the presence of lower molecular weight carbohydrates (as some non-digestible oligosaccharides) and some non-carbohydrate substances. In 2002, the Panel on the Definition of Dietary Fiber of the National Academy of Science (USA) differentiated three concepts on fiber definition: **Dietary Fiber** (non-digestible carbohydrates and lignin) that are intrinsic and intact in plants; **Functional Fiber,** as isolated, non-digestible carbohydrates which have beneficial physiological effects in humans; and **Total Fiber** consisting on the sum of Dietary Fiber and Functional Fiber [2].

In 2009, the Codex Alimentarius Commission defined dietary fiber as carbohydrate polymers[a] with ten or more monomeric units[b], which are not hydrolyzed by the endogenous enzymes in the small intestine of humans and belong to the following categories:

- Edible Carbohydrate polymers naturally occurring in the food as consumed,
- Carbohydrate polymers, which have been obtained from food raw material by physical, enzymatic or chemical means and which have been shown to have a physiological effect of benefit to health as demonstrated by generally accepted scientific evidence to competent authorities,
- Synthetic carbohydrate polymers which have been shown to have a physiological effect of benefit to health as demonstrated by generally accepted scientific evidence to competent authorities.
  a) When derived from a plant origin, dietary fiber may include fractions of lignin and/or other compounds when associated with polysaccharides in the plant cell walls and if these compounds are quantified by the AOAC gravimetric analytical method for dietary fibre analysis: Fractions of lignin and the other compounds

(proteic fractions, phenolic compounds, waxes, saponins, phytates, cutin, phytosterols, etc.) intimately "associated" with plant polysaccharides in the AOAC 991.43 method. These substances are included in the definition of fiber insofar as they are actually associated with the poly- or oligo-saccharidic fraction of fiber. However when extracted or even re-introduced in to a food containing non digestible polysaccharides, they cannot be defined as dietary fiber. When combined with polysaccharides, these associated substances may provide additional beneficial effects (pending adoption of Section on Methods of Analysis and Sampling).

b) Decision on whether to include carbohydrates of 3 to 9 monomeric units should be left up to national authorities (http://www.codexalimentarius.net/download/report/710 /al32_26e.pdf).

After these changes in the fiber concept, dietary fiber is now recognized as food material, particularly plant material, that is not hydrolyzed by enzymes secreted by the human digestive tract, but that may be digested by gut microflora. Plant components that fall within this definition include non-starch polysaccharides (NSP) such as celluloses, some hemi-celluloses, gums and pectins, as well as lignin, resistant dextrins and resistant starches [3]. In this sense, fiber involves a complex concept, including not only non-starch polysaccharides, but also other non-digestible carbohydrates and associated non-carbohydrate compounds. According to the criteria followed by EFSA, the concept of dietary fiber should be referred to as non-digestible carbohydrates plus lignin (table 1), including non-starch polysaccharides (NSP) – cellulose, hemicelluloses, pectins, hydrocolloids (i.e., gums, mucilages, β-glucans), resistant oligosaccharides – fructo-oligosaccharides (FOS), galacto-oligosaccharides (GOS), other resistant oligosaccharides, resistant starch – consisting of physically enclosed starch, some types of raw starch granules, retrograded amylose, chemically and/or physically modified starches, and lignin associated with the dietary fibre polysaccharides [4]. Examples of oligosaccharides that fall under the category of dietary fiber are those that are normally constituents of a dietary fiber source, such as raffinose, stachyose, and verbascose in legumes, and the low molecular weight fructans in foods, such as Jerusalem artichoke and onions. Also some non-carbohydrate fiber-associated substances such as waxes, cutin, and some polyphenols (which have recently raised great interest due to their *in situ* antioxidant potential at the gut level), have been suggested to take part in this non-digestible fraction of plant foods, and responsible of some of the beneficial effects of fiber [5].

Due to the complexity of the fiber components, several classifications of dietary fiber are considered, based on their chemical, analytical, and physiological characteristics, among others. One of the most accepted classification of fiber components are related to their solubility in water system at physiological conditions, into insoluble fiber (IF) and soluble fiber (SF).

The main components of insoluble dietary fiber are cellulose, some hemicelluloses, resistant starch, and lignin. Cellulose is a polymer of β-(1-4)-D-Glucose, while hemicelluloses consist of long chains of cellulose bound to other monosaccharides including hexoses, pentoses as arabinose or rhamnose, and uronic acids. The digestibility of an insoluble dietary fiber depends upon its amount and ratio of cellulose to hemicellulose, which affects the physical structure of the fiber and its transit time in the gut. Lignin is a non-carbohydrate polymer with a complex structure including phenylpropane molecules with methoxyl and other functional groups, intimately formed with the cellulose of plant cell walls, and very resistant to digestion [6].

Table 1. Total, insoluble and soluble fiber content in legume pulses (g/100g dwb)

Common name	Latin name	IDF	SDF	TDF	Country of origin	Reference
Chickpea	*Cicer arietinum* L.	23.95	2.05	26.00	Spain	[7]
	Cicer arietinum L.	10.00	8.4	18.40	USA	[8]
	Cicer arietinum L.	-	-	25.50 (NDF); 9.53 (ADF)	Pakistan	[9]
	Cicer arietinum L.	19.54	1.51	21.05	Spain	[10]
	Cicer arietinum L.	24.90	1.30	26.20	Philippines	[11]
	Cicer arietinum L.	20.48	0.96	21.44	Spain	[12]
	Cicer arietinum L.	15.37	0.75	16.12	USA	[13]
Flat pod pea vine	*Lathyrus cicera* L.	-	-	6.7 – 7.3*	Various	[14]
Chickling pea, Grass pea	*Lathyrus sativus* L.	-	-	5.3 – 10.0*	Various	[14]
Lentils	*Lens* sp.	-	-	6.75*	Egypt	[15]
	Lens sp.	-	-	13.50 – 15.00*	Canada	[16]
	Lens esculenta L.	-	-	29.80 (NDF); 9.52 (ADF)	Pakistan	[9]
	Lens esculenta L.	15.30	0.71	16.7	India	[17]
	Lens culinaris Med.	-	-	6.83*	Brazil	[18]
	Lens culinaris L.	10.36	0.47	15.642	USA	[13]
White Lupin	*Lupinus albus* L.	-	-	14.50*	Unspec.	[19]
	Lupinus albus L.	4.02	0.52	4.54	Unspec.	[20]
	Lupinus albus L.	-	-	17.10*	Africa	[21]
Yellow Lupin	*Lupinus luteus* L.	-	-	20.20*	S. Africa	[21]
	Lupinus madrensis Seem.	-	-	15.40*	Mexico	[22]
French Beans	*Phaseolus vulgaris* L.	15.5	3.7	19.2	Spain	[23]
Haricot bean	*Phaseolus vulgaris* L.	16.13	4.22	20.35	Spain	[7]
Red kidney bean	*Phaseolus vulgaris* L.	-	-	27.60 (NDF); 7.43 (ADF)	Pakistan	[9]
White kidney bean	*Phaseolus vulgaris* L.	-	-	24.30 (NDF); 9.32 (ADF)	Pakistan	[9]
	Phaseolus vulgaris L.	20.35	2.37	21.50	India	[17]
	Phaseolus vulgaris L.	-	-	22.20*	Unspec.	[24]
	Phaseolus vulgaris L.	20.6-24.5	5.95-9.96	11.16-17.13	Polan	[25]
	Phaseolus vulgaris L.	22.14	5.82	26.96	Spain	[12]
Pink-mottled cream bean	*Phaseolus vulgaris* L.	16.37	5.43	21.81	Spain	[12]
Dry Green Peas	*Pisum sativum L.*	-	-	3.31*	Spain	[26]
	Pisum sativum L.	11.30	8.70	20.00	USA	[8]
	Pisum sativum L.	-	-	5.50*	Canada	[27]

Common name	Latin name	IDF	SDF	TDF	Country of origin	Reference
	Pisum sativum L.	-	-	4.25	Egypt	[15]
	Pisum sativum L.	-	-	8.30*	S. Africa	[21]
	Pisum sativum L.	-	-	10.4*	Brazil	[28]
	Pisum sativum L.	20.30	1.73	22.03	Brazil	[18]
	Pisum sativum L.	28	2.10	29.70	Philippines	[11]
	Pisum sativum L.	10.88	0.27	11.14	USA	[13]
Broad bean	*Vicia faba L.*	-	-	15.40*	India	[29]
	Vicia faba L.	27.68	1.37	28.50	India	[17]
	Vicia faba L.	-	-	13.30*	S.Africa	[21]
	Vicia faba L.	-	-	5.67*	Pakistan	[30]
Black gram	*Vigna mungo L.*	26.23	4.69	30.92	India	[7]
	Vigna mungo L.	-	-	23.4 (NDF); 11.7 (ADF)	Pakistan	[9]
	Vigna mungo L.	11.4	-	11.40*	Unspec.	[24]
Green gram	*Vigna radiata L..*	19.26	1.22	20.49	India	[7]
	Vigna radiata L.	-	-	20.00*	Unspec.	[6]
Moth bean	*Vigna aconitifolia Jacq.*	17.33	1.21	18.54	India	[7]
	Phaseolus aconitifolins	-	-	16.30*	Unspec.	[24]
Horse gram	*Macrotyloma uniflorum Lam.*	21.61	0.86	22.47	India	[7]

IDF: Insoluble dietary fiber); SDF: Soluble dietary fiber; TDF: Total dietary fiber; *Crude Fiber according to AOAC (enzymatic-gravimetric method); dwb: dry weight basis.

On the other hand, the main components of soluble dietary fiber are pectins, some hemicelluloses, gums, mucilages, inulin, as well as non-digestible oligosaccharides. Pectic substances are considered to be a mixture of arabinogalactans, branched polymers with a galactouronic acid. Polymers of soluble fiber dissolve readily in water to form a gel and maximize the viscosity of the food matrix, slowing down digestion; it also helps to lower postprandial blood glucose, insulin and serum cholesterol levels [31]. Non-digestible oligosaccharides are other components of soluble fiber, which are also involved in some physiological mechanisms responsible for the health benefits of dietary fiber

2. Fiber Components in Leguminous Seeds

Dietary fiber in legume pulses is composed of cellulose, which represents 16-30 % of the total dietary fiber content depending on the degree of seeds maturation; hemicellulosic polysaccharides (3-26 %) and pectic substances (pectin, galactan and araban). These components have important physiological roles, which have direct beneficial effects on lipidemia by inducing a reduction in plasma cholesterol and the risk of atherosclerosis in animal and human studies [32]. Legumes have also been used as coadyuvants for diabetes treatment, due to the reduction on glucose intestinal absorption induced by fiber polysaccharides, as well as the presence of amylase inhibitors which slow starch hydrolysis [32]. Human studies where people were fed diets containing lupin kernel fiber have shown that this fiber has potential to bulk and dilute the contents of the upper gastrointestinal tract, delay gastric emptying and slow orocecal transit, thus prolonging the intestinal absorption of nutrients. These gastrointestinal effects may lead to benefits for the dietary control of non-insulin-dependent diabetes mellitus [33] and obesity [34]. From this point on, the terms "legume", "pulses", and/or "legume pulses" will be used interchangeable throughout this chapter.

Dietary fiber polysaccharides are more abundant in the hull than in the cotyledons of legume seeds. The cell walls of the cotyledons are non-lignified and contain a range of polysaccharides, including 55 percent pectic substances, 9 percent cellulose, and non-starch non-cellulosic glucans (in the range 5–12%); while the seed coat contains large quantities of cellulose (ranging from 35 to 57%) and smaller amounts of hemicelluloses and pectins [35]. This difference in dietary fiber composition lead to changes in important physicochemical properties; as for example, the higher water-retention capacity of the cotyledon fiber fractions compared to the lower oil-binding capacity of this fiber fractions, which is in the same range as that of hull fibers [36-37; 13; 38].

Carbohydrate fraction of pulses is composed mainly by starch, soluble sugars and the components of dietary fiber, with total dietary fiber contents of about 3-30 % (dry weight basis). This wide variation in TDF is due to either genetic or environmental factors, or the effect of some food processing conditions [10; 39]. Table 2 shows the total dietary fiber (TDF), insoluble dietary fiber (IDF) and soluble dietary fiber (SDF) values in different raw legume seeds. The TDF content in the seed of chickpea, lentils, common beans, and dry green peas range between 16.12 – 26.20 %, 6.5 – 29.80 %, 20.35 – 27.60%, and 3.31 – 29.70%, respectively. Additionally, the TDF content in the seed of other less common legumes such as lupins, grass peas, and black gram range between 4.54 – 30.92 %, 5.3 - 10.0 %, and 11.40 – 30.92%, respectively.

The IDF fraction in legume seeds range widely between 4 to 28 % dry weight basis. This wide range in IDF content is due to the effect of environmental conditions, seed maturity, and other factors (table 2). The IDF constitutes the major portion of legumes´ dietary fiber. In the case of haricot bean, the IDF fraction constitutes around 79.26% of TDF; while for broad beans and chickpea these values are between 95 – 97% of their TDF, respectively. Table 2 present important published data on the fiber fractions of main commercial pulses, including resistant starch. Different studies highlight that hemicellulose, and in lesser extent cellulose, constitute the major components of the IDF fraction of pulse's seeds. Rehman & Shah [9] reported hemicellulose contents of 16.0, 20.2 and 20.3 mg/100g, in chickpea, red kidney and white kidney bean, respectively; the cellulose fraction represented only a content of 8.45, 6.22 and 8.00 mg/100g, respectively, in the same legume seeds. Ligin content was found in even lower values of IDF in legume seeds, ranging in values between 1.08 mg/100g in chickpea to 6.60 mg/100g in black grams. Also, some difference in lignin content was observed within the same legume species; e.g., in the case of *Phaseolus vulgaris,* the lignin content ranged between 1.32 to 2.38 mg/100g and in chickpea it ranged between 1.08 to 2.49 mg/100g.

Resistant starch (RS) is the portion of starch and starch products that resist digestion as they pass through the small intestinal tract of healthy individuals. The EURESTA definition describes RS in terms of its physiological functionality. A wide number of beneficial physiological effects have been attributed to RS, such as the improvement of glycaemic and insulinaemic response, bowel health, blood lipid profile, and micronutrient absorption among other effects. The RS that reach the large intestine are fermented into different products as carbon dioxide, methane, hydrogen, organic acids (e.g. lactic acid) and short chain fatty acids

(SCFA) [40]. A unique characteristic of RS is that it can deliver some of the benefit of IDF as well as those of SDF. To support EURESTA's physiological definition of RS and take into account the physical characteristics of the various resistant starches, four types of RS have been categorized.

These resistant starches include the physically protected form of starch found naturally in legume seeds and unprocessed whole grains (RS1), starch that occurs in granular form (RS2), retrograded starch obtained by heating and cooling (RS3), and modified starches obtained by chemical treatments (RS4). Table 2 presents the RS content of some main commercial pulses. As observed, the values of RS varied in the range of 2.45 % for dry pea to 8.31 % for red kidney beans. The RS values of all other pulses were between the above indicated values.

Table 2. Contribution of different fiber components present in legume pulses (g/100g dwb)

Common name	Latin name	Cellulose	Hemicellulose	Lignin	Resistant starch	Pectin	Country of origin	References
Chickpea	*Cicer arietinum L.*			2.49*	4.17		Spain	[7]
	Cicer arietinum L.	8.45	16.0	1.08			Pakistan	[9]
	Cicer arietinum L.				3.39		Brazil	[18]
	Cicer arietinum L.			2.24*			Spain	[7]
	Cicer arietinum L.	6.5	5.5	2.1		2.7	Pakistan	[41]
	Cicer arietinum L.	3.70	2.69	2.17		4.21		[42]
Lentils	Lens sp.				3.25			[18]
	Lens sp.			4.12*			Spain	[10]
	Lens esculenta	8.10	20.3	1.42			Pakistan	[9]
	Lens esculenta	2.9	1.0	2.5		2.9	Pakistan	[42]
Red kidney bean	*Phaseolus vulgaris* L.			2.38*	8.31		Spain	[7]
White kidney bean	*Phaseolus vulgaris* L.	6.22	20.2	1.21			Pakistan	[9]
Haricot bean	*Phaseolus vulgaris* L.	8.00	14.9	1.32			Pakistan	[9]
	Phaseolus vulgaris L.				3.7		Brazil	[18]
	Phaseolus vulgaris L.	5.9	2.0	1.4		2.3	Pakistan	[42]
Black grams	*Vigna aconitifolia* Jacq.			4.57*	3.90		Spain	[7]
	Vigna mungo			6.60*			Spain	[7]
Moth bean	*Vigna mungo*	9.70	12.0	1.70	3.40		Pakistan	[9]
Green gram	*Vigna radiata* L.			4.40*	4.18		Spain	[7]
	Vigna unguiculeta	6.0	3.9	2.0		1.8	Pakistan	[42]
Horse gram	*Macrotyloma uniflorum* Lam.			5.61*	5.21		Spain	[7]
Pea	*Pisum sativum*				2.45		Spain	[10]
	Pisum sativum	2.4	1.0	2.5		2.4	Pakistan	[42]

* Klason lignin Method; dwb: dry weight basis.

As previously mentioned, the SDF is not the predominant dietary fiber fraction in legumes. However, the RS content in legumes is associated to important health benefits for the human body. Khatoon and Prakash [17] reported that the SDF fraction in lentils and French beans ranged from 0.61% to 2.37%. Similarly, Berrios et al. [13] reported that the concentration of SDF fraction in lentils, chickpeas and dry peas was significantly different ($P < 0.05$) with values in the range of 0.27 to 0.75%.

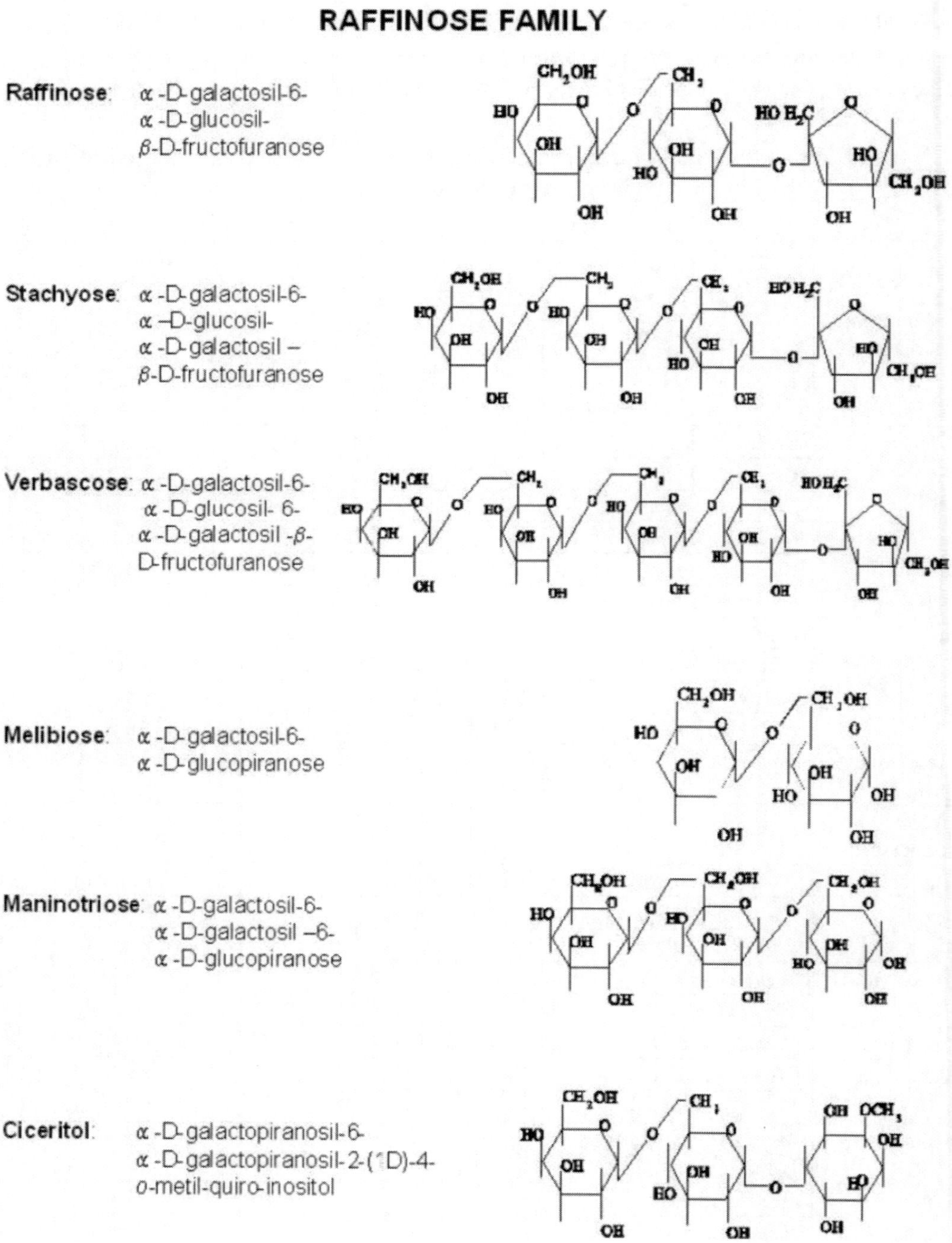

Figure 1. Chemical structures of main non-digestible oligosaccharides present in legume pulses.

The SDF fraction present in legume pulses contains high concentration of a very important group of non digestible oligosaccharides. These compounds are mainly α-galactosides of the raffinose family (containing galactose with α-D-1,6-linkages), present in levels varying between 1.7-7.9 % [43; 44]. Galactosides derived from sucrose, such as raffinose, stachyose, and verbascose represent the most studied sugars in legumes pulses (figure 1). The non digestible oligosaccharides content in legume pulses are presented in table 3. It is observed that the concentration of stachyose is consistently higher than raffinose, which is a physiological characteristic of the legume pulse seeds of *Fabaceae* family. The pentasaccharide verbascose has been identified in chickpeas (0.19-0.72%), as well as in different types of beans, lentils, and dry peas in levels ranging from 0.24 to 2.19 % [43; 45; 46; 47].

Another group of α-galactosides on legume pulses are the glucose galactosides (melibiose and manninotriose), and inositol galactosides (galactinol, galactopinitol, and ciceritol). Ciceritol, a galactoside derived from pinitol (α-D-galactopiranosil-6-α-D-galactopiranosil-2-(1D)-4-O-methyl-*chiro*-inositol) was previously wrongly reported as manninotriose and identified in chickpeas, from which it take his name (*Cicer arietinum* L.). Ciceritol is the α-galactoside most abundant in legume pulses (4 g/100 g). It is even more abundant than stachyose in some pulse cultivars [48] and it has been also identified in lentil and white lupin [54-56].

The presence of α-galactosides is associated to various important functions in seeds of many plants. According to García & Mendoza [52], some of these functions are: being storage and transport mechanism for carbohydrates; protecting vegetables against toxic effects of lectins and other toxins (as well as against desiccation); acting as germination inhibitors in dry conditions; and they can also play a role in cold acclimation of many plants.

The biosynthesis of α-galactoside takes place during seed maturation, as reserve carbohydrates, because they can be hydrolyzed by plant α-galactosidase, which activity increases during seed germination to allow for the utilization of galactose by the embryo [53]. Several studies have been carried out on the determination of oligosaccharides in legume pulses. These studies have reported a wide variation in the soluble sugar composition of different legumes, even from the same botanical species and variety (table 3). These differences are generally attributed to genetic and environmental factors.

2.1. Effect of Processing on Legume Fiber Components

Legume fibers can be modified by food-processing operations (Table 4), such as grinding or extrusion, to improve their functional properties. These functional modifications could be attributed to physicochemical changes induced by the variation on insoluble/soluble fiber ratios, due to processing treatments [54].

Culinary or technological processes, such as soaking, cooking, fermentation and extrusion, may also affect α-galactoside content in legumes by dilution in the medium and/or transformations in the carbohydrate fraction that may include hydrolysis of α-galactosides. Several studies done on beans, lentils, chickpeas, and other legume seeds have reported the differences on α-galactosides transformations depending on the conditions applied during processing [12; 55; 56]. One of these studies analyzed the liquids obtained after soaking and cooking of chickpeas. The study reported that the soaking process did not extract appreciable amounts of raffinose and stachyose, but cooking processing could extract up to 26-47 % of α-galactosid

Table 3. Non Oligosaccharides content in main commercial legume pulses (g/100g dwb)

Common name	Latin name	Raffinose	Ciceritol	Stachyose	Verbascose	Total galactosides	Country of origin	References
Chickpea	*Cicer arietinum L.*	0.57-0.63	2.51-2.79	0.74-1.17			Spain	[48]
	Cicer arietinum L.	0.63-2.79	0.86-3.93	0.74-1.17			Spain	[55]
	Cicer arietinum L.	1.45		2.56	0.19			[45]
	Cicer arietinum L.	0.39	2.03	1.4		3.82		[10]
	Cicer arietinum L.	0.65	3.83	2.29		2.94	Spain	[43]
	Cicer arietinum L.	0.86		1.91	0.72	3.49	India	[46]
Lentils	Lens sp.	0.73-0.79	1.35-1.82	1.63-1.71			Spain	[48]
	Lens sp.	0.22	1.05	1.64		2.91		[10]
	Lens culinaris L.	0.27	2.30	2.24	1.65	4.17	USA	[57]
	Lens culinaris L.	0.29	1.28	2.26	2.19	4.75	Spain	[43]
Lupins	*Lupinus albus*	0.46		6.30	0.36		Spain	[58]
	Lupinus albus	0.63	0.59	6.13	1.17	7.93	Spain	[43]
	Lupinus angustifolius	0.98		4.61	1.51		Spain	[58]
	Lupinus luteus	0.59		7.36	3.09		Spain	[88]
White bean	*Phaseolus vulgaris* L.	0.27-0.51		1.65-2.49			Spain	[48]
Pinto bean	*Phaseolus vulgaris* L.	0.25-0.34		1.96-2.06			Spain	[48]
	Phaseolus vulgaris L.	0.64		4.14	0.24		Spain	[47]
	Phaseolus vulgaris L.	0.35		1.41		1.76	USA	[44]
	Phaseolus vulgaris L.	0.41		1.49				[59]
Dry Pea	*Pisum sativum*	0.48		0.64			Spain	[48]
	Pisum sativum	0.92		2.30	1.40		Spain	[47]
	Pisum sativum	1.14		2.36	0.60	4.10		[60]
Cowpea	*Vigna unguiculata* L. Walp	1.03		1.78	0.36	3.17	India	[46]
Horse gram	*Macrotyloma uniflorum* Lam	0.71		1.56	0.41	2.68	India	[46]

Table 4. Effect of thermal and pressure treatments on total dietary fiber (TDF) content in legume pulses

Common name	Latin name	Traditional soaking	Traditional cooking	High pressure	Autoclaving	Extrusion	Microwave
Chickpea	*Cicer arietinum* L.	No significant effect after treatment [61]	↓33.35% [63]; ↑ 17-21%[15]; ↑20.94%[45];	↓37.36% [63]	↓41.36% [63]; ↑29.84% [45]	↓37.58% [63]	↑1.2% [64]; ↓2% [63]; ↑26.17%[45]; ↓2.5%[13]
Flat pod pea vine	*Lathyrus cicera* L.						
Lentils	*Lens culinaris* L.	No significant effect after treatment [61]; ↑10-26% [61]					↓15.40%[13]
White Lupin	*Lupinus albus* L.		↑ 26% [15];				
Common beans	*Phaseolus vulgaris* L.	No significant effect after treatment [61]	↑ 1.4% [64];				↑ 2.2 %[63]
Dry Green Peas	*Pisum sativum* L.						↑16.15%[13]
Broad bean	*Vicia faba* L.						
Black gram	*Vigna mungo* L.	No significant effect after treatment [61]	↓6% [9];				
Green gram	*Vigna radiata* L.		↓22.88% [63]	↓22,52% [63]	↓20.05% [63]	↓25.72% [63]	
Cowpea	*Vigna unguiculata* L. Walp	↑20.51% [62]	↑30.03% [62]	↓26%[17]			

Values in table are expressed as the TDF percent reduction after treatment (g/100g, dwb).

3. PHYSIOLOGICAL EFFECTS OF DIETARY FIBER

Although there is still no official consensus on what constitutes a functional food [65-67], in 1994 the Institute of Medicine's Food and Nutrition Board defined functional foods as "*any food or food ingredient that may provide a health benefit beyond the traditional nutrients it contains*" [68].

The main physiological effects of dietary fiber on human body take place in the digestive tract. At stomach level, some of its components are involved in gel formation, causing gastric distension and slowing gastric emptying and satiety perception; with the consequent influence on weight gain control. However, more complex mechanisms take place at the intestinal level (figure 2), where the intake of adequate amounts of fiber has shown beneficial effects on the regulation of intestinal transit and prevention of related digestive disorders including diverticular disease, haemorrhoids, constipation and colon cancer prevention. Dietary fiber has also shown a positive effect on innate immune responses of the gut mucosa both directly and indirectly, by mechanisms involving the gut-associated lymphoid tissue, with the result of an improvement on intestinal defense systems against pathogens [38; 69]. In addition, soluble fiber has an influence on lipidemia, and the risk of atherosclerosis and heart attacks, as well as on the improvement of glucose tolerance and insulin response in diabetics [38].

Many of these properties are related to water interaction leading to a complex behavior of fibers involving water binding, viscosity, gel formation, and other physicochemical processes of interest from either a technological or a physiological point of view. In this way, the ratio of insoluble to soluble dietary fiber in the seeds may play an important role in human nutrition and potential health benefits [12; 31].

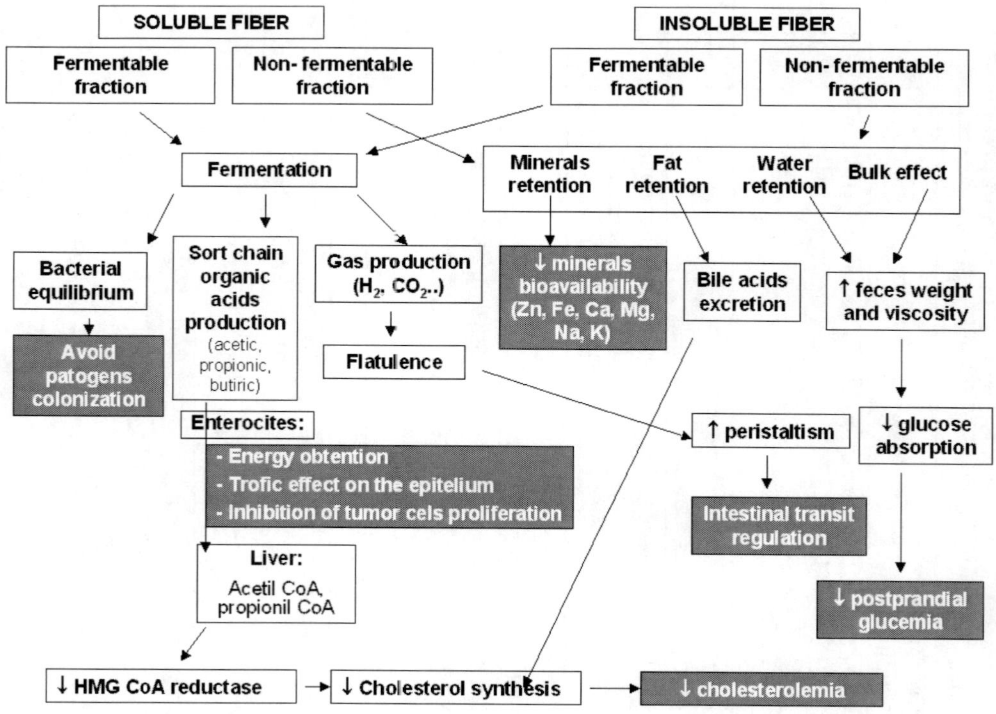

Figure 2. Main intestinal effects of dietary fiber functions in the human body.

Also, many of the health benefits linked to dietary fiber consumption are due to products of the fermentation processes that take place at gut level, leading to positive effects on beneficial gut microflora development and short chain fatty acid production, which contribute to the metabolism of gut epithelium cells. Most fiber types are at least partially fermented; however differences in fermentation patterns can be found. Generally, well-fermented fibers are soluble in water, while partially or poorly fermented fibers are insoluble [2].

The physiological functions of most dietary fibers, including legume fibers, are often attributed to their physiochemical properties, water-holding capacity, swelling, rheological, fat-binding properties and susceptibility to bacterial degradation or fermentation. In fact, the different components of dietary fiber are involved in different mechanisms of physiological effects. Some of the health benefits of SDF, such as lowering cholesterol effect and the rate of glucose absorption and postprandial plasma glucose concentration, have been associated with its ability to increase viscosity of the food undergoing digestion. This increase in viscosity due to SDF, which reduces glucose and/or fat absorption, is usually dependent on the physical characteristics of the fiber rather than its chemical composition; whereas, the susceptibility of SDF to fermentation in the colon is dependent on its chemical structure [38; 70-72].

Raffinose-family oligosaccharides, are particular components of special relevance in legume seeds that tend to normalize bowel functions, increase lactobacilli and bifidobacteria (decreasing enterobacteria in intestinal microflora), and reduce potentially carcinogenic N-nitroso compounds levels in the gut [73-74]. On the other hand, the human digestive system lacks the enzyme α-galactosidase so α-galactosides cannot be hydrolyzed. Therefore, these oligosaccharides accumulate in the lower intestine and undergo anaerobic fermentation by bacterian α-galactosidase, with the production of H_2 and CO_2, and traces of CH_4, causing the flatus effect and sometimes diarrhea and abdominal pain [56; 75]. There are some differences between flatulence inducing potential of α-galactosides. Quemener & Brillouet [50] reported that *in vitro* hydrolysis rate of ciceritol by α-D-galactosidase is much lower than for raffinose, stachyose and verbascose. These authors considered the presence of pinitol as a factor that reduces the sensitivity of the α-galactoside to hydrolysis by α-galactosidase. *In vivo* studies have shown the ability of raffinose, stachyose and verbascose to produce flatulence, while there is no evidence of ciceritol acting as a flatogenic sugar. This fact has been attributed to the differences in their sensitivity to α-galactoside hydrolysis, which is higher for raffinose-family oligosaccharides (mainly stachyose and verbascose), than for pinitol galactosides [56; 75].

Besides the negative flatulence production, laxative effects, and orosmotic diarrhea production caused by the intake of raffinose-family oligosaccharides, these complex sugars have been related to important beneficial effect, such as reduction of glycemia and cholesterol absorption, reduction of the risk of some chronic diseases as some cardiovascular alterations, and reduction of colon carcinogenesis [76]. Due to their physiological effects and actions on colonic microbiota, these compounds are being studied as potential prebiotic substances.

In fact, some fiber components have demonstrated a prebiotic effect, frequently remarked in food labelling as "bifidus effect". The concept of prebiotic was firstly defined by Gibson & Robertfroid in 1995, and reviewed in 2004, as *"a selectively fermented ingredient that allows specific changes, both in the composition and/or activity in the gastrointestinal microflora that confers benefits upon host wellbeing and health"* [77]. According to this definition, a prebiotic compound should meet the three following criteria: 1) resistance to gastric acidity, hydrolysis by mammalian enzymes and gastrointestinal absorption; (2) being fermented by the intestinal microflora; (3) selective stimulation of the growth and/or activity of intestinal bacteria associated with health and wellbeing. There is enough scientific evidence to affirm that compounds with prebiotic effect induce health benefits at different levels: gastrointestinal and immune functions, mineral absorption, lipid metabolism, inflammatory bowel disease and colorectal cancer prevention [69].

In this sense, prebiotic value cannot be attributed to all the components of dietary fiber, but all prebiotics can be included in dietary fiber definition, as the resistance to digestive enzymes and the utilization by intestinal microflora are two of the requirements of prebiotic concept. Currently, although many fiber compounds have been candidates to be classified as prebiotics, only fructans, galactooligosaccharides and lactulose have clearly demonstrated to meet the previously mentioned requirements [69; 78-80]. Fructans include inuline, either native (purified from cichorium root or other plant sources) or synthetic, both with an average degree of polimerization higher than 12; the oligofructose (OF), obtained by hydrolysis of inuline (2-7 units); and fructooligosacarides (FOS), as kestose (GF_2), nistose (GF_3) or fructosyl-nistose (GF_4), naturally present in many vegetables as onions, artichoke, among others, that can also be synthesized by enzymatic techniques. Prebiotic galactoologi-saccharides are not present in plant tissues, but only present in human milk.

As it has been previously mentioned, research has been conducted on pulse α-galactosides to show their potential prebiotic effect; these studies have been conducted mainly in soybean products and their by-products. Espinosa-Martos et al. [80] and Inoguchi et al. [81] studied fermentation of soybean oligosaccharides by *Bifidobacterium bifidum* and *Lactobacillus acidophilus*. They found different pattern of fermentation in different soybean products. These studies concluded on the fermentation of soybean oligosaccharides by bifidobacteria, and non-fermentation by pathogens as *Escherichia coli* and *Clostridium perfringens*. These oligosaccharides are nowadays considered under fiber definition, and their prebiotic effect has been suggested, but more scientific evidence is necessary to affirm this effect according to the concept proposed by Gibson et al. [77].

3.1. Fiber Intake Recommendations

It is known that insufficient consumption of fiber in Western societies is closely linked to certain health problems, such as the so-called developed society's diseases, which include constipation, cardiovascular disease, and cancer, among others [82-84].

Daily dietary fiber intake recommended by international agencies is estimated at 25-30 grams containing 75% SDF and 25% IDF. This recommended intake can be satisfied by consuming foods naturally rich in dietary fiber, such as fruits and vegetables, legumes, nuts, and whole grains; or eating foods with added fiber as a functional ingredient [85].

Whole grain cereals, pulses, fruit, vegetables and potatoes are the main food sources of dietary fiber. Average dietary fiber intakes varied from 10 to 20 g per day in young children (<10 to 12 years), from 15 to 30 g per day in adolescents, and from 16 to 29 g per day in adults [86]. Average intakes of dietary fiber per energy units ranged from 1.7 to 2.5 g per MJ in young children and from 1.8 to 2.9 g per MJ in adults [4].

3.2. Fiber Health Claims

The regulatory instrument governing nutrition and health claims on foods in the European Union is Regulation (EC) No 1924/2006 [87] of The European Parliament and of the Council, December 20, 2006. This regulation aims to protect the right of consumers to receive truthful information about the food they purchase, increase their protection against exaggerated or misleading information, and establishes the following types of statements: "nutrition" or "content", "health claims", "statements of reducing the risk of disease".

The authorization of these claims corresponds to the Authority, which issues the relevant opinions on the basis of "Panel on Dietetic Products, Nutrition and Allergies (NDA)" scientific opinion.

In addition, some of these foods and functional ingredients can be considered as "Novel food" in accordance with Regulation (Ec) No 258/97 [88] of the European Parliament and of the Council, January 27, 1997, concerning novel foods and novel food ingredients. Claims that have EFSA approval could be used in food labeling as an added value for consumer health, which may also improve the market success of the final product(s). Currently, according to European Regulations, pulse seeds can be labeled as "High Fiber Content", since they surpassed 6g of fiber per 100g of product.

The denominations "High Fiber Content", or “Source of Fiber” could be also used for products using pulses as ingredients, providing 6 g/100 g of fiber or 3 g/100 g of fiber, respectively.

References

[1] Hipsley, E. H. (1953). Dietary fibre and pregnancy Toxaemia. *British Medical Journal*, *22*, 420-422.

[2] Tungland, B. & Meyer, D. (2002). Nondigestible oligo and polysaccharides (dietary fiber): Their physiology and role in human health and food. Comprehensive Reviews. *Food Science and Food Safety*, *3*, 73-92.

[3] IFST, The Institute of Food Science & Technology, through its Public Affairs and Technical & Legislative Committees, has authorised the following Information Statement, dated April 2007, updating that of February 2007.

[4] EEFSA Panel on Dietetic Products, Nutrition, and Allergies (NDA); Scientific Opinion on Dietary Reference Values for carbohydrates and dietary fibre. *EFSA Journal*, 2010; *8*(3), 1462 [77]. doi:10.2903/j.efsa.2010.1462. Available online: www.efsa.europa.eu

[5] Saura-Calixto, F. García-Alonso, A. & Goñi I. Bravo, L. (2000). *In vitro* determination of the indigestible fraction in foods: An alternative to dietary fiber analysis. *Journal of Agricultural and Food Chemistry*, *48*, 3342-3347.

[6] Ghavidel, R.A. & Prakash, J. (2007). The impact of germination and dehulling on nutrients, antinutrients, in vitro iron and calcium bioavailability and in vitro starch and protein digestibility of some legume seeds. *LWT*, *40*, 1292-1299.

[7] Bravo, L. Siddhuraju, P. & SauraCalixto, F. (1999). Composition of underexploited Indian pulses. Comparison with common legumes. *FoodChemistry*, *64*, 185192

[8] Dalgetty D. D & Baik, B. K. (2003). Isolation and characterization of composition cotyledon fibers from peas, lentils, and chickpeas. *Cereal Chemistry*, *80*, 310 315.

[9] Rehman, Z. U. & Shah, W. H. (2004). Domestic processing effects on some insoluble dietary fibre components of various food legumes. *Food Chem.*, *87*, 613 617.

[10] Martín-Cabrejas, M. A. Aguilera, Y. Benítez, V. Mollá, E. López-Andréu, F. J. & Esteban, R. M. (2006). Effect of industrial dehydration on the soluble carbohydrates and dietary fibre fractions in legumes. *Journal of Agricultural and Food Chemistry*, *54*, 7652–7657.

[11] Mallillin, A. C. Trinidad, T. P. Raterta, R. Dagbay, K. & Loyola, A. S. (2008). Dietary fiber and fermentability characteristics of root crops and legumes. *British Journal of Nutrition*, *100*, 485–488.

[12] Aguilera, Y. Martín-Cabrejas, M. A. Benítez, V. Mollá, E. López-Andréu, F. J. & Esteban, R. M. (2009). Changes in carbohydrate fraction during dehydration process of common legumes. *Journal of Food Composition and Analyses*, *22*, 678–683.

[13] Berrios, J. Morales, P. Cámara, M. & Sánchez Mata, M. C. (2010). Carbohydrate Composition of Raw and Extruded Pulse Flours. *Journal of Food Research International*, *40*, 531- 536

[14] Hanbury, C. D. White, C. L. Mullan, B. P. & Siddique, K. H. M. (2000). A review of the potential of Lathyrus sativus L. and L. cicera L. for use as animal feed. *Anim. Feed Science and Technoogy*, *87*, 1-27.

[15] El-Adawy, T. A. Rahma, E. H. El-Bedawey, A. A. & El-Beltagy A. E. (2003). Nutritional potential and functional properties of germinated mung bean, pea and lentil seeds. *Plant Foods for Humuman Nutrition*, *58*, 1-13.

[16] Wang, N. & Daun, J. K. (2006). Effect of variety and crude protein content on nutrients and anti-nutrients in lentils (*Lens culinaris*). *Food Chemistry*, *95*, 493-502.

[17] Khatoon, N. & Prakash, J. (2004). Nutritional quality of microwave-cooked and pressure-cooked legumes. *International Journal of Food Science and Nutrition*, *55*, 441–448.

[18] De Almeida Costa, G. E. Queiroz-Monici, K. d. S. Reis, S. M. P. M. & de Oliveira, A. C. (2006). Chemical composition, dietary fiber and resistant starch contents of raw and cooked pea, common bean, chickpea and lentil legumes. *Food Chemistry*, *94*, 327–330.

[19] Wiryawan, I. K. G. (1997). Grain Legumes for Poultry. University of Queensland, Australia: PhD thesis.

[20] Farrell, D. J. & Mannion, P. F. (1997). Final report to Rural Industry. RAD Council Egg Industry and Chicken Meat Programs. Available from: www.rirdc.gov. au/reports/ CME/DAQ-28E.doc

[21] Brand, T. S. Brandt, D.A. & Cruywagen C. W. (2004). Chemical composition, true metabolisable energy content and amino acid availability of grain legumes for poultry. *South African Journal of Animal Science*, *34*, 116-122.

[22] Ruiz, M.A. & Sotelo, A. (2001). Chemical Composition, Nutritive Value, and Toxicology Evaluation of Mexican wild Lupins. *Journal of Agricultural and Food Chemistry*, *49*, 5336-5339.

[23] Martín-Cabrejas, M. A. Jaime, L. Karanja, C. Downie, A. J. Parker, M. L. Lopez-Andreu, F. J. Maina, G. Esteban, R. M. Smith, A. C. & Waldron, K. W. (1999) Modifications to physicochemical and nutritional properties of hard-to-cook beans

(*Phaseolus Vulgaris* L.) by extrusion cooking. *Journal of Agricultural and Food Chemistry*, *47*, 1174-1182.

[24] Kahlon, T. S. Smith G. E. & Shao Q. (2005). In vitro binding of bile acids by kidney bean (*Phaseolus vulgaris*), black gram (*Vigna mungo*), Bengal gram (*Cicer arietinum*) and moth bean (*Phaseolus aconitifolins*). *Food Chemistry*, *90*, 241-246

[25] Korus, J. Gumul, D. & Czechowska, K. (2007). Effect of extrusion on the phenolic composition and antioxidant activity of dry beans of *Phaseolus vulgaris*. *Food Technology and Biotechnology*, *45*, 139-146.

[26] Alonso, R. Grant, G. Dewey, P. & Marzo, F. (2000). Nutritional assessment in Vitro and in Vivo of Raw and Extruded Peas (*Pisum sativum* L.). *J. Agric. Food Chem*, *48*, 2286- 2290

[27] Hickling, D. (2003). Canadian Feed Peas Industry Guide. 1-35. Manitoba, Canada: Pulse Canada: Winnipeg.

[28] Mesquita, V. L. V. & Giada, M. L. R. (2005). Organic-mineral content of the main Brazilian cereals and legumes. *Nutrition and Food Science*, *35*, 264-270.

[29] Saharan, K. Khetarpaul, N. & Bishnoi, S. (2002). Variability in Physico-chemical Properties and Nutrient Composition of Newly Released Ricebean and Fababean Cultivars. *Journal of Food Composition and Analyses*, *15*, 159-167.

[30] Hossein, M. S. & Mortuza, M. G. (2006). Chemical composition of Kalimatar, a locally grown strain of faba bean. Pakist. *Journal of Bioogical Science*, *9*, 1817-1822.

[31] Jenkins, D. J. A. Kendall, C. W. C. & Vuksan, V. (2000). Viscous fibers, health claims, and strategies to reduce cardiovascular disease risk. *The American Journal of Clinical Nutrition*, *71*, 401–402.

[32] Schneeman, B.O. (1987). Soluble vs, Insoluble Fiber-Different Physiological Responses. *Food Technology*, *41*, 81-86.

[33] Brand Miller, J. & Foster-Powell, K. (1999). Diets with a low glycemic index: from theory to practice. *Nutrition Today*, *34*, 64–72.

[34] Burton-Freeman, B. (2000). Dietary fiber and energy regulation. *Journal of Nutrition*, *130*, S272–S274.

[35] Van Laar, H. Tamminga, S. Williams, B. A. Verstegen, M. W. A. & En-gels, M. (1999). Fermentation characteristics of cell-wall sugars from soya bean meal, and from separated endosperm and hulls of soya beans. *Anim. Feed Science and Technology*, *79*, 179–193.

[36] Pfoertner, H. N. & Fischer, J. (2001). Dietary fibers of lupins and other grain legumes. In: McCleary, B.V. Prosky, L. (Eds.), Advanced dietary fiber technology. 361–366 Oxford, England: Blackwell Science.

[37] Guillon, F. & Champ, M. M. J. (2002). Carbohydrate fractions of legumes: uses in human nutrition and potential for health. *British Journal of Nutrition*, *88*, S293–S306.

[38] Tiwari, U. & Cummins, E. (2011). Functional and physicochemical properties of legume fibers. In B. K. Tiwari, A. Gowen, B. McKenna (Ed.) Pulse Foods: Processing, Quality and Nutraceutical Applications. (Food Science and Technology International Series) Academic Press. London Burlington, MA. 121-156.

[39] Wang, N. Lewis, M. J. Brennam, J. G. & Westby, A. (1997). Effect of processing methods on nutrients and anti nutritional factors in cowpea. *Food Chemistry*, *58*, 59–68.

[40] Christl, S. U. Murgatroyd, P. R. Gibson, G. R. & Cummings, J. H. (1992). Production, metabolism, and excretion of hydrogen in the large intestine. *Gastroenterology 102*, 1269–1277.

[41] Riaz-Khan, A. Alam, S. Ali, S. Bibi, S. & Khalil, I.A. (2007) Dietary fiber profile of food legumes. *Sarhad Journal of Agriculture*, *23*, 763-766

[42] Vasishtha, H. & Srivastava, R. P. (2011) Effect of soaking and cooking on dietary fibre components of different type of chickpea genotypes. *Journal of Food Science and Technology*, DOI 10.1007/s13197-011-0366-4

[43] Pedrosa, M. M. Cuadrado, C. Burbano, C. Allaf, K. Haddad, J. Gelencser, E. Takács, K. Guillamón, E. & Muzquiz, M. (2012). Effect of instant controlled pressure drop on the oligosaccharides, inositol phosphates, trypsin inhibitors and lectins contents of different legumes. *Food Chemistry*, *131*, 862-868.

[44] Berrios, J. De J. Camara, M. Torija, M. E. & Alonso, M. (2002). Effect of extrusion cooking and sodium bicarbonate addition on the carbohydrate composition of black bean flours. *Journal of Food Processing and Preservation*, *26*, 113– 128.

[45] Alajaji, S. A. & El-Adawy, T. A. (2006). Nutritional composition of chickpea (*Cicer arietinum* L.) as affected by microwave cooking and other traditional cooking methods. *Journal of Food Composition and Analyses*, *19*, 806-812.

[46] Sreerama, Y. N. Sashikala, V. B. Pratape, W. M. & Singh, V. (2012) Nutrients and antinutrients in cowpea and horse gram flours in comparison to chickpea flour: Evaluation of their flour functionality. *Food Chemistry*, *131*, 462–468.

[47] Alonso, R. Rubio, L. A. Muzquiz, M. & Marzo, F. (2001). The effect of extrusion cooking on mineral bioavailability in pea and kidney bean seed meals. *Animal Feed Science and Technology*, 94, 1-13.

[48] Sanchez-Mata, M. C. Penuela-Teruel, M. J. Camara-Hurtado, M. Diez-Marques, C. & Torija-Isasa, M. E. (1998). Determination of mono-, di-, and oligosaccharides in legumes by high-performance liquid chromatography using an amino-bonded silica column. *Journal of Agricultural and Food Chemistry*, *46*, 3648–3652.

[49] Sosulki, F. W. Elkowicz, L. Reichert, R. D. (1982). Oligosaccharides in eleven legumes and their air-classified protein and starch fractions. *Journal of Food Science*, *47*, 498-502.

[50] Quemener, B. & Brillouet, J. M. (1983). Ciceritol, a Pinitol Digalactoside from Seeds of Chickpea, Lentil and White Lupin. *Phytochemistry*, *22*, 1745-1751.

[51] Bernabé, M. Fenwick, R. Frias, J. Jiménez-Barbero, J. Price, K. Valverde, S. & Vidal-Valverde, C. (1993). Determination, by NMR Spectroscopy, of the Structure of Ciceritol, a pseudotrisaccharide isolated from Lentils. *Journal of Agricultural and Food Chemistry*, *1993*, *41*, 870-872.

[52] García, V. V. & Mendoza, E. M. T. (1992). Raffinose oligosaccharides in legumes. *Comments Agric & Food Chemistry*, *2*, 302-320.

[53] Kuo, T.M. Van Middlesworth, J. F. Wolf, W. J. (1988). Content of raffinose oligosaccharides and sucrose in various plant seeds. *Journal of Agricultural and Food Chemistry*, *36*, 32-36.

[54] Martín-Cabrejas, M. A. Ariza, N. Esteban, R. Mollá, E. Waldron, K. López-Andréu, F. J. (2003). Effect of germination on the carbohy- drate composition of the dietary fiber of peas (*Pisum sativum* L.). *Journal of Agricultural and Food Chemistry*, *51*, 1254–1259.

[55] Sánchez-Mata, M. C. Cámara Hurtado, M. M. & Díez-Marqués, C. (1999). Effect of domestic processes and water hardness on soluble sugars content of chickpeas (Cicer arietinum L.). *Food Chemistry, 65*, 331–338.

[56] Reddy, N. R. Salunkhed, D. K. & Sharma, R.P. (1980). Flatulence in rats following ingestion of cooked and germinates black gram and a fermented product of black gram and rice blend. *Journal of Food Science, 45*, 1161-1164.

[57] Morales, P. Berrios, J. De J. Varela, A. Burbano, C. Cuadrado, C. Muzquiz, M. & Pedrosa, M. M. (2013). Extrusion effect on the oligosaccharides, inositol phosphates, tripsin inhibitiors and lectins content in different lentil flours fiber rich formulation. *Journal of Agricultural and Food Chemistry* (*In press*).

[58] Martínez-Villaluenga, C. Frías, J. & Vidal-Valverde, C. (2005). Raffinose family oligosaccharides and sucrose contents in 13 Spanish lupin cultivars. *Food Chemistry, 91*, 645-649.

[59] Mubarak, A.E. (2005). Nutritional composition and antinutritional factors of mung bean seeds (Phaseolus aureus) as affected by some home traditional processes. *Food Chemistry, 89*, 489–495.

[60] Frias, J. Vidal Valverde, C. Sotomayor, C. Diaz Polla, C. & Urbano, G. (2000). Influence of processing on available treatments carbohydrate content and antinutritional factors of chickpeas. *European Food Research and Technology, 210*, 340-345.

[61] Wang, N. Hatcher, D. W. Tyler, R. T. Toews, R. & Gawalko, E. J. (2010). Effect of cooking on the composition of beans (*Phaseolus vulgaris* L.) and chickpeas (*Cicer arietinum* L.) *Food Research International, 43*, 589-594.

[62] Ragab, H. I. Kijora, C. Abdel Ati, K. A. Danier, J. (2010). Effect of Traditional Processing on the Nutritional Value of Some Legumes Seeds Produced in Sudan for Poultry Feeding. *International Journal of Poultry Science, 9*, 198-204.

[63] Mahadevamma. S. & Tharanathan, R. N. (2004). Processing of legumes: resistant starch and dietary fiber contents. *Journal of Food Quality, 27*, 289-303.

[64] Marconi, E. Ruggeri, S. Capelloni, M. Leonardi, D. & Carnovale, E. (2000). Physicochemical, nutritional, and microstructural characteristics of chickpeas (*Cicer arietinum* L.) and common beans (*Phaseolus vulgaris* L.) following microwave cooking. *Journal of Agricultural and Food Chemistry, 48*, 5986-5994.

[65] Prosky, L. Asp, N.G. Furda, I. De Vries, J. W. Schweizer, T. F. & Harland, B. (1985). The determination of total dietary fiber in food, and food products: Collaborative study. *Journal of the Association of Official Analytical Chemists, 68*, 677 - 679.

[66] ILSI. Conceptos sobre los Alimentos Funcionales. (2004). Belgium: ILSI (internacional Life Science Intitute).

[67] Mazza, G. (2001). Alimentos Funcionales. Ed. Acribia, Zaragoza, Spain.

[68] Institute of Medicine's Food and Nutrition Board (IOM/FNB, 1994). (2010). *EFSA Journal, 8(12)*,1885 [15].

[69] Gibson, G.R. & Roberfroid, M.B. (2008). Handbook of prebiotics. Boca Raton, FL (USA): CRC Press.

[70] Morris, E. R. (2001). Assembly and rheology of non-starch polysaccha-rides. In: McCleary, B.V. Prosky, L. (Eds.), Advanced dietary fiber technology. 30–41 Oxford, England: Blackwell Science.

[71] Tosh, S. M. & Yada, S. (2010). Dietary fibers in pulse seeds and fractions: characterization, functional attributes, and applications. *Food Research International*, *43*, 450–460.

[72] Dikeman, C. L. & Fahey, Jr. G. C. (2006). Viscosity as related to dietary fiber: a review. *Critical Review of Food Science and Nutrition*, *46*, 649–663.

[73] Van Loo, J.A, Cummings, J. Delzenne, N. Englyst, H. Frank, A. & Hpkins, M (1999). Functional food properties of non-digestible oligosaccharides: a consensus report from ENDO project (DGXII AIRII-CT94-1095). *British Journal of Nutrition*, *81*, 121-132.

[74] Rowland, I. Rummey, C. Silvi, S. Cresci, A. & Pool-Zobel, B. (1998). Effect of non-digestible oligosaccharides and resistant starch on biochemical and mucosal biomarkers of colonic neoplasia in rats. Proceedings of Profibre. Lisbon, Portugal.

[75] Fleming, S.E. (1981). A study of relationships between flatus potential and carbohydrate distribution in legume seeds. *Journal of Food Science*, *46*, 794-803.

[76] Cummings, J. H. Englyst, H. N. & Wiggins, H. S. (1986). The role of carbohydrates in lower gut function. En "The Nutritional reemergence of starchy foods. Suppl. *Naeringsforsk*, *30*, 17-21.

[77] Gibson, G. Probert, H.M. Van Loo, J. Rastall, R.A. Roberfroid, M.B. (2004). Dietary modulation of the human colonic microbiota: updating the concept of prebiotics. *Nutrition Research Reviews*, *17*, 259–275.

[78] Delzenne, N. M. (2003). Oligosaccharides: state of the art. *Proceedings of the Nutrition Society*, *62*, 177–182.

[79] Van Loo, J. (1998). Non-digestible oligosaccharides are prebiotic functional food ingredients with promising health benefits. *Proceedings of Profibre*. Lisbon, Porgugal.

[80] Espinosa-Martos, I. & Rupérez. P. (2009). Indigestible fraction of okara from soybean: composition, physicochemical properties and in vitro fermentability by pure cultures of Lactobacillus acidophilus and Bifidobacterium bifidum. *European Food Research and Technology*, *228*, 685-693.

[81] Inoguchi, S. Ohashi, Y. Narai-Kanayama, A. Aso, K. Nakagaki, T. & Fujisawa, T. (2012). Effects of non-fermented and fermented soybean milk intake on faecal microbiota and faecal metabolites in humans. *International Journal of Food Sciences and nutrition*, *63*, 402-410.

[82] World Health Organization Study Group. (2003). Diet, nutrition and the prevention of chronic diseases. WHO Technical Report Series, No. 916 (TRS 916), Geneva: World Health Organization.

[83] Flamm, G. Glinsmann, W. Kritchevsky, D. Prosky, L & Roberfroid, M. (2001). Inulin and Oligofructose as Dietary Fiber: A Review of the Evidence. *Critical Reviews in Food Science and Nutrition*, *41*, 353-362.

[84] Schneeman, BO. (2009) Fiber, Inulin and Oligofructose: Similarities and Differences Nutritional and Health Benefits of Inulin and Oligofructose. *Journal of Nutrition*, *4*, 1424S-1426S.

[85] Madrigal, L. & Sangronis, E. (2007). La inulina y derivados como ingredientes claves en alimentos funcionales. Archivos latinoamericanos de nutrición. Organo Oficial de la Sociedad Latinoamericana de Nutrición, *57(4)*, 387-396.

[86] Trumbo, P. Schlicker, S. Yates, A. A. & Poos, M. (2002). Dietary References intakes for energy, carbohydrate, fiber, fat, fatty acids, cholesterol, protein and amino acids. *Journal of the American Dietetic Association*, *102*, 1621-1630.

[87] Regulation (EC) No 1924/2006 of the European Parliament and of the Council of 20 December 2006 on nutrition and health claims made on foods.

[88] Regulation (EC) No 258/97 of the european parliament and of the council of 27 January 1997, concerning novel foods and novel food ingredientsEFSA Panel on Dietetic Products, Nutrition and Allergies (NDA); Scientific Opinion on the safety of 'phosphated distarch phosphate' as a novel food ingredient. *EFSA Journal 2010*, *8*(9), 1772. [21].doi:10.2903/j.efsa.2010.1772. Available online: www.efsa.europa.eu/efsajournal.htm

In: Seeds as Functional Foods and Nutraceuticals
Editors: R. Mora-Escobedo, J. De J. Berrios et al.
ISBN: 978-1-62808-489-4

Chapter 5

NUTRACEUTICAL POTENTIAL OF PEANUT (*ARACHIS HYPOGAEA*) SEEDS

María del Carmen Robles-Ramírez[1*], Laura Almazán-Rodríguez[1] and Rosalva Mora-Escobedo[2]

[1]Departamento de Ingeniería Bioquímica,
Escuela Nacional de Ciencias Biológicas, Instituto Politécnico Nacional,
Prol. Carpio y Plan de Ayala S/N, Col. Sto. Tomás, México

ABSTRACT

In the global production of oilseeds, some seeds have had a greater importance than others, such as soybean, canola and sunflower. However, over time, new uses and better nutritive properties have been identified for oilseeds other than soy, canola, and sunflower. One such seed is the peanut seed, which has gained great importance in countries such as Mexico, the United States, India, China, and Russia, among others. Most of the peanut crop is processed into peanut butter, salted peanuts, candies, and snacks. In other areas of the world, such as Africa, India and China, where the majority of the world's peanut crop is grown, the peanut crop is also converted into oil and cake. The peanut (*Arachis hypogaea*) is an oilseed (also a legume) that contains a considerable amount of protein (approximately 30%), which is more than other true nuts. Peanuts also have a high oil content (almost 50%) and are rich in oleic acid, a monounsaturated fatty acid that has been found to decrease LDL cholesterol levels without reducing the good HDL cholesterol levels.

Moreover, peanuts are a rich source of other nutrients, such as fiber, niacin, folate, vitamin E, magnesium, manganese and phosphorous, as well as phytochemicals including phytosterols and phenolic compounds, which have important beneficial effects on human health. Several studies have associated peanut consumption with a reduced risk of cardiovascular disease. This fact has been linked with the high contents of dietary fiber, folate, alpha tocopherol, copper, magnesium, arginine and oleic acid in peanuts. On the other hand, peanuts are one of the few foods that contain resveratrol, a phytoalexin that has been demonstrated to have anticancer, neuroprotective, anti-inflammatory, blood

* Corresponding author: maricarmenrobles58@yahoo.com.mx Tel. 525557296000 Ext. 62462.

sugar-lowering and cardioprotective effects in animal experiments. In this chapter we will review the origin, taxonomy, morphology, market trends, chemical composition and nutraceutical potential of this interesting seed.

ORIGIN AND TAXONOMY OF THE PEANUT

Peanuts or groundnuts (*Arachis hypogaea L.)* are the world's most widely distributed edible legume. The peanut is an annual legume, and a valuable cash crop for millions of small-scale farmers living mostly in the semi-arid tropics, and it makes an important contribution to human nutrition due to its high oil and protein contents (Rao-Jagana, 2012).

As shown in Figure 1, the peanut plant is unusual because its pods develop underground. For this reason, peanuts are often called groundnuts. Other common names for peanuts include earthnut, monkeynut, pinder, mani or goober pea (FAO, 2013).

Plants from the genus *Arachis*, family *Fabaceae*, are native to South America, must likely from a region that includes Brazil, Bolivia, Paraguay, Argentina and Uruguay. Freitas et al., (2007) studied the genetic variability of the Brazilian Indian landraces of *Arachis hypogaea* L. and reported that the Xingu Indigenous Park (located west central Brazil, in the Mato Grosso State) proved to be an important center of peanut diversity. There are 69 species described in the genus, which have been assembled into nine groups, according to their morphology, geographic distribution and crossability (Krapovicas and Gregory 1994; Singh and Simpson 1994). The archeological record indicates that the peanut was domesticated by indigenous peoples at least 3500 years ago (Singh and Simpson 1994). Peanut specimens before this domestication consisted only of wild species. At the time of the explorations by the Spanish and Portuguese in the 16th century, the peanut was cultivated in many parts of South America, as well as in the Caribbean and Mexico. As a result of explorations, peanut cultivation spread quickly from the Americas to Africa and Asia.

Some of these species have been used for forage in South America, but the most important species is the cultivated peanut, *Arachis hypogaea* L. This crop is widely grown in more than 80 countries in the Americas, Asia and Africa and is used for both human consumption and as a source of oil.

Plants of the cultivated peanut species, *Arachis hypogaea* L. are annuals and grow either erect (up to 24" tall) or prostrate (usually under 12" tall). Inflorescences consist of approximately three perfect flowers that usually appear several days apart and occur in the axils of foliage leaves.

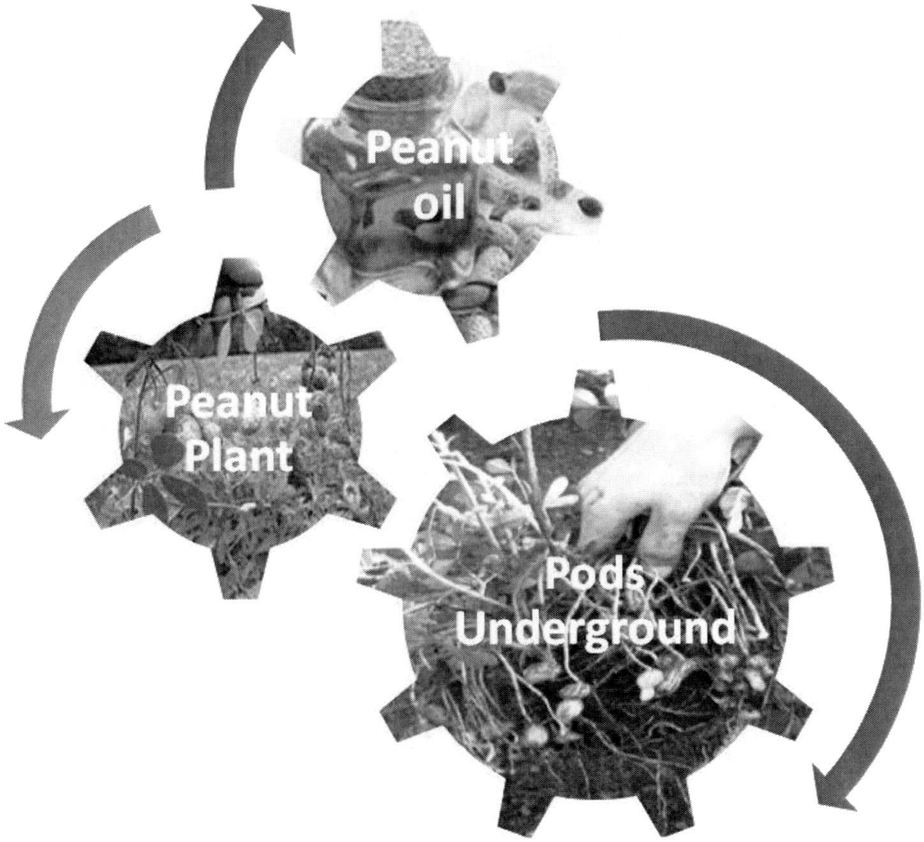

Figure 1. Peanut (*Arachis hypogaea L.)* plant.

Peanuts differ from most angiosperms in that the fertilized ovary is carried in a peg (gynophores) into the ground, where the development of the pods begins and the peanuts are produced. Depending on the genotype and environment, the peanuts mature 2.5 to 6 months after sowing.

There are two basic types of peanut plants, runner peanuts and bush peanuts. Runner plants spread out like vines, with peanuts developing from the main horizontal branches. Bush peanut plants look similar to pea plants and grow to a height between 18" and 30" with the peanut nodules developing closer to the roots. Both varieties require about five months before the peanuts are ready to be harvested.

Arachis hypogaea is classified, based on the presence or absence of flowers on the main axis, into two subspecies, A. *hypogaea hypogeal* and A. *hypogaea fastigiata* Waldron. The subspecies hypogaea is divided into two botanical varieties, hypogaea and hirsute Köhler, while the subspecies fastigiata is divided into the varieties *fastigiata vulgaris* Harz, *aequatoriana* Krapov & W.C.Greg, and *peruviana* Krapov & W.C. Greg (Krapovickas and Gregory, 1994).

The commonly cultivated botanical varieties are listed in Table 1 and they are as follows: Spanish (Sub esp. *fastigiata* var. *vulgaris*), Valencia (Sub esp. *fastigiata* var. *fastigiata*) and Virginia (Sub esp. *hypogaea* var. *hypogaea*). These varieties differ in their agronomic and morphologic characteristics but have some overlapping characteristics due to the suspected introgression between inter and intra-subspecific levels (Bhagat et al., 1991; Freitas et al., 2007).

Production

Peanuts are grown on nearly 21.45 million ha worldwide with total production of 36 million Tons and an average yield of 1.68 metric tons/ha in 2011 (USDA, 2013). China, India, Nigeria, the USA and Myanmar are the major groundnut growing countries.

With an annual production of 37.7 million of tons (MT) covering 24.1 million of ha (M-ha) achieving an average yield of 1564 kg/ha during 2010, this crop is cultivated in more than 100 countries and stands as the fourth largest oilseed crop in the world (FAO, 2012). The largest producers of groundnuts include China and India followed by other countries in Sub-Saharan Africa and the Americas. Production is concentrated in Asia (50% of the global area and 64% of the global production) and Africa (46% of the global area and 28% of the global production), where the crop is mostly grown by small farmers under rainfed conditions with limited inputs. Between 2000 and 2009, the annual global production increased marginally by 0.4%, the area by 0.3% and the yield by 0.1% (ICRISAT, 2013)

Table 1. Commonly cultivated botanical varieties of *Arachis hypogaea*

Common name	Botanical varieties	Characteristics
Spanish	Sub esp. *fastigiata* var *vulgaris*	Smaller kernels covered with a reddish-brown skin. Used predominantly in peanut candy, with significant quantities being used for salted nuts and peanut butter. Higher oil content than the other types of peanuts which is advantageous when crushing them for oil.
Valencia	Sub esp. *hypogaea* var. *hypogaea*	Have three or more small kernels in a pod. Very sweet peanuts that are usually roasted and sold in the shell. Excellent for fresh use as boiled peanuts.
Virginia	Sub esp. *hypogaea* var. *hypogaea*	Have the largest kernels and account for most of the peanuts roasted and eaten "in shells". When shelled, the larger kernels are sold as salted peanuts.

According to the USDA´s Foreign Agricultural Service (FAS), the United States is ranked third in the world in terms of the production of peanuts. China is the largest producer of peanuts and has approximately doubled the production of India, the world´s second largest peanut producer. However, while the world yields are only approximately 800 to 900 pounds per acre, USA yields average 2500 to 2600 pounds per acre resulting in 10 to 12% of the total world production (Godsey, 2011).

Physical and Chemical Characteristics

Peanuts and their products have been a component of the world's diet for years. The geographical location, cultivar type and cultivation conditions influence the nutritional profile of the nuts. Peanut kernels contain approximately 16.2 to 36% protein. These proteins are classified into albumin (water soluble), globulins (salt soluble) and glutelins (acid/alkaline soluble); the globulins constitute approximately 87% and consist of arachin and conarachin. Peanuts are known to be rich in acidic amino acids; however, they are deficient in the essential amino acids lysine, methionine and threonine (Salunkhe et al., 1992).

The oil content of peanuts is 47-50%, of which approximately 76–80% is unsaturated fatty acids. Oleic acid makes up 40–45%, and linoleic acid makes up 30–35% of the composition of the unsaturated fatty acids. Peanuts are 18% carbohydrates and have a starch content of 0.5–5%, a sucrose content of 4–7%, and ash content of 3%. The ash content is composed of 26 inorganic constituents of which the phosphorus, potassium, magnesium and sulfur contents are high and are virtually unaffected by heat; the remaining 22, though also heat stable, are deficient from a nutritional viewpoint.

The vitamin profile of peanuts includes riboflavin, thiamin (which is destroyed to a great degree by roasting and blanching), nicotinic acid and vitamin E, as well as appreciable amounts of B complex vitamins and vitamin K, but peanuts contain practically no vitamins A, C or D (Woodroof, 1983).

Grosso et al. (1999) studied the composition of 9 aboriginal *Arachis hypogaea* subsp. *fastigiata* var. *vulgaris* cultivars originating from Uruguay. They showed high protein levels, like other varieties of the subspecies *fastigiata* (these protein percentages were higher than varieties of the subspecies *hypogaea*). Palmitic (16:0), stearic (18:0), oleic (18:1), linoleic (18:2), arachidic (20:0), eicosenoic (20:1), behenic (22:0), and lignoceric (24:0) acids were detected. The chemical quality of these seeds was characterized by higher protein and oil levels with lower oleic/linoleic (O/L) ratios and higher iodine values.

The biochemical compositions of the peanut seeds collected from six landrace accessions of *Arachis hypogaea* var. *hirsuta* cultivated in Mexico were investigated. This study examined the oil content, oil stability, sugar composition and amino acid composition of the six lines of the "hairy" peanut. The oil content and oil stability characteristics were generally not as good when compared with two standard U.S. grown cultivars; however, two of the "hairy" lines contained less than 37 % oil compared to common values of ca. 49%. This result may have an application in breeding low oil content cultivars, considering the current trend for consumers to prefere for low fat products (Grimm et al., 1996).

Peanut Products with Added Value

Many options exist for adding value to commodities or specialty crops and are limited only by one's imagination. Peanuts are an excellent cash crop, and there are multipurpose uses for each plant part in direct consumption, confectionary preparations, cooking oil and as a rich source of protein feed for animals (Padley et al., 2012). Peanuts have been marketed as a cheap source of protein compared to cheese and red meat and as a good source of essential vitamins and minerals; therefore, industry has been developing many foods with this legume. For example, it is possible to taste a bar made with peanuts and chocolate. Food scientists (Halthorn and Sanders, 2012) reported that it is possible to add peanut skin to peanut butter and peanut paste to increase the nutritional value and antioxidant capacities of these products while only subtly changing the taste. The majority of peanut consumption in the United States can be attributed to peanuts as a food source, as opposed to as oil or meal, and peanut butter has been and remains the largest source of consumption.

Peanut Flour

Made from raw peanuts which have been cleaned, blanched and electronically sorted to select the highest quality peanuts, the nuts are then roasted and naturally processed to obtain low fat peanut flour with a strong roasted peanut flavor. Chefs and manufacturers have turned to roasted peanut flour for years to boost of flavor, texture and nutrition for a variety of menu items and products.

This healthy flour is a great thickener for soups, a flavorful and aromatic ingredient for a variety of breads and pastries, and a creative and flavorful coating for meats, fish and other center of the plate proteins. It is low in fat and contains over twice the protein of all purpose flour. In addition, it is nutrient rich, providing the following percentages of the daily recommended intake in a quarter cup: 8% fiber, 35% manganese, 20% niacin, 15% copper, 10% folate and 10% phosphorus (U.S. Department of Agricultural, 2008).

Peanut flour is used in confectionery products, seasoning blends, bakery mixes, frostings, fillings, cereal bars and nutritional bars. Because the flour is partially defatted, it works well as a fat binder in applications such as confection centers. Using peanut flour at a level of 4-8% in a formulation has been found to extend the shelf life of confections and can contribute a peanut flavor to the product. Because of its high protein content of 45-50% (USDA, 2008), peanut flour is a good protein source, in addition to its function as a flavoring agent.

Peanut Oil

Peanut oil is extracted from shelled and crushed peanuts by one or a combination of the following methods: hydraulic pressing, expeller pressing, or solvent extraction. Refined peanut oil, like all processed vegetable oils, has been refined, bleached and deodorized. This process removes the protein component of the oil that can cause an allergic reaction, making it non-allergenic. Refined peanut oil is the main type of oil utilized in major US fast-food chains. Highly aromatic 100% peanut oil and peanut extract are also available. These products have a strong roasted peanut flavor and aroma. Suggested applications for these products include flavoring compounds, confections, sauces and baked goods.

Peanut oil is one of the healthiest oils. It is a vegetable oil that is naturally free of trans-fats and cholesterol, and is low in saturated fats. Peanut oil is high in unsaturated fats, especially monounsaturated fats, similar to olive oil. It is also a source of antioxidants vitamin E and phytosterols, which benefit heart-health. Vassiliou et al., (2009) studied the effect of oleic acid and peanut oil high in oleic acid in reversing the inhibitory effect of the inflammatory cytokine TNF-α on the insulin production in both *in vitro* and *in vivo* systems. They found that type II diabetic mice that were administered a diet derived from peanut oil high in oleic acid had decreased glucose levels compared to animals administered a high fat diet with no oleic acid. Oleic acid was found to be effective in reversing the inhibitory effect on insulin production of the inflammatory cytokine TNF-α.

Roasted Peanuts

Roasted peanuts are available in several different packages and roasting variations. Different coatings can be applied to the peanuts prior to and after roasting to provide a variety of products with different flavors, including honey, smoked, sweet, hot and spicy, and salty. Coatings applied to peanuts can enhance the oxygen barrier to reduce rancidity. Ultrasonication has been used to remove lipids from the peanuts before coating application to improve the attachment of the coating onto the peanut's surface. The treatment was an effective method in delaying the formation of oxidative volatile compounds and hence inhibiting the rancidity of roasted peanuts. (Wambura et al, 2010).

Peanut Butter

Peanut butter was first developed in the United States during the 1890s as a high protein health food. Dr. John Harvey Kellogs initially developed peanut butter as a meat alternative for his patients and was the first to patent the product. It is used as a cooking ingredient throughout the world.

It has intense flavor, therefore only a small amount is needed to enliven sauces, soups and other savory dishes. Medical science has revealed that peanut butter, like olive oil, contains monounsaturated fats and is therefore good for people. It has been found that monounsaturated fats really help to reduce cholesterol levels and improve heart function. It is convenient, inexpensive and a nutritional power-house, which provides fiber, niacin, magnesium, B-complex vitamins, copper, and folate in addition to protein (Robertson, 2006).

A variety of different peanut butter products are currently available. Peanuts are roasted, blanched and sorted before being ground into a creamy consistency. Peanut butter produced in the U.S. contains a minimum of 90% peanuts; sweeteners and salt can be added to enhance the flavor while small amounts of stabilizers are used to prevent oil separation.

The small amount of stabilizer added does not contribute trans fatty acids to the product. Peanut pieces can be added to provide a crunchy texture and custom formulations can also be developed to modify the texture or sweetness or to add flavoring. Reduced fat peanut butters are also available, which provide a fat reduction of at least 25%. Several different varieties are sold for both consumer and industrial use with varying peanut content depending on the flavor and consistency of the product needed. Other modified formulations of peanut butter, peanut spreads and peanut paste are available from most manufacturers.

Peanut Milk

Bensmira and Jiang (2012) used peanut milk as the main raw material for kefir preparation. The rheological characteristics, textural properties, mineral elements and amino acid composition of kefir made from peanut milk (PMK) and 7:3 peanut:skim milk (70% PMK), were investigated, using whole milk kefir (WMK) as a control. The authors found that rheological and textural properties of 70% PMK (peanut milk Kefir) were not significantly different from those of WMK. The mineral composition, including the K, Na, and Ca contents, was higher in both 70% PMK and WMK compared to in the PMK. In general, the amino acid composition of PMK was lower than that of both 70% PMK and WMK.

Feed Supplements for Ruminants

Peanut husks have been shown to be a good substrate when treated with white rot fungi (*Pleurotus ostreatus and Pleurotus pulmonarius*) for solid state fermentation through a biotechnological approach, which leads to an effective and economic way to utilize this otherwise unused residue and improves the economy of the peanut industry (Akinfemi, 2010).

Protective Components of Peanuts

Peanuts are a rich source of nutrients such as healthy fats, fiber, arginine, niacin, folate, vitamin E, magnesium, manganese and phosphorous as well as bioactive compounds such as phytosterols and phenolic compounds, which have been proven to have important effects on human health. Extensive research has demonstrated that peanuts can prevent chronic diseases including cardiovascular disease (CVD), diabetes, cancer and obesity (Table 2).

Healthy Fats

Peanuts are rich in unsaturated fatty acids, which compose between 75 and 85% of peanut oil (Campos-Mondragón *et al.,* 2009; Sabaté, 2003). These "healthy fats" contribute to the association of peanut intake with health benefits, especially the prevention of coronary heart disease. Thirty grams (a single serving) of peanuts provides 28 and 38% of the dietary reference values (DRV) of monounsaturated (MUFA) and polyunsaturated fatty acids (PUFA), respectively (Higgs, 2003).

Peanuts have a high content of MUFA, mainly in the form of oleic acid, which represents approximately 51% of the seed oil (Sabaté, 2003). The evidence indicates that the substitution of MUFA instead of carbohydrate for saturated fatty acid calories may positively influence the risk of CVD (Kris-Hetherton, 1999). Monounsaturated fatty acids were identified as a key nutrient of the Mediterranean diet.

The Mediterranean population has low prevalence of CVD despite the consumption of a diet high in total fat. This low prevalence of CVD has been attributed to the high consumption of olive oil (high in MUFAs) in that region. Some studies suggest that MUFAs may decrease platelet aggregation, increase bleeding time, and increase fibrinolysis, which protect against thrombogenesis (Kris-Etherton, 1999). Studies have shown that both a high-fat diet rich in MUFA and low in SFA and a low-fat but carbohydrate-rich diet diminish the total and low density lipoprotein (LDL) cholesterol; however, a diet rich in carbohydrates increases the tryglicerides and decreases the HDL cholesterol whereas a MUFA-rich diet does not.

Table 2. Bioactive components or properties of peanuts which prevent disease

Disease	Bioactive component or condition
Coronary heart disease	Polyunsaturated fatty acids, Monounsaturated fatty acids Arginine Vitamin E Folate Magnesium, copper, potassium Fiber Resveratrol Antioxidant compounds β-sitosterol Bioactive peptides
Diabetes	Fiber Arginine Monounsaturated fatty acids Antioxidant compounds α-amylase inhibitors
Obesity	Fiber α-amylase inhibitors High satiety properties Poor fat digestibility and absorption
Cancer	Resveratrol Other antioxidant compounds Fiber

On the other hand, peanuts have a good proportion of PUFA, and approximately 35% of these types of fatty acids (mainly linoleic acid) have been reported (Sabaté, 2003). Linoleic acid (a fatty acid with 18 carbons, two double bonds, and omega 6) is one of the nutrients most effective for the reduction of cholesterol levels (Sacks and Campos, 2006). Studies have indicated that higher plasma levels of linoleic acid lead to decreased CVD risk through its beneficial effects on inflammatory markers. Polyunsaturated fatty acids reduce the activation of nuclear factor-κB (NF- κB), a transcription factor of proinflammatory genes. Thus, linoleic acid inhibits the endothelial cells production of adhesion molecules, interleukins and chemokines, key mediators that promote the atherosclerosis process.

Dietary Fiber

Peanuts have a total dietary fiber content between 9 and 13% (Jonnala *et al.,* 2005). The American Association of Cereal Chemists defines dietary fiber as the edible parts of plants or analogous carbohydrates that are resistant to digestion and absorption in the human small intestine with complete or partial fermentation in the large intestine. Fiber is an important component of food that, in recent decades, has attracted the attention of food specialists due to epidemiological studies that have related dietary fiber deficiency with diseases such as gastrointestinal disorders (constipation, diverticulitis and colon cancer), obesity, cardiovascular diseases and diabetes. However, over the years, evidence indicating that high-fiber foods can regulate digestion and absorption processes has accumulated. The physical and chemical properties of the fiber sources (water retention capability, viscosity, organic molecule absorption, cation exchange and large intestine fermentation) are important in the determination of the physiological responses generated, such as the volume and weight of feces, the digestion control of nutrients, the absorption velocity, the reduction of cholesterol in plasma and the attenuation of postprandial glucose levels (Tungland and Meyer, 2002).

Arginine

Arginine is one of the most versatile amino acids from a metabolic and physiological point of view. In addition to participating in the synthesis of proteins and bioactive peptides, arginine participates in the detoxification of ammonia via the synthesis of urea, the liberation of hormones, and the synthesis of pyrimidine bases. Arginine is also the common substrate of two enzymes, arginase and nitric oxide synthase (Figure 2) (Robles-Ramírez *et al.,* 2011). Thus, it is the precursor for the syntheses of nitric oxide, ornithine, proline, polyamines, creatine and agmatine (Wu and Meininger, 2000). Dietary arginine augments the synthesis of nitric oxide (NO) because it is the substrate nitric oxide synthase. Nitric oxide inhibits endothelin-1 release; leukocyte adhesion; platelet adhesion, activation and aggregation; superoxide generation, the expression of vascular cell adhesion molecules and monocyte chemotactic peptides; and the smooth muscle cell proliferation. Nitric oxide also induces muscle relaxation in smooth muscle cells (vasodilation) diverting the endothelial dysfunction associated with cardiovascular risk factors (De Lorgeril, 1998; Wu and Meininger, 2000; Higgs, 2003).

Arginine has also been implicated in cancer prevention. The nitric oxide molecule is a free radical, which is relevant to understanding its high reactivity. Nitric oxide freely diffuses through membranes and interacts with specific molecular receptors, often iron, present in certain proteins with heme groups and SH-Fe complexes, and it can activate or inhibit enzymatic systems. This seems to be the case in the regulation of the activity of ornithine decarboxylase by NO. Ornithine decarboxylase is the key enzyme in the biosynthesis of polyamines, and its inhibition may explain a portion of NO's cytostatic effect because it may decrease the synthesis of the polyamines necessary for growth and cell proliferation (Teixeira et al., 2003).

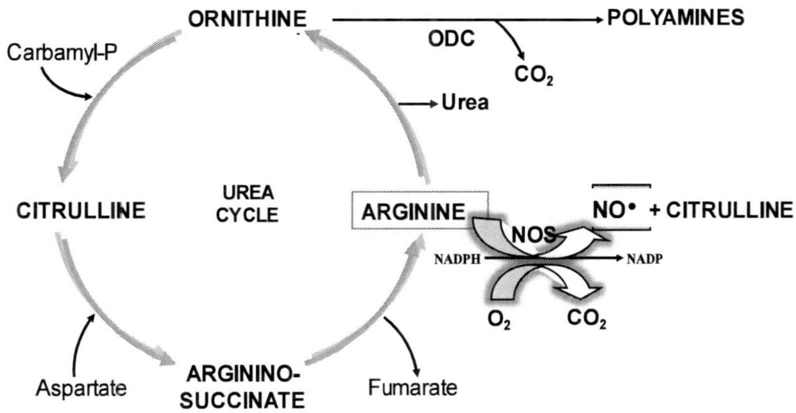

NO: nitric oxide; NOS: nitric oxide synthase; ODC: ornithine decarboxilase

Figure 2. Arginine metabolism.

Vitamin E

Considerable evidence indicates that reactive free radicals contribute to many diseases, including heart disease and cancer. Cells contain many potentially oxidizable substrates such as polyunsaturated fatty acids, proteins, and DNA. Thus, a complex antioxidant defense system usually protects cells from the deleterious effects of endogenous and exogenous free radical production. When exposure to free radicals exceeds the protective capacity of the antioxidant defense system, damage to biological molecules may occur.

There is considerable evidence indicating that disease causes an increase in oxidative stress; therefore, consumption of foods rich in antioxidants, which are potentially able to neutralize excess radicals, may play an important role in modifying the development of such diseases.

Vitamin E (mainly as α-tocopherol) is the main lipid-soluble antioxidant in the cell's antioxidant defense system and is exclusively obtained from the diet. The major biologic role of vitamin E is to protect PUFAs, other components of cell membranes and low-density lipoprotein (LDL) from oxidation by free radicals (FAO, 2004). High doses of vitamin E have been associated with a reduced risk of CVD via the inhibition of LDL oxidation (Kris-Etherton, 1999; Higgs, 2003). The daily consumption of peanuts can contribute to increased vitamin E ingestion as this legume is a rich source of this vitamin.

Folate

Peanuts provide 110 μg folate/100 g and are one of the best sources of this vitamin. Folate reduces the levels of homocysteine, an amino acid that is related to vascular toxicity due to its inhibition of NO synthesis. Peanuts have high contents of arginine and folate and a low level of methionine, and all of these conditions may help to prevent CVD through their effects in reducing the homocystein level (Higgs, 2003).

Magnesium, Copper and Potassium

Peanuts, like other legumes, are good sources of magnesium, copper and potassium and are low in sodium. These conditions may help to prevent CVD. Whereas magnesium is important in maintaining the correct calcium/potassium balance, which helps to prevent dysrhythmias and myocardial infarction, copper is related to positive changes in lipids, glucose tolerance and blood pressure (Higgs, 2003).

Phytochemicals

As in other plants, peanuts contain several bioactive phytochemicals synthesized in the plant to protect it against predators and diseases. These phytochemicals may provide different health benefits. Several studies have shown their beneficial effects on obesity, diabetes, cancer and cardiovascular disease. Peanuts are one of the few foods that contain resveratrol (Figure 3). This phytoalexin is a polyphenol that has been associated with reduced cancer risk (Aggarwal et al., 2004), as well as reduced risks of atherosclerosis, hypertension, ischemia/reperfusion, and heart failure, along with other degenerative diseases such as diabetes, obesity and aging (Petrovski et al., 2011). Phytochemicals act by different mechanisms that will be described later.

Peanuts are also a rich source of phytochemicals with antioxidant properties, such as p-hydroxybenzoic acid, p-coumaric acid, quercetin and kaempferol, which can prevent chronic-degenerative diseases by neutralizing the free radicals generated by oxidative stress.

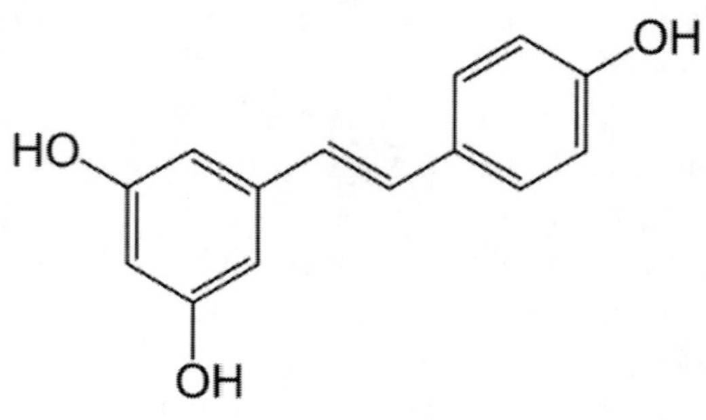

Figure 3. Resveratrol structure.

Peanuts are also one of the main dietary sources of phytosterols. These are cholesterol-like molecules found in plant foods (Figure 4). Phytosterols are absorbed only in small amounts by the intestine but inhibit the absorption of intestinal and biliary cholesterol, a key step in the elimination of cholesterol (Ostlund, 2002; Brufau et al., 2008).

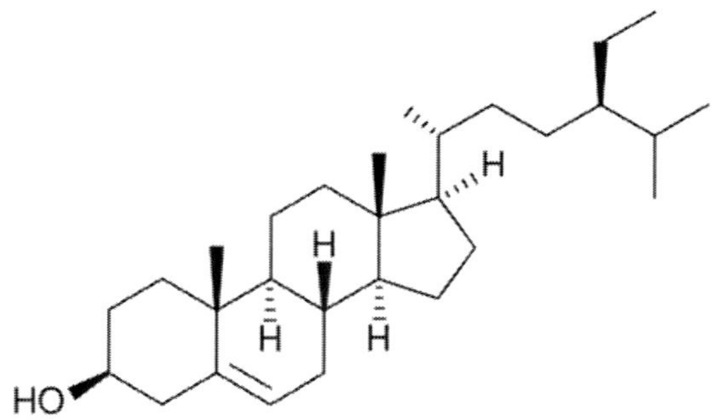

Figure 4. β-sitosterol structure.

Some studies have revealed that phytosterols can protect against prostate, colon and breast cancers by reducing tumor cell adhesion, migration and invasion and, by its effects on the membrane structures of tumor cells, affect the signal transduction pathways that regulate tumor growth, apoptosis, and the immune response of the host (Awad and Fink, 2000). The sitosterol content in peanuts can be up to 180mg/100g mainly in the form of β-sitosterol (Higgs, 2003; Tasan et al., 2006)

Peanuts and Disease

Coronary Heart Disease (CHD)

Evidence suggests the important role of dietary factors in the modulation of endothelial function. In particular, Ω3 fatty acids, antioxidant vitamins, folic acid, and L-arginine appear to have beneficial effects on vascular endothelial function, either by decreasing endothelial activation or by improving endothelium-dependent vasodilation in patients at a high risk for cardiovascular disease (CVD) as well as in healthy subjects (Brown and Hu, 2001). Epidemiologic studies and several clinical trials have shown that these nutrients may reduce the risk of CVD. Peanuts are an especially rich source of all of these nutrients.

Epidemiological and metabolic studies have demonstrated that the type of fat rather than the total fat content affects cholesterol levels. Despite their high fat content, the consumption of nuts has been constantly linked with a reduced risk for coronary heart disease (Kris-Etherton et al., 1999; Kris-Etherton et al., 2008; Sabaté et al., 2010; Jenkins et al., 2011), which has been mainly attributed to their fatty acid composition. Nuts are low in saturated fatty acids and high in monounsaturated and polyunsaturated fatty acids.

A pooled analysis of 25 intervention trials conducted with 583 men and women in seven countries showed that, with a mean daily consumption of 67 g of diverse nuts, mean reductions of 5.1% in the cholesterol concentrations were obtained, 7.4% reduction in LDL-cholesterol, 8.3% in the LDL/HDL cholesterol ratio, and 10.3% in trygliceride levels (in subjects with triglyceride levels greater than 150 mg/dL). Different types of nuts had similar effects on blood lipid levels (Sabaté et al., 2010).

When the results of studies that have examined the effects of diets containing nuts on blood lipids and lipoproteins were collected and compared with the results predicted from regression equations using changes in dietary fatty acid profiles, nuts were shown to have a stronger lipid-lowering effect than expected. The results from this comparison showed that when subjects consumed test diets including nuts, there was about a 25% greater cholesterol-lowering response than was predicted by the equations. These results suggested that there were non–fatty acid constituents in nuts with additional cholesterol-lowering and therefore cardio protective effects (Kris-Etherton et al., 1999). Some of these are the amino acid arginine, vitamin E, folate, magnesium, dietary fiber and the phytoalexin resveratrol.

Peanuts are rich in the amino acid arginine. Dietary L-arginine is the precursor of endothelial nitric oxide (NO) which plays an important role in maintaining active vasodilatation in healthy blood vessels. The biological activity of NO is severely impaired in hypercholesterolemia and atherosclerosis, leading to reduced endothelium-dependent vasodilation and internal thickening. In studies with cholesterol-fed rabbits, dietary L-arginine improved the NO-dependent vasodilator function and completely blocked the progression of plaques via the restoration of NO synthase substrate availability and the reduction of vascular oxidative stress (Böger et al., 1997). It has been suggested that the high content of arginine in peanuts could be responsible for its anti-atherogenic effect (Kris-Etherton et al., 1999; Hu and Stampfer, 1999).

Vitamin E in doses higher than 100 UI/d reduces the risk of coronary heart disease through the inhibition of LDL oxidation (Kris-Etherton et al., 1999). Cardiovascular disease is largely attributable to atherogenesis, and lipid peroxidation plays a central role in atherogenesis. Therefore, it is believed that antioxidants like vitamin E may protect against cardiovascular disease (Schürks et al., 2010). Some observational studies have suggested this protective effect, although these results should be taken with caution because a meta-analysis of nine trials investigating the effect of vitamin E on the incidence of stroke (118765 participants) showed that vitamin E reduced the risk of an ischaemic stroke by 10% but increased the risk of hemorrhagic stroke by 22% (Schrüks *et al.,* 2010). A serving of 42 g of peanuts provides approximately 20% of the recommended dietary allowance of vitamin E (King et al., 2008). The amounts needed to obtain the CVD risk reduction are greater; however, peanut consumption can contribute to an increase in the vitamin intake and its antioxidant effect.

Peanuts are also a good source of folate; 42 g of peanuts provide 16% of the RDA (King et al., 2008). This vitamin acts as a coenzyme in the synthesis of methionine from homocysteine (Hcy) lowering the concentration of the latter. High concentrations of Hcy in the plasma have been linked to an increased risk of atherosclerosis because this amino acid causes increased oxidative stress, the inhibition of nitric oxide sythase, the proliferation of the smooth muscle cells and endothelial dysfunction (Held et al., 2008). It has been observed that supplementation with folic acid (or a combination of folic acid and vitamins B6 and B12) lowered the Hcy levels preventing against cardiovascular disease (Toole et al., 2004; Held et al., 2008). The vascular toxicity of homocysteine is in part related to its inhibition of NO. So, the high contents of arginine and folate in peanuts could have synergic effects due to their effects on the Hcy levels (Higgs, 2003). On the other hand, the low content of methionine in peanuts also contributes to the reduction of plasma homocysteine levels, as homocysteine synthesis requires that amino acid.

Peanuts are a rich source of copper and magnesium (King et al., 2008). These micronutrients may also contribute to the cardioprotective effects of peanuts. Increased Cu intake may reduce the risk of CVD and atherosclerosis in women through the promotion of an improved fibrinolytic capacity. In a study with sixteen women, supplementation with 3 mg Cu/d significantly increased the serum Cu concentration and the activity of erythrocyte superoxide dismutase. In addition, the concentration of the fibrinolytic factor plasminogen activator inhibitor type 1 was significantly reduced by approximately 30% after supplementation with 6 mg Cu/d (Bügel et al., 2005). Magnesium improves insulin resistance and the serum lipid profiles, as well as lowering inflammation, endothelial dysfunction, oxidative stress, and platelet aggregation. It is involved in energy metabolism, fatty acid synthesis, glucose utilization, ATPase functions, neurotransmitter release, and endothelial cell function (Bo and Pisu, 2008).

A 42 g portion of peanuts provides approximately10% of the RDA for fiber (Higgs, 2003; King et al., 2008). Epidemiological studies have shown that dietary fiber may reduce the risk of cardiovascular disease by improving the blood lipid profile, blood pressure, body-weight, and insulin sensitivity (Pereira et al., 2004; Streppel et al., 2008). The analysis of the data from 10 prospective cohort studies from the USA and Europe showed that the consumption of dietary fiber in cereals and fruits was inversely associated with the risk of CHD. Each 10g/d increment of total dietary fiber was associated with a 14% decrease in the risk of all coronary events and a 27% decrease in the risk of coronary death (Pereira et al., 2004). Likewise, a cohort study with 1373 men was carried out to investigate the effects of recent and long-term dietary fiber intake on CHD and all-cause mortality. This study showed that every additional 10 g of recent dietary fiber intake per day reduced CHD mortality by 17% (Streppel et al., 2008).

Peanuts contain between 0.02 and 1.79 μg resveratrol/g. It has been proposed that this phytoalexin is responsible for the epidemiological phenomenon observed in the French population that links the moderate consumption of red wine with lower risk of CVD. Resveratrol can exert an adaptive stress response and force the expression of cardioprotective genes and proteins such as heat shock and antioxidant proteins (Petrovski et al., 2011).

In addition to all of the bioactive peanut components mentioned above, the protein in peanuts could also contribute to their health benefits. The production of peptides with antihypertensive activity from the simulated gastric digestion of peanut protein has been demonstrated (Jimsheena and Gowda, 2011); this suggests that digestive enzymes could release those peptides after the consumption of peanuts, and they could contribute to a reduced CVD risk.

Diabetes

Peanuts have a very low glycemic index (GI) of approximately 13, as compared to a GI of 100 for glucose (Foster-Powel et al., 2002). The GI classifies carbohydrate-containing foods according to how they affect blood glucose levels. Because peanuts are absorbed slowly through the intestinal wall, they do not cause a rapid increase in blood sugar. Foods with low GI have been proven to help in the control of non-insulin dependent diabetes mellitus. In a study, two diets (high and low-GI, respectively) were randomly assigned to sixteen subjects with well controlled diabetes mellitus, for a period of 12 weeks for each diet. The mean glycosylated hemoglobin and the plasma glucose profile at the end of the low-GI diet were lower than at the end of the high-GI diet (Brand et al., 1991). In a prospective cohort study of 83818 women aged 34 to 59 years, it was found that the consumption of nuts and peanut butter was inversely associated with the risk of type II diabetes. The relative risk of developing diabetes was reduced by 27 and 21% in women who consumed nuts or peanut butter, respectively, 5 or more times per week compared with those who rarely or almost never ate nuts (Jiang et al., 2002). Another population-based study, in which approximately 64,000 Shanghai women were followed for 4.6 years, showed an inverse association between peanut consumption and the risk of diabetes in pre- and post-menopausal women (Villegas et al., 2008). In that study, a remarkable risk reduction of 20% was observed between the first quintile (0.1 g/d) and the upper quintile (3.1 g/d) of peanut consumption.

Johnston and Buller (2005) evaluated the effect of peanut products as complementary food for the reduction of postprandial protein. Eleven healthy individuals (ten women and one man) who were 27.9 ± 2.9 years of age and had BMI values of 22.7 ± 1.0 participated in this study. These participants ate two test meals (a bagel and juice or chicken and rice), with and without the addition of peanuts, in a randomized crossover experimental design. It was observed that the consumption of peanuts reduced the postprandial blood glucose. This effect was attributed to the high arginine content of this legume, which is a potent secretagogue of insulin. However, it must also be considered that peanuts contain other components such as monounsaturated fatty acids and fiber, which could contribute to that activity. These compounds may influence the glycemic response by slowing gastric emptying and/or activating the incretin system with the consequence of a reduction of the glucose uptake. Furthermore, it is believed that the physical form of peanuts impedes the complete breaking or grinding of the grains by chewing, which affects the digestive process. The lack of rupture of the fibrous cell walls affects the enzyme bioaccessibility (Esteves et al., 2011).

Irshad and Sharma (1981) reported a heat resistant α-amylase protein-inhibitor from peanuts. This inhibitor showed highly specific inhibitory activity towards hog pancreatic and human salivary α-amylase. This could prevent the release of glucose from starch, which would contribute to a lower glycemic response.

The beneficial effects of peanut in the control of diabetes mellitus were also attributed to its fatty acid composition. In a study with 117 type 2 diabetic subjects who were randomized to one of the three treatments for 3 months (mixed nuts, muffins or half portions of both), it was found that two ounces of nuts daily, as a replacement for foods high in carbohydrates, improved both the glycemic control and the serum lipid levels in type 2 diabetics. This result was attributed to the MUFA and PUFA composition of nuts (Jenkins et al., 2011).

Obesity

Obesity is associated with a number of chronic diseases such as type II diabetes, cardiovascular disease and cancer. Despite their high fat content, it has been demonstrated that peanut consumption does not increase body weight (Mattes et al., 2008). This has been attributed to the high satiety of peanuts (Mattes and Dreher, 2010), as well as the limited efficiency of energy absorption due to their reduced bioaccessibility and the elevation of fecal fat loss (Traoret et al., 2008).

Moreover, it has been observed that the chronic consumption of peanuts leads to a high resting energy expenditure resulting in the dissipation of an additional portion of the energy that they supply (Alper and Mattes, 2002). In a review examining the impact of low-glycemic load diets on weight loss, it was concluded that weight loss was superior in overweight people given low glycemic load diets than in people given higher glycemic load diets and conventional weight loss diets.

Likewise, the total fat mass loss and the decrease in the body mass index (BMI) were significantly greater in the group receiving a low glycemic load diet (Thomas et al., 2009). As was mentioned before, peanuts have a very low GI. Using data from the continuing Survey of Food Intakes by Individuals, Sabaté (2003) and Griel et al. (2004) reported that female peanut consumers had a significantly lower BMI than nonusers and that there was also a tendency for a lower BMI in male peanut consumers.

Cancer

There are relatively few studies related to the effects of peanut consumption on cancer, and those that exist are not conclusive (Davis *et al.,* 2008; Sabaté and Ang, 2009). Therefore, more research concerning this topic is required. However, peanuts contain different bioactive compounds that might have anticarcinogenic activities. Boiled peanuts are a rich source of resveratrol, a phytoalexin that has the ability to suppress the proliferation of a broad variety of tumor cells through different mechanisms, such as cell-cycle arrest; the up-regulation of p21, p53 and Bax; the down-regulation of cyclin D1, cyclin E, Bcl2, Bcl-xL and c-IAPs; and the activation of caspases. Moreover, resveratrol suppresses the activation of several transcription factors, which inhibit different protein kinases and down-regulate products of genes, all of which account for the suppression of angiogenesis (Aggarwal et al., 2004). Peanuts contain resveratrol in a similar level to that found in red wines.

An analysis from the USDA Agricultural Research Service in Raleigh, North Carolina (Sanders et al., 2000), showed that peanuts have 0.02 to 1.79 μg of resveratrol/g, compared to 0.7 μg in a glass of red grape juice or 0.6 to 8.0 μg/mL in red wine. On the other hand, peanuts have other phenolic compounds such as p-hydroxybenzoic acid, p-coumaric acid, quercetin, kaempferol and tocopherol, which may prevent cancer by reducing oxidative stress (Talcott et al., 2005; Duncan et al., 2006; Win et al., 2011).

A prospective cohort study was carried out in Taiwan to examine whether peanut consumption was associated with a reduced risk of colorectal cancer. That study suggested that the frequent intake of peanuts may reduce colorectal cancer in women (Yeh et al., 2006). Several studies have demonstrated the strong protective effects of dietary fiber against colon and rectal cancers.

The mechanisms proposed are as follows: reduction of the colonic pH affecting pH-dependent enzymatic reactions such as secondary bile acid formation; reduction of carcinogenic substances by the absorption of fiber to the cell walls of microbiota; reduction of intestinal transit time; increase in the colonic contents, thereby diluting carcinogens; and inhibition of the initiation and promotion stages of colon cancer formation due to the production of short chain fatty acids, mainly butyric acid (Tungland and Meyer, 2002).

CONCLUSION

Peanuts (*Arachis hypogaea*) are approximately 30% protein and approximately 50% oil rich in oleic acid, a monounsaturated fatty acid that has been found to decrease LDL cholesterol levels without reducing the good HDL cholesterol. Moreover, peanuts are a rich source of other nutrients such as fiber, niacin, folate, vitamin E, magnesium, manganese and phosphorous, as well as phytochemicals including phytosterols and phenolic compounds which have important beneficial effects on human health.

This chapter shows that several studies have associated peanut consumption with a reduced risk of cardiovascular disease. Peanuts are one of the few foods that contain resveratrol, a phytoalexin that has been demonstrated to have anticancer, neuroprotective, anti-inflammatory, blood sugar-lowering and cardioprotective effects in animal experiments. Additionaly, a high cholesterol level is associated with the formation of gallstones, and eating peanuts may help to prevent these. Peanuts have a low content of sodium, thus making them a healthy snack choice. Peanuts are also rich in dietary fiber which has been linked to preventing colon cancer and improving digestion and bowel movement.

REFERENCES

Aggarwal, BB; Bhardwaj, A; Aggarwal, RS; Seeram, NP; Shishodia, S; Takada, Y. Role of resveratrol in prevention and therapy of cancer: preclinical and clinical studies. *Anticancer Res.*, 2004, 24, 2783-2840.

Akinfemi, A. Bioconversion of peanut husk with rot fungi: *Pleurotus ostreatus* and *Pleurotus pulmonarius. Livestock Research for Rural Development*, 2010, 22,3.

Alper, CM; Mattes, RD. The effects of chronic peanut consumption on energy balance and hedonics. *Int. J. Obes.*, 2002, 26, 1129–1137.

Awad, AB; Fink, CS. Phytosterols as anticancer dietary components: evidence and mechanism of action. *J. Nutr.*, 2000, 130, 2127-2130.

Bensmira, M. and Jiang, B. Rheological characteristics and nutritional aspects of novel peanut based kefir beverages and whole milk kefir. *Inter. Food Research J.* 2012, 19,647-650.

Bhagat, NR; Rajgopal K; Ghetia, NR. and Bhalodia P.K. NRCG Valencia peanut germplasm evaluation catalogue based over four years. *National Research Centre for Groundnut.* India 1991, 74.

Bo, S; Pisu, E. Role of dietary magnesium in cardiovascular disease prevention, insulin sensitivity and diabetes. *Curr Opin Lipidol*, 2008, 19, 50-56.

Bogër, RH; Bode-Böger, SM; Brandes, RP; Phivthong-ngam, L; Böhme, M; Nafe, R; Mügge, A; Frölich, JC. Dietary L-arginine reduces the progression of atherosclerosis in cholesterol-fed rabbits comparison with lovastatin. *Circulation*, 1997, 96, 1282-1290.

Brand, JC; Colagiuri, S; Crossman, S; Allen, A; Roberts, DC; Truswell, AS. Low-glycemic index foods improve long-term glycemic control in NIDDM. *Diabetes Care*, 1991, 14, 95-101.

Brown, AA; Hu, FB. Dietary modulation of endothelial function: implications for cardiovascular disease. *Am. J. Clin. Nutr.*, 2001, 73, 673-681.

Brufau, G; Canela, MA; Rafecas, M. Phytosterols: physiologic and metabolic aspects related to cholesterol-lowering properties. *Nutr. Res.,* 2008, 28, 217-225.

Bügel, S; Harper, A; Rock, E; O'Connor, JM; Bonham, MP; Strain, JJ. Effect of copper supplementation on indices of copper status and certain CVD risk markers in young healthy women. *Br. J. Nutr.*, 2005, 94, 231-236.

Campos-Mondragón, MG; Calderón de la Barca, AM; Durán-Prado, A; Campos-Reyes, LC; Oliart-Ros, RM; Ortega-García, J; Medina-Juárez, LA; Angulo, O. Nutritional composition of new peanut (Arachis hypogaea) cultivars. *Grasas y Aceites*, 2009, 60, 161-167.

Davis, PA; Jenab, M; Vanden, JP; Furlong, T; Taylor, S. Tree nut and peanut consumption in relation to chronic and metabolic diseases including allergy. *J. Nutr.*, 2008, 138, 1757S-1762S.

De Lorgeril, M. Dietary arginine and the prevention of cardiovascular diseases. *Cardiovascular. Res.*, 1998, 37, 560-563.

Duncan, CE; Gorbet, DW; Talcott, ST. Phytochemical content and antioxidant capacity of water-soluble isolates from peanuts (*Arachis hypogaea* L.). *Food Res. Int.*, 2006, 39, 898-904.

Esteves, OFC; Pinheiro VAC; Alfenas RC. Impact of different protein sources in the glycemic and insulinemic responses. *Nutr Hosp*, 2011, 26, 669-676.

FAO. Vitamin E. In: FAO/WHO, editor. *Vitamin and mineral requirements in human nutrition.* Bangkok, Thailand: FAO/WHO expert consultation on human vitamin and mineral requirements; 2004; 121-131.

FAO. 2013. [http://faostat.fao.org/faostat/collections?subset=agriculture] *website*

Foster-Powell, K; Holt, SH; Brand-Miller, JC. International table of glycemic index and glycemic load values. *Am. J. Clin. Nutr.*, 2002, 76, 5-56.

Freitas, F.O; Moretzsohn, M.C; Valls, J.F.M. Genetic variability of Brazilian Indian landraces of Arachis hypogaea L. *Genet. Mol. Res.* 2007, 6, 675-684

Godsey, Ch. Peanut production guide. 2011

Griel, AE; Eissenstat, B; Juturu, V; Hsieh, G; Kris-Etherton, PM. Improved diet quality with peanut consumption. *J. Am. Coll. Nutr.*, 2004, 23, 660–668.

Grimm, DT; Sanders,TH; Pattee HE; Williams, DE. and Sanchez-Dominguez, S. Chemical Composition of *Arachis hypogaea L.* Subsp. *hypogaea Var. hirsuta* Peanuts1. *Peanut Science,* 1996, 23,111-116.

Grosso, NR; Lucini,EI; López, AG. and Guzmán, CA. Chemical composition of aboriginal peanut (Arachis hypogaea L.) seeds from Uruguay. Grasas y Aceites. 1999, 50, 203-207

Halthorn CS., Sanders TH. Flavor and antioxidant capacity of peanut paste and peanut butter supplemented with peanut skins. *J. Food Sci.* 2012, 77,407-411.

Held, C; Sumner, G; Sheridan, P; McQueen, M; Smith, S; Dagenais, G; Yusuf, S; Lonn, E. Correlations between plasma homocysteine and folate concentrations and carotid atherosclerosis in high-risk individuals: baseline data from the homocysteine and atherosclerosis reduction trial (HART). *Vascular Med*, 2008, 13:245-253.

Higgs, J. The beneficial role of peanuts in the diet. Part 2. *Nutrition and Food Science*, 2003, 33, 56-64.

Hu, FB; Stampfer, MJ. Nut consumption and risk of coronary heart disease: a review of epidemiologic evidence. *Current Atherosclerosis Reports,* 1999, 1, 204-209.

ICRISAT International Crops Research Institute for the Semi-Arid tropics, January, 2013.

Irshad, M; Sharma, CB. Purification and properties of an α-amylase protein-inhibitor from *Arachis hypogaea* seeds. *Biochimica et Biophysica Acta*, 1981, 659, 326-333.

Jenkins, DJA; Kendall, CWC; Banach, MS; Srichaikul, K; Vidgen, E; Mitchell, S; Parker, T; Nishi, S; Bashyam, B; de Souza, R; Ireland, C; Josse, RG. Nuts as a replacement for carbohydrates in the diabetic diet. *Diabetes Care*, 2011, 34, 1706-1711.

Jiang, R; Manson, JE; Stampfer, MJ; Liu, S; Willett, WC; Hu, FB. Nut and peanut butter consumption and risk of type 2 diabetes in women. *JAMA*, 2002, 288, 2554-2560.

Jimsheena, VK; Gowda LR. Angiotensin 1-converting enzyme (ACE) inhibitory peptides derived from arachin by simulated gastric digestion. *Food Chem.*, 2011, 125, 561-569.

Johnston, CS; Buller, AJ. Vinegar and peanut products as complementary foods to reduce postprandial glycemia. *American Diabetic Association,* 2005, 105, 1939-1942.

Jonnala, RS; Dunford, NT; Chenault, K. Nutritional composition of genetically modified peanut varieties. *J Food Sci*, 2005, 70, S254-S256.

King, JC; Blumberg, J; Ingwersen, L; Jenab, M; Tucker, KL. Tree nuts and peanuts as components of a healthy diet. *J. Nutr.*, 2008, 138, 1736S-1740S.

Krapovickas, A. and Gregory W.C. Taxonomía del género Arachis (Leguminosae). Bonplandia, 1994, 8:1–186.

Kris-Etherton, PM. Monounsaturated fatty acids and risk of cardiovascular disease. *Circulation*, 1999, 100, 1253-1258.

Kris-Etherton, PM; Yu-Poth, S; Sabaté, J; Ratcliffe, HE; Zhao, G; Etherton, TD. Nuts and their bioactive constituents: effects on serum lipids and other factors that affect disease risk. *Am. J. Clin. Nutr.*, 1999, 70, 504S-511

Kris-Etherton, PM; Hu, FB; Ros, E; Sabaté, J. The role of tree nuts and peanuts in the prevention of coronary heart disease: multiple potential mechanisms. *J. Nutr.*, 2008, 138, 1746S-1751S.

Mattes, RD; Dreher, ML. Nuts and healthy body weight maintenance mechanisms. *Asia Pac. J. Clin. Nutr.*, 2010, 19, 137-141.

Mattes, RD; Kris-Etherton, PM; Foster, GD. Impact of peanuts and tree nuts on body weight and healthy weight loss in adults. *J. Nutr.*, 2008, 138, 1741S-1745S.

Ostlund, R.E. Phytosterols in human nutrition. *Annu. Rev. Nutr.*, 2002, 22, 533-549.

Pandey, M.K. , Monyo, E., Ozias-Akins, P., Liang, X., Guimarães, P., Nigam S.N. *et al.* Advances in *Arachis genomics* for peanut improvement. *Biotechnol. Adv., 2012, 30,* 639–651

Pereira, MA; O'Reilly, E; Augustsson K; Fraser, GE; Goldbourt, U; Heitmann, BL; Hallmans, G; Knekt, P; Liu, S; Pietinen, P; Spiegelman, D; Stevens, J; Virtamo, J; Willet, WC; Ascheiro, A. Dietary fiber and risk of coronary heart disease. A pooled analysis of cohort studies. *Arch. Intern. Med.*, 2004, 164, 370-376.

Petrovski, G; Gurusamy, N; Das, DK. Resveratrol in cardiovascular health and disease. *Ann NY Acad Sci*, 2011, 1215, 22-33.

Rao Jagana S; Vadez V; Bhatnagar-Mathur, P; Lakshmi Narasu, M; Sharma K. K. Better root:shoot ratio conferred enhanced harvest index in transgenic groundnut overexpressing the rd29A:DREB1A gene under intermittent drought stress in an outdoor lysimetric dry-down trial. 2012. An Open Access Journal published by ICRISAT. (10)1-7. ejournal.crisat.org/Volumen 10/Groundnut/KKS.pdf.

Robertson, R. Peanut butter planet: around the world in 80 recipes, from starters to main dishes to desserts. 2006. Ed. Noltzbrink Pu. Pag. 2

Robles- Ramírez MC; Ramón-Gallegos, E; Mora-Escobedo R. Soybeans germination and cancer disease. In: Maxwell JE, editor. *Soybeans: Cultivation, Uses and Nutrition.* New York: Nova Science Publishers Inc; 2011; 223-250.

Sabaté, J. Nut consumption and body weight. *Am. J. Clin. Nutr.*, 2003, 78, 647S-500S

Sabaté, J; Ang, Y. Nuts and health outcomes: new epidemiologic evidence. *Am. J. Clin. Nutr.*, 2009, 89, 1S-6S.

Sabaté, J; Oda, K; Ros, E. Nut consumption and blood lipid levels: a pooled analysis of 25 intervention trials. *Arch. Intern. Med.*, 2010, 170, 821-827.

Sacks, FM; Campos, H. Polyunsaturated fatty acids, inflammation and cardiovascular disease: time to widen our view of the mechanisms. *J. Clin. Endocrinol. Metab.*, 2006, 91, 398-400.

Salunkhe, D.K; Chavan, J.K; Adsule, R. N. and Kadam, S.S. World Oil Seeds: Chemistry, Technology and Utilization. Van Nostrand Reinhold, New York, USA. 1992, 112-6

Sanders, TH; McMichael, RW; Hendrix, KW. Occurrence of resveratrol in edible peanuts. *J. Agric. Food. Chem.*, 2000, 48:1243-1246.

Schürks, M; Glynn, RJ; Rist, PM; Tzourio, C; Kurth, T. Effects of vitamin E on stroke subtypes: meta-analysis of randomized controlled trials. *BMJ*, 2010, 341, c5702

Singh U, Singh B. Tropical grain legumes as important human foods. *Econ. Bot.*, 1992, 46:310-321.

Singh, A.K and Simpson C.E. Biosystematics and genetic resources. In: The groundnut crop: A scientific basis for improvement (Smartt J, ed.). London, UK: Chapman &Hall.Stalker ST. 1994, 96-137

Streppel, MT; Ocké, MC; Boshuizen, HC; Kok, FJ; Kromhout, D. Dietary fiber intake in relation to coronary heart disease and all-cause mortality over 40 y: the Zutphen study. *Am. J. Clin. Nutr.*, 2008, 88, 1119-1125.

Talcott, ST; Passeretti, S; Duncan, CE; Gorbet, DW. Polyphenolic content and sensory properties of normal and high oleic acid peanuts. *Food Chem*, 2005, 90, 379-388.

Tasan, M; Bilgin, B; Gecgel, U; Demirci, AS. Phytosterols as functional food ingredients. *J. Tekirdag. Agr. Fac.*, 2006, 3, 153-159.

Teixeira, D; Santaolaria, L; Iglesias, E. La arginina en su contexto metabólico y fisiológico. *Acta Bioquim. Clin. Latinoam*, 2003, 37, 165-179.

Thomas, D; Elliott, EJ; Baur, L. Low glycaemic index or low glycaemic load diets for overweight and obesity (Review). *Cochrane Database Syst. Rev.*, 2009, 3, 1-38.

Toole, JF; Malinow, MR; Chambless, LE; Spence, JD; Pettigrew, LC; Howard, VJ; Sides, EG; Wang, C; Stampfer, M. Lowering homocysteine in patients with ischemic stroke to prevent recurrent stroke, myocardial infarction, and death. The vitamin intervention for stroke prevention (VISP) randomized controlled trial. *JAMA*, 2004, 291, 565-575.

Traoret, C; Lokko, P; Cruz, ACRF; Oliveira, CG; Costa, NMB; Bressan, J; Alfenas, G; Mattes, RD. Peanut digestion and energy balance. *Int. J. Obes*, 2008, 32, 322–328.

Tungland, B.C; Meyer, D. Non digestible oligo and polysaccharides (dietary fiber): their physiology and role in human health and food. *CRFSFS*, 2002, 1, 90-109.

USDA. 2008. USDA National Nutrient Database for Standard Reference, Release 21. USDA Nutrient Data Laboratory website: http://www.ars.usda.gov/nutrientdata

USDA United States Department of Agriculture. 2013. http://www.fas.usda.gov.

Varshney, R.K.; Mohan, S.M; Gaur, P.M; Gangareo, N.V.P.R. et al. Achievements and prospects of genomics-assisted breeding in three legume crops of the semi-arid tropics. *Biotech Adv*. 2013, In press. Available on line http://www.sciencedirect.com/science/article/pii/S0734975013000037.

Vassiliou, EK; Gonzalez A; Garcia C; Tadros JH; Chakraborty, G. and Toney JH. Oleic acid and peanut oil high in oleic acid reverse the inhibitory effect of insulin production of the inflammatory cytokine TNF-α both *in vitro* and *in vivo* systems. *Lipids Health Dis*. 2009, 8: 25.

Villegas, R; Gao, Y-T; Yang, G; Li, Hong-Lan; Elasy, TA; Zheng, W; Shu, XO. Legume and soy food intake and the incidence of type 2 diabetes in the Shanghai Women Health Study. *Am. J. Clin. Nutr.*, 2008, 87, 162–167.

Wambura, P; Yang,W. and Mwakatage N. Reduction of roasted peanut lipid oxidative rancidity by power ultrasound and edible coatings containing natural extracts. *J.Food Process Engineering* 2010, 33, 883-898

Win, MM; Abdul-Hamid, A; Baharin, BS; Anwar, F; Sabu, MC; Pak-Dek, MS. Phenolic compounds and antioxidant activity of peanut's skin, hull, raw kernel and roasted kernel flour. *Pak. J. Bot.*, 2011, 43, 1635-1642.

Woodroof J.G. Peanuts.Production, Processing,Products, 3th ed, AVI Publishing Company Inc.,Westport, CT, USA, 1983,

Wu, Guoyao; Meininger, CJ. Arginine nutrition and cardiovascular function. *J. Nutr.*, 2000, 130, 2626-2629.

Yeh, CC; You, SL; Chen, CJ; Sung, FC. Peanut consumption and reduced risk of colorectal cancer in women: A prospective study in Taiwan. *World J. Gastroenterol,* 2006, 12, 222-227.

In: Seeds as Functional Foods and Nutraceuticals
Editors: R. Mora-Escobedo, J. De J. Berrios et al.
ISBN: 978-1-62808-489-4

Chapter 6

ANTIOXIDANT AND ANGIOTENSIN-CONVERTING ENZYME INHIBITORY ACTIVITIES IN COMMON BEANS (*PHASEOLUS VULGARIS* L.)

Anaberta Cardador-Martínez[1*], Cristian Jiménez-Martinez[2], Silvia L. Amaya-Llano[3], Mercedes M. Pedrosa[4], Carmen Cuadrado[4] and Mercedes Muzquiz[4]

[1]Tecnológico de Monterrey Campus Querétaro. Epigmenio González No. 500. Fracc. San Pablo. Querétaro, Qro., México

[2]Escuela Nacional de Ciencias Biológicas. IPN. Prolongación de Carpio y Plan de Ayala s/n, Col. Santo Tomas, Delegación Miguel Hidalgo México, D.F.

[3]Programa de Posgrado en Alimentos del Centro de la República (PROPAC), Facultad de Química, Universidad Autónoma de Querétaro, Centro Universitario Cerro de las Campanas, s/n, Querétaro, Qro, México

[4]Instituto Nacional de Investigación y Tecnología Agraria y Alimentaria (INIA). Madrid, España

ABSTRACT

Beans have been recognized as nutraceutical foods due to their content of phytochemicals such as phenolic compounds, tannins, anthocyanins, oligosaccharides, and phytates, among others. Biological activities attributed to beans include anti-inflammatory, antioxidant, and anticarcinogen. This study considers two aspects in the study of beans. In the first one, the contents of phenolic compounds, tannins, phytates, oligosaccharides and haemagglutinins, as well as the scavenging free radicals capacity (DPPH and ABTS), were evaluated in 25 varieties of beans. In the second part, 2 varieties of beans were used to produce peptides using Alcalase enzyme immobilized in a membrane bioreactor. The peptides produced were evaluated for their anti-hypertensive (Angiotensin converting enzyme inhibition) activity. The results showed that either

* Corresponding author: Email: mcardador@itesm.mx.

phytochemicals or peptides have biological activities, making beans a source of compounds that could benefit the health of consumers.

INTRODUCTION

Common bean is a crop that has been part of the culture and food of the Mexican people for many years. It is a widely accepted food in the diet and provides significant advantages, not only for its high nutritional content, but for the nutraceutical potential and the benefits it provides to health.

Beans are a very valuable food, considered the second largest source of vegetable protein in the country, they are an excellent source of complex carbohydrates (dietary fiber, starch, resistant starch and oligosaccharides), are a major source of minerals (Fe, Ca and Zn), B vitamins, and contain about 3% fat, of which up to 70% are polyunsaturated fatty acids [1, 2]. In addition to their nutritional, beans are considered as nutraceutical food, they contain components that prevent or reduce the risk of chronic degenerative diseases, such components have different biological activity: function as antioxidants, antimutagenic, anticancer enzyme modulators, lowering cholesterol blood, impart glucose tolerance, reduced intestinal transit time and promote good digestion [3-6]. Bioassays have been conducted that show the importance of bean consumption on health [7].

ANTIOXIDANTS IN BEANS

Polyphenols are a large group of compounds, which includes a large number of potential chemopreventive agents. They are present in all plants, and therefore many of them are in the diet, but the concentration varies depending on the type of food [8]. Chemically, the phenolic compounds can be defined as substances that possess an aromatic ring to which are attached one or more hydroxyl groups, including functional derivatives. The main polyphenols in legumes and cereals are flavonoids and tannins. Its main feature, important for health, is that most of them show a significant antioxidant activity [9]. The consumption of phenolic compounds may be beneficial to health, because of their potential antioxidant anticancer, antimicrobial, and anti-inflammatory activities [10-14]

Many cellular processes result in the production of free radicals, which lose an electron molecules tending to be more reactive. The accumulation of these free radicals, oxidative stress produced in the cell may cause damage to biomolecules such as lipids, proteins and DNA, leading to increased cancer risk [15]. Therefore, the importance of phenolic compounds as free radical scavengers is increasing.

PROTEINS AND PEPTIDES IN BEANS

Beans are an important source of protein in the diet of the people of Latin America, where its consumption is between 10 and 25 kg per person per year [1]. Protein bean content is within the range of 18% to 32% of dried grains [16]. These proteins are of lower quality in comparison with other proteins such as egg and milk, which are deficient in methionine (Met) and cysteine (Cys). The majority of proteins are in bean seed cotyledons, a small part is in the embryonic axes, and in the shell are present only in minimal quantities [17].

Peptides are short proteins formed by a chain consisting of 50 or less amino acids. The term bioactive peptides is used to identify molecules of peptidic nature or origin which display a biological behavior or activity which can be developed at the industrial level for pharmaceutical, diagnostic, chemical and agro-food applications [18].

In living organisms, endogenous peptides often function as hormones and neurotransmitters and play important physiological roles such as regulation of metabolism, and hormones. In the past several decades, researchers have found that bioactive peptides can also be derived from dietary proteins. They may be present as independent entities or encrypted in the parent protein. It is known that during gastrointestinal digestion or food processing, these peptides are released from the parent protein and act as regulatory compounds with hormone-like activities [19, 20].

Hypertension is a growing undesired symptom that damages health and threatens mostly the developed societies. It is one among the major independent risk factors for atherosclerosis, stroke, myocardial infarction, and end stage renal disease. Antihypertension has been reported to be mediated through inhibition of angiotensin-converting enzyme (ACE; EC 3.4.15.1). ACE is a Zn–metallopeptidase and plays an important role in regulating blood pressure. ACE catalyzes the conversion of an inactive form, decapeptide (Angiotensin I) to a potent vasoconstrictor octapeptide (Angiotensin II) and also inactivates the antihypertensive vasodilator bradykinin. Many ACE inhibitory peptides have been derived from not only milk proteins but also from other food proteins [21, 22].

POLYPHENOSL AND FLAVONOIDS IN 42 BEAN CULTIVARS

42 cultivars of beans were analyzed for their content of total phenolics [23] and total flavonoids [24] in methanolic bean flour extracts. **Table 1** shows the results obtained, as well as the color of hulls for each bean variety. Of the 42 varieties tested, 39 belong to the genus *Phaseolus vulgaris*, while 2 more belong to *Phaseolus coccineus* (Ayocote amarillo and morado) and one cultivar was *Vigna radiata* (Mungo bean).

Table 1. Total phenolic content and total flavonoid content in 42 beans cultivar

Bean cultivar	Hull colour	TPC[1]		TFC[2]	
Alubia grande	White	1138.9	E	491.3	E
Alubia pequeña	White	1219.3	D	543.5	E
Ayocote Amarillo	Yellow	3465.1	B	1281.2	B
Ayocote Morado	Purple	2556.1	C	906.4	C
Bayo	Yellow	2437.5	C	595.4	E
Bayo Chilar	Yellow	1620.3	D	847.1	C
Bayo gordo	Yellow	2250.4	C	615.2	E
Bayo INIFAP	Yellow	2624.5	C	1513.6	B

Table 1. (Continued)

Bean cultivar	Hull colour	TPC[1]		TFC[2]	
Bayo INIFAP Santa Lucía	Yellow	2000.0	C	846.7	C
Bayo Mecentral	Yellow	2631.6	C	1222.4	B
Bayo Mecentral Santa Lucía	Yellow	2049.8	C	812.6	C
Blanco de Villaflores	White	592.8	F	543.9	E
Botí Comitán	Red	4660.2	A	983.8	C
Cacahuate	Light brown with dark brown spots	3019.2	B	1189.1	B
Cacahuate Importado	Light brown with dark brown spots	2431.9	C	994.6	C
Cacahuate Nacional	Light brown with dark brown spots	2735.9	C	1236.6	B
Flor de Durazno Santa Lucía	Creamy with pink spots	2418.9	C	932.0	C
Flor de Mayo Santa Lucía	Creamy with pink spots	3134.9	B	841.6	C
Mantequilla	Yellow	3300.2	B	847.2	C
Moro	Gray with black spots	1956.6	C	805.7	C
Mungo 2010	Green	3851.0	B	2026.4	A
Negro 8025	Black	3377.5	B	766.4	D
Negro Americano	Black	3828.7	B	1216.5	B
Negro Americano Don Pepe	Black	3286.2	B	885.1	C
Negro Bola	Black	3138.2	B	1271.2	B
Negro Tuxtla	Black	4097.1	A	1415.8	B
Patashete	Light with brown black spots	1556.7	D	619.3	E
Patashete 2010	Light with brown black spots	1135.0	E	677.4	D
Peruano	Yellow	730.0	F	591.2	E
Peruano DF	Yellow	645.9	F	645.9	E
Pinto	Creamy with brown spots	3194.7	B	690.8	D
Pinto Bill	Creamy with brown spots	2193.3	C	1331.6	B
Pinto Saltillo	Creamy with brown spots	2021.4	C	964.3	C
Querétaro bola	Black	2106.7	C	956.0	C
Rojo Comitán	Red	2539.7	C	690.9	D
Rojo Villaflores	Red	3987.4	B	1061.8	B
Vaquita Café Chiapas	White with brown spots	1205.5	E	812.9	C
Vaquita Café DF	White with brown spots	1296.3	D	803.1	C
Vaquita Café DF2010	White with brown spots	1278.0	D	673.6	D
Vaquita Negra Chiapas	White with black spots	1605.8	D	680.8	D
Vaquita Negro DF	White with black spots	2495.3	C	743.6	D
Vaquita Negro DF2010	White with black spots	1613.2	D	567.3	E

[1] TPC= Total phenolic content expressed as mg Gallic Acid equivalent per 100 g of dry beans

[2] TFC= Total flavonoid content expressed as mg Rutin equivalent per 100 g of dry beans

Results are the mean of two independent experiments with three replicates. Same letter in the same column indicate samples are in the same group. Samples were grouped after Tukey´s comparison (α= 0.05)

Table 2. Phenolic content in common beans, analyzed by hull color

Hull color	Total phenolics[1]	Flavonoids[2]
Black	3299.4 ^{a,b}	1084.5 ^{b,c}
Creamy brown spots	3100.8 ^{b,c}	662.7 ^{e}
Creamy pink spots	2664.0 ^{c}	882.8 ^{c}
Gray black spots	1877.1 ^{e}	739.5 ^{d,e}
Green	3813.6 ^{a}	2049.7 ^{a}
Light brown dark brown spots	2674.9 ^{c}	1112.2 ^{b}
Light brown black spots	1322.8 ^{f}	652.0 ^{e}
Purple	2558.0 ^{c,d}	881.9 ^{c,e}
Red	3739.8 ^{a}	901.3 ^{c}
White	970.5 ^{g}	521.5 ^{f}
White black spots	1810.7 ^{e}	668.0 ^{e}
White brown spots	1187.7 ^{f,g}	762.0 ^{d}
Yellow	2195.6 ^{d}	893.4 ^{c}
Yellow brown spots	2009.5 ^{d,e}	1117.5 ^{b}

[1] Total phenolics expressed as μg Gallic Acid equivalent per 100 g of dry beans

[2] Flavonoids expressed as μg Rutin equivalent per 100 g of dry beans

Results are the mean of two independent experiments with three replicates. Different letters in the same column indicate statistical differences (Tukey, α=0.05).

According to the total phenolic content of 42 varieties tested, they were grouped into six categories. The first is comprised of two bean varieties, black and red ones with values of 4660.2 μg equivalent Gallic Acid/100 g of dry bean and 4097 μg /100 g, respectively. In the second group composed of eight varieties, the average content of total phenolics was 3417 mg/100 g, with different hull colors.

In the third group, with 16 varieties phenolics content varied between 1956 and 2700 μg /100 g. Seven varieties were comprised in group four, with an average of 1456 μg /100 g. Category 5 consisted of three varieties showed a value of only 300 μg / 100 g (around 1000 μg /100 g) less than the previous group. Finally, the latter group showed half the content of group 5 and approximately 5 times lower than Group 1 (590 μg/100g). Given the variety of colors in each group of total phenolics content, there was no correlation between color and total phenol content. This only happened in the first and the last group, comprising dark colors and light colors respectively. Similar results for total phenolic content have been observed for other authors as analyzing several beans with varied hull colors ([14, 25, 26].

Total flavonoids are shown in Table 1, Mungo beans, with green hull, is the cultivar with the highest content. Additionally, there are other 4 groups, the first of which showed an average content of flavonoids 1271 μg Rutin equivalents per 100 g of dry bean. The color of hulls in this group ranged from yellow to black and red, and it is composed of a total of 10 varieties. The second group with colors from yellow to purple (15 varieties) showed a content of 883 μg /100 g. The third group composed of eight varieties, all dark in color, but one of Pinto (Creamy with brown spots), had a content between 673 and 766 μg / g, an average of 200 μg less than the previous group. Finally, the group with the lowest flavonoid content was composed of nine varieties containing flavonoids between 490 and 645 μg /100 g. Boateng et al [27]studied four legumes (including kidney bean and black eyed bean) founding flavonoids contents up to 1000 μg/g. In general, the highest flavonoid content was showed by Mungo beans, 4 times greater than the white beans flavonoids content. Due to the variability in the flavonoid content compared to the hull color, it was not allowed to establish a color-flavonoid content correlation.

Univariated analysis using hull color as independent variable, showed that the group with dark colors (green, red, and black) had the highest values for total phenolic content, followed by those with dark spots in their coloration such as Vaquita café and negra, whereas the white varieties show lower total content of total phenolics. In this case, a correlation was established between color and total phenols, ie. darker color, higher the total phenols content (**Table 2**).

ANTIOXIDANT CAPACITY OF METHANOLIC 42 BEAN EXTRACTS

The antioxidant capacity of bean extracts was measured by two methods. The first one consisted in the scavenging of free radical DPPH [28]. 42 bean cultivars can be grouped into three groups, the first and largest with 26 varieties shows a range between 65 and 87%, and an average of 74% DPPH discoloration. The second group with 8 varieties and an average of 59% discoloration, was 7% discoloration over the third group, consisting of eight varieties in the range of 32-55% (**Table 3**).

Rocha Guzman et al. [14] found a 50 % antioxidant capacity in several beans, although they used methanolic extracts of hulls, not the whole bean. The second measurement of antioxidant ability was expressed as μM Trolox (TEAC, Trolox Equivalent Antioxidant Capacity).

In this case 4 groups could be separated, the first one with 7 varieties with yellow and white with black spots hulls, showed the highest TEAC value (464-623 μM). The second group of cultivars with a total of 26 varieties had an average value of 379 μM followed by a group of four varieties (248 μM) and the fourth group with 4 varieties and 151 μM. Notably, one of the varieties identified as Peruano (yellow hull) had a TEAC value as little as 8 μM, 100 times less than the variety with a higher TEAC (Mantequilla, yellow hull, 623 μM).

Table 3. Antioxidant capacity of beans related to bean cultivar

Bean cultivar	ARA[1]		TEAC[2]	
Alubia grande	54.4	C	152.1	D
Alubia pequeña	32.3	C	186.8	D
Ayocote Amarillo	79.5	A	464.0	A
Ayocote Morado	67.7	A	358.4	B
Bayo	82.2	A	493.2	A
Bayo Chilar	65.1	A	433.0	B
Bayo gordo	72.6	A	426.2	B
Bayo INIFAP	73.3	A	289.2	C
Bayo INIFAP Santa Lucía	62.8	B	323.9	B
Bayo Mecentral	71.8	A	370.5	B
Bayo Mecentral Santa Lucía	60.0	B	423.4	B

Bean cultivar	ARA[1]		TEAC[2]	
Blanco de Villaflores	40.6	C	128.5	D
Botí Comitán	84.6	A	554.9	A
Cacahuate	78.4	A	396.5	B
Cacahuate Importado	54.7	C	311.4	B
Cacahuate Nacional	58.8	B	294.7	B
Flor de Durazno Santa Lucía	75.1	A	320.5	B
Flor de Mayo Santa Lucía	70.1	A	349.9	B
Mantequilla	87.6	A	623.2	A
Moro	61.6	B	372.4	B
Mungo 2010	58.3	B	297.7	B
Negro 8025	64.6	A	387.7	B
Negro Americano	65.0	A	365.7	B
Negro Americano Don Pepe	64.5	A	453.7	B
Negro Bola	79.2	A	433.0	B
Negro Tuxtla	84.6	A	523.9	A
Patashete	74.5	A	475.5	A
Patashete 2010	65.3	A	412.7	B
Peruano	55.0	C	8.2	E
Peruano DF	37.9	C	214.4	C
Pinto	59.4	B	374.5	B
Pinto Bill	69.5	A	385.7	B
Pinto Saltillo	67.9	A	344.7	B
Querétaro bola	76.7	A	440.5	B
Rojo Comitán	75.2	A	391.9	B
Rojo Villaflores	71.7	A	449.5	B
Vaquita Café Chiapas	46.5	C	134.5	D
Vaquita Café DF	53.7	C	247.7	C
Vaquita Café DF2010	57.6	B	242.4	C
Vaquita Negra Chiapas	82.7	A	509.9	A
Vaquita Negro DF	67.5	A	374.7	B
Vaquita Negro DF2010	57.2	B	349.0	B

[1] Antiradical activity expressed as percent DPPH discoloration.[2] Trolox Equivalent Antioxidant Capacity expressed as μM Trolox equivalent.Results are the mean of two independent experiments with three replicates. Same letter in the same column indicate samples are in the same group. Samples were grouped after Tukey´s comparison (α= 0.05)

The univariated analysis considering the color of the hull as an independent variable, with 13 levels (13 hull colors), showed that black and red varieties have the greatest ability to scavenge DPPH free radicals, followed by beans with yellow and white with brown spots hulls; and finally white hull colors showed the lowest antioxidant capacity (**Table 4**). TEAC values in relation to the color of the hull followed a trend similar to the capacity to catch the DPPH free radical. These results indicated that when considering only the color of the husk, bean antioxidant capacity is directly related to the intensity of the color, implying that the darker the color, the better antioxidant capacity.

Table 4. Antioxidant capacity in beans related to hull color

Hull color	ARA[1]	TEAC[2]
Black	72.9^{a}	432.9^{b}
Creamy with brown spots	59.4^{c}	374.5^{b,c}
Creamy with pink spots	72.6^{a,b}	335.2^{d}
Gray with black spots	61.6^{c}	372.4^{b,c}
Green	58.3^{c}	297.7^{d,e}
Light brown with dark brown spots	64.0^{c}	334.2^{c}
Light Brown with black spots	69.9^{b}	444.1^{a,b}
Purple	67.7^{b,c}	358.4^{c}
Red	77.2^{a}	465.4^{a}
White	42.5^{e}	155.8^{f}
White with black spots	69.1^{b}	411.2^{b}
White with brown spots	52.6^{d}	208.2^{e}
Yellow	68.0^{b,c}	369.9^{b,c}
Yellow with brown spots	68.7^{b}	365.2^{c}

[1] Antiradical activity expressed as percent DPPH discoloration.
[2] Trolox Equivalent Antioxidant Capacity expressed as μM Trolox equivalent.
Results are the mean of two independent experiments with three replicates. Different letters in the same column indicate statistical differences (Tukey, α=0.05).

PEPTIDES WITH ACE INHIBITORY ACTIVITY IN 2 BEANS CULTIVARS

To evaluate the antihypertensive activity of peptides only two varieties were used, a black one (Negro 8025) and a light-colored (Pinto, Cream with brown spots). Protein extracts prepared by solubilization at alkaline pH, were hydrolyzed with Alcalase enzyme. The degree of hydrolysis was determined by the method of o-phthaldialdehyde (OPA). In both bean cultivars, a high degree of hydrolysis was achieved (**Figure 1**). Similar results were obtained by Oseguera [29] who worked with Pinto Durango beans and obtained a degree of hydrolysis of about 60% at 180 minutes, higher than those shown by Torruco-Uco [30] with 45% for Jamapa beans and 35 % for Lima bean at 90 minutes of enzymatic hydrolysis.

Regarding the black beans, the higher degree of hydrolysis was reached after 180 minutes with a total of 87% (**Figure 1**). Yeboa et al [31] mentioned that phaseolin hydrolysis, the main bean protein, reaches values of up to 82% with heat treatment followed by enzymatic hydrolysis with Alcalase.

In pinto bean hydrolysates, the higher degree of ACE inhibition was obtained after 60 minutes (35% hydrolysis, ACE inhibition of 80%). Thereafter, although the degree of hydrolysis was increased to 60%, the percentage inhibition of ACE remained constant. Black bean hydrolysate showed inhibition rates of 70 - 90%, greater than pinto beans, with a degree of hydrolysis also higher. Bean antihypertensive activity has been reported by Torruco et al [30] who stated that the alcalase can be used to obtain bioactive peptides with ACE-inhibitory activity, because this enzyme has a serine group in the active site and hydrolyzes peptide bonds involving phenylalanine, tyrosine, tryptophan, leucine, isoleucine, proline and methionine (**Figure 2**). Moreover, the preferred ACE substrates or competitive inhibitors possess hydrophobic amino acids in their structure.

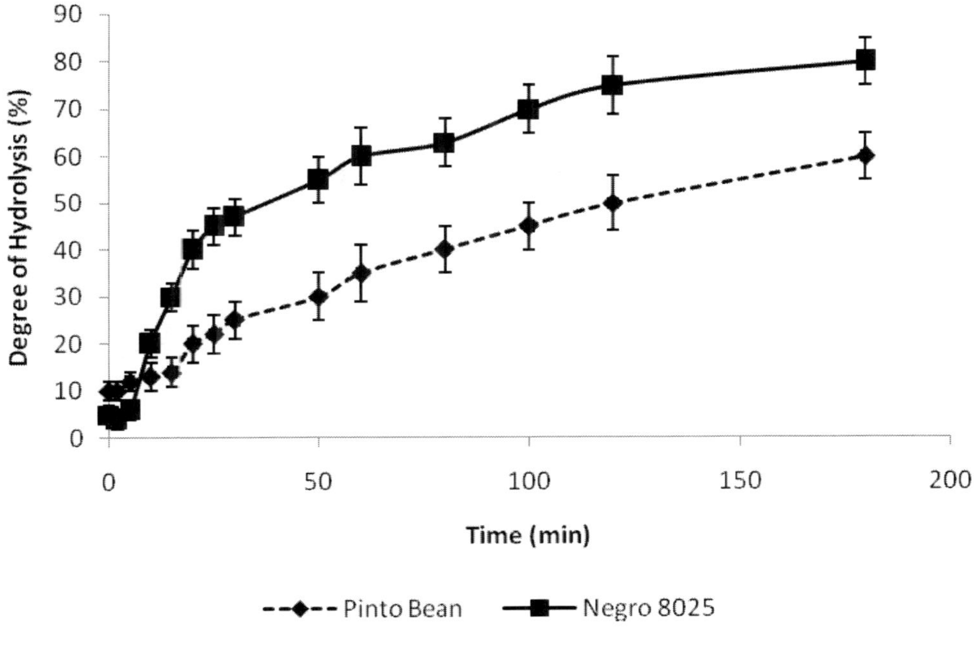

Figure 1. Degree of hydrolysis of protein extracts of two bean cultivars hydrolyzed with Alcalase enzyme. Results are the mean ± SD of two independent experiments with three repetitions.

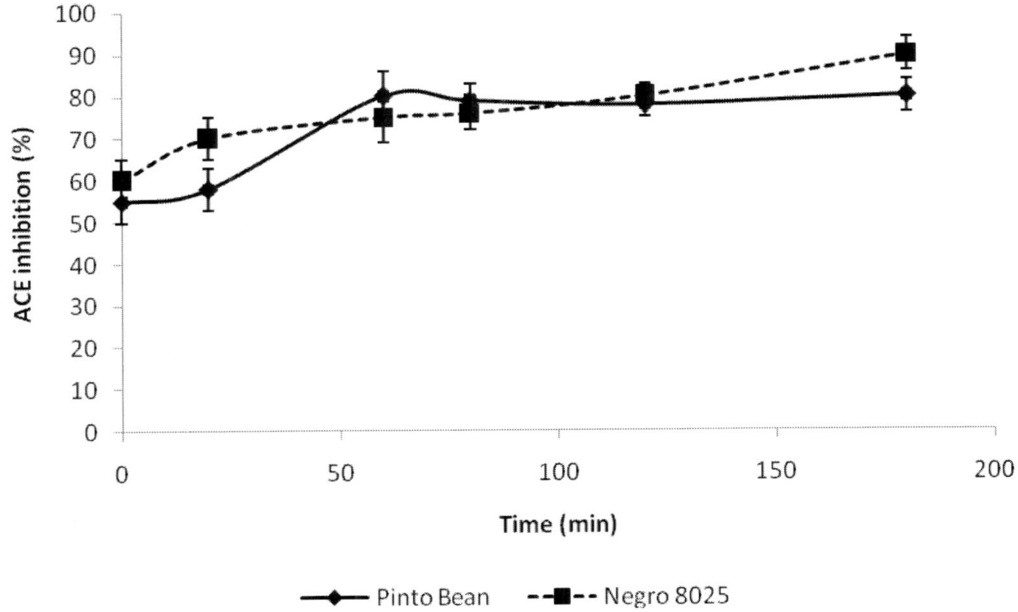

Figure 2. ACE inhibitory activity of bean protein hydrolysates of two bean cultivars. Results are the mean ± SD of two independent experiments with three repetitions.

Conclusion

Beans are a good source of nutritious factors, but also of phytochemicals such as phenolic compounds, flavonoids, and peptides that have biological activities (Antioxidant, ACE inhibition) making beans a source of compounds that could benefit the health of consumers.

Acknowledgment

We thank Foundación Carolina for granting a short postdoctoral scholarship to the first author (May – July, 2009).

References

[1] Guzmán Maldonado, SH; Paredes López, O. Pérdidas irreversibles de un rico legado de la naturaleza para la humanidad. El caso del frijol común en México. *Perspectivas*, 1998. 49(4), 5-13.

[2] Guevara-Lara, F; Espinosa-Alonso, LG; Valverde, ME; Lygin, A; Widholm, J; Paredes-López, O. Phenolics, flavonoids and oher nutraceuticals in Mexican wild common beans (*Phaseolus vulgaris*), in *Hispanic Foods*, 2006, American Chemical Society: Washington, DC. 77-88.

[3] Madhujith, T; Shahidi, F. Beans: A source of natural antioxidants, in *Phenolic compounds in foods and natural health products*, 2005, American Chemical Society: Washington, DC. 83-93.

[4] Aguilera, Y; Estrella, I; Benitez, V; Esteban, RM; Martín-Cabrejas, MA. Bioactive phenolic compounds and functional properties of dehydrated bean flours. *Food Research International*, 2011, 44(3), 774-780.

[5] Azevedo, L; Gomes, JC; Stringheta, PC; Gontijo, ÁMMC; Padovani, CR; Ribeiro, LR; Salvadori, DMF. Black bean (*Phaseolus vulgaris* L.) as a protective agent against DNA damage in mice. *Food and Chemical Toxicology*, 2003, 41(12), 1671-1676.

[6] Oomah, BD; Luc, Gv; Leprelle, C; Drover, JCG; Harrison, JE; Olson, M. Phenolics, phytic acid, and phytase in canadian-grown low-tannin faba bean (*Vicia faba* L.) genotypes. *Journal of Agricultural and Food Chemistry*, 2011, 59(8), 3763-3771.

[7] de Mejía, EG; Castaño-Tostado, E; Loarca-Piña G. Antimutagenic effects of natural phenolic compounds in beans. *Mutation Research/Genetic Toxicology and Environmental Mutagenesis*, 2002, 441(1), 1-9.

[8] Balaban, DP; Petcu, L; Busuricu, F; Zaharia, A; Bucur, LA; Basa, M; Amariei, C. Antioxidant activity of certain vegetal active principles compared to that of citrus juices and wines. *Romanian Biotechnological Letters*, 2008, 13(6), 43-48.

[9] Balasundram, N; Sundram, K; Samman, S. Phenolic compounds in plants and agri-industrial by-products: Antioxidant activity, occurrence, and potential uses. *Food Chemistry*, 2006, 99(1), 191-203.

[10] Cardador-Martinez, A; Loarca-Pina, G; Oomah, BD. Antioxidant activity in common beans (*Phaseolus vulgaris* L.). *Journal of Agricultural and Food Chemistry*, 2002, 50(24), 6975-6980.

[11] Beninger, CW; Hosfield, GL. Antioxidant activity of extracts, condensed tannin fractions, and pure flavonoids from *Phaseolus vulgaris* L. seed coat color genotypes. *Journal of Agricultural and Food Chemistry*, 2003, 51(27), 7879-7883.

[12] Dueñas, M; Hernández, T; Estrella, I. Assessment of in vitro antioxidant capacity of the seed coat and the cotyledon of legumes in relation to their phenolic contents. *Food Chemistry*, 2006, 98(1), 95-103.

[13] Moneam, NMA. Effects of presoaking on faba bean enzyme inhibitors and polyphenols after cooking. *Journal of Agricultural and Food Chemistry*, 1990, 38(7), 1479-1482.

[14] Rocha-Guzmán, NE; Herzog, A; González-Laredo, RF; Ibarra-Pérez, FJ; Zambrano-Galván, G; Gallegos-Infante, JA. Antioxidant and antimutagenic activity of phenolic compounds in three different colour groups of common bean cultivars (*Phaseolus vulgaris*). *Food Chemistry*, 2007, 103(2), 521-527.

[15] Goyoaga, C; Burbano, C; Cuadrado, C; Romero, C; Guillamón, E; Varela, A; Pedrosa, MM; Muzquiz, M. Content and distribution of protein, sugars and inositol phosphates during the germination and seedling growth of two cultivars of *Vicia faba*. *Journal of Food Composition and Analysis*, 2011, 24(3), 391-397.

[16] Granito, M; Guinand, J; Pérez, D; Pérez, S. Valor nutricional y propiedades funcionales de *Phaseolus vulgaris* procesada: un ingrediente potencial para alimentos. *Interciencia*, 2009, 34(1), 64-70.

[17] Chel-Guerrero, L; Pérez-Flores, V; Betancur-Ancona, D; Dávila-Ortiz, G. Functional properties of flours and protein isolates from *Phaseolus lunatus* and *Canavalia ensiformis* seeds. *Journal of Agricultural and Food Chemistry*, 2001, 50(3), 584-591.

[18] Korhonen, H; Pihlanto, A. Food-derived bioactive peptides - opportunities for designing future foods. *Current Pharmaceutical Design*, 2003, 9(16), 1297-1308.

[19] Vioque, J; Sanchez-Vioque, R; Clemente, A; Pedroche, J; Yust, MM; Millan, F. Bioactive peptides in storage proteins. *Grasas y Aceites*, 2000, 51(5), 361-365.

[20] Wang, W; Gonzalez de Mejia, E. A new frontier in soy bioactive peptides that may prevent age-related chronic diseases. *Comprehensive Reviews in Food Science and Food Safety*, 2005, 4(4), 63-78.

[21] Muir, AD. Natural peptides in blood pressure control. A review. *Agro Food Industry Hi Tech*, 2005, 16(5), 15-17, 26 ref.

[22] Donkor, O; Henrikson, A; Todor, V; Nagendra, S. Probiotic strains as starter cultures improve Angiotensin-Converting Enzyme inhibitory activity in soy yogurt. *Journal of Food Science*, 2005,70(8), M375-M380.

[23] Singleton, VL; Orthofer, R; Lamuela-Raventós, RM; Lester, P. Analysis of total phenols and other oxidation substrates and antioxidants by means of Folin-Ciocalteu reagent, in *Methods in Enzymology* 1999, Academic Press. 152-178.

[24] El Hariri, B; Sallé, G; Andary, C. Involvement of flavonoids in the resistance of two poplar cultivars to mistletoe (*Viscum album* L.). *Protoplasma*, 1991, 162(1), 20-26.

[25] Aparicio-Fernandez, X; Yousef, GG; Loarca-Pina, G; de Mejia, E; Lila, MA. Characterization of polyphenolics in the seed coat of black Jamapa bean (*Phaseolus vulgaris* L.). *Journal of Agricultural and Food Chemistry*, 2005, 53(11), 4615-4622.

[26] Lin, L.-Z; Harnly, JM; Pastor-Corrales, MS; Luthria, DL. The polyphenolic profiles of common bean (*Phaseolus vulgaris* L.). *Food Chemistry*, 2008,107(1), 399-410.

[27] Boateng, J; Verghese, M; Walker, LT; Ogutu, S. Effect of processing on antioxidant contents in selected dry beans (*Phaseolus* spp. L.). *LWT - Food Science and Technology*, 2008, 41(9), 1541-1547.

[28] Burda S; Oleszek W. Antioxidant and antiradical activities of flavonoids. *Journal of Agricultural and Food Chemistry*, 2001, 49(6), 2774-2779.

[29] Sharma A; Sehgal S. Effect of domestic processing, cooking and germination on the trypsin inhibitor activity and tannin content of faba bean (*Vicia faba*). *Plant Foods for Human Nutrition*, 1992,42(2), 127-133.

[30] Youssef, MM; Abd El-Aal, MH; Shekib, LAE; Ziena, HM. Effects of dehulling, soaking and germination on chemical composition, mineral elements and protein patterns of faba beans (*Vicia faba* L.). *Food Chemistry*, 1987, 23(2), 129-138.

[31] López, A; El-Naggar, T; Dueñas, M; Ortega, T; Estrella, I; Hernández, T; Gómez-Serranillos, MP; Palomino, OM; Carretero, ME. Effect of cooking and germination on phenolic composition and biological properties of dark beans (*Phaseolus vulgaris* L.). *Food Chemistry*, 2013, 138(1), 547-555.

In: Seeds as Functional Foods and Nutraceuticals
Editors: R. Mora-Escobedo, J. De J. Berrios et al.
ISBN: 978-1-62808-489-4

Chapter 7

BIOLOGICALLY ACTIVE COMPOUNDS IN THE COMMON BEAN (*PHASEOLUS VULGARIS* L.) AND THEIR HEALTH BENEFITS

J. Carrasco Castilla, A. Rojas Ávila, A. Cabrera Orozco, C. Jiménez Martínez and G. Dávila Ortiz*

Graduados e Investigación en Alimentos, Depto Ingenieria Bioquimica, Escuela Nacional de Ciencias Biológicas, Instituto Politécnico Nacional, México D.F., México

ABSTRACT

Among dry beans, common bean (*Phaseolus vulgaris* L.) seed is widely consumed throughout the world, especially in many African, Central and South American countries. Beans are recognized for their high protein and carbohydrate contents, and as good sources of minerals, vitamins, and polyunsaturated fatty acids. In addition to its nutritional value, common bean is gaining attention as a functional or nutraceutical food. Several studies have shown a correlation between consumption of beans and a reduced risk of suffering chronic and degenerative diseases such as cancer, cardiovascular diseases, obesity and diabetes. This chapter describes these health benefits attributed to carbohydrates (fiber and oligosaccharides), proteins (lectins, α-amylase inhibitors, trypsin inhibitors) and bioactive peptides.

INTRODUCTION

Common beans (*Phaseolus vulgaris L.*) represent the major staple of eastern and southern Africa and Central and South America, where people's diets lack animal protein are largely based on beans and legumes to supply vegetable protein [1]. In addition, beans are low in fat and rich in vitamins, complex carbohydrates and minerals.

* Corresponding author: E- mail: gdavilao@yahoo.com, Graduados e Investigación en Alimentos.Departamento de Ingenieria Bioquimica, Escuela Nacional de Ciencias Biológicas. Instituto Politécnico Nacional. México D.F. C.P. 11340. México. Tel. 52 5557296000 ext 62462 fax 62463.

The actual bean cultivars produced around the world come from the wild ancestors of America. Two centers of origin have been identified according to their archeological morphological, seed protein evidence, agronomic and allozyme patterns: the highland regions of Mexico and Andean South America [2]. They were further divided into Races based on geographic origin and genetic lineage (Mesoamerican, Nueva Granada, Jalisco, Durango) [3, 4].

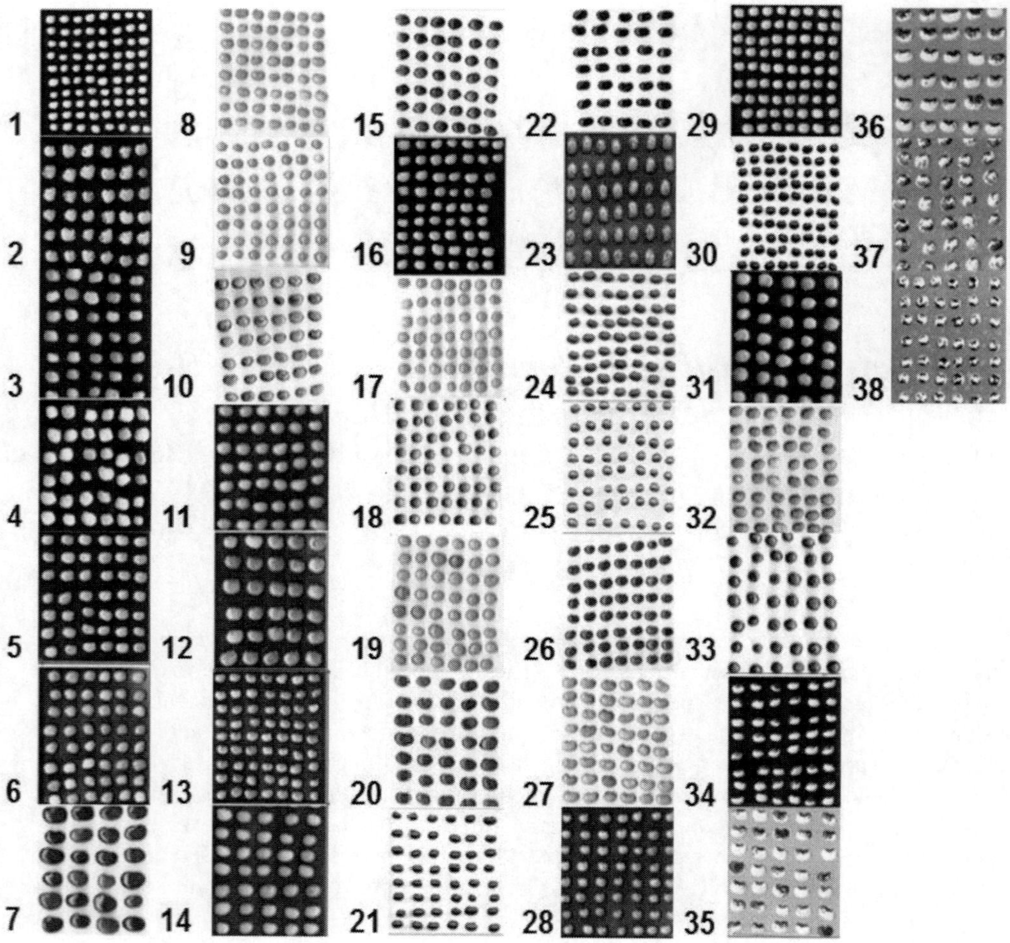

Figure 1. Common bean varieties cultivated in Mexico: 1) Alubia, 2) Batacheta, 3) Bayo, 4) Bayo gordo, 5) Bayo INIFAP, 6) Bayo Mecentral, 7) Boti, 8) Café, 9) Café bola, 10) Café rayado, 11) Era 270308, 12) Flor de Durazno, 13) Flor de Mayo, 14) Garbancillo, 15) Gris pinto, 16) Lluvia alubia, 17) Mantequilla amarillo, 18) Mantequilla bayo, 19) Mantequilla dorado, 20) Morado, 21) Negro Querétaro, 22) Negro Zacatecas, 23) Pintillo, 24) Rojo, 25) Rojo Babro, 26) Rojo varón, 27) Rosa, 28) Rosado, 29) Santa Lucía Bajo, 30) Santa Lucía Jamapa, 31) Super bayo, 32) Tepe Flor de Mayo, 33) Tigre, 34) Vaquita café, 35) Vaquita grande, 36) Vaquita grande café, 37) Vaquita morada, 38) Vaquita negra.

Mexico is a major center in domestication and genetic diversity of the common bean, counts with a high diversity of wild populations, land races or traditional cultivars and improved cultivars [4]. The wide collection of commercial cultivars of common beans from Mexico, include Azufrado, Bayo, Flor de Mayo, Flor de Junio, Pinto, black bean types, as well as some endemic varieties (such as Peruano) [5] (**Figure 1**). The second center of diversity for common bean is Africa [6]. Genetic diversity is kept at the Instituto Nacional de Investigaciones Forestales, Agricolas y Pecuarias (INIFAP) Germplasm Bank. In addition, the International Center for Tropical Agriculture in Colombia (CIAT) is the main repository for the dry bean germplasm.

Total production of beans exceeds 23.1 million tons, of which 9.9 million tons are produced in Latin America and Africa. The six major producing countries are India (4.9 million of tons), Brazil (3.2 million of tons), Myanmar (3.0 million of tons), United States of America (1.4 million of tons), China (1.63million of tons), and Mexico (1.2 million of tons) according to 2010 data released by FAO [7].

In Mexico, per capita bean consumption of common bean is 10.3 kg/year (2009) [7]. Every region of the country has developed preferential habits for consuming the beans and there are specific features that define these habits in each region. These features are related to the appearance of the grain (shape, size, and coat color) in the market, as well as to the way of processing them and the appearance of the cooked beans. In the North West region, consumers prefer the "Azufrado" types (sulphur yellow); in the North East, consumers prefer "pinto" (beige with brown mottles) and "Bayo" (cream) types; in the South, the "Black" type is preferred; and in the Center of the country, all commercial classes are consumed [8]. In Mexico there are a number of varieties of common bean; however, there is still a lack of information related to less commercial bean varieties (Mantequilla, Flor de Durazno, Rosado, Vaquita).

Beans have recently gained increasing attention as a functional food item because they contain a number of bioactive compounds such as phytochemicals, fiber, enzyme inhibitors, lectins, oligosaccharides and bioactive peptides that may play metabolic roles in humans or animals that frequently consume this food [9] . Consumption of common beans has been linked to reduced risk of diabetes, obesity, heart disease and colon cancer [10, 11]. These beneficial effects have been studied in common varieties of beans (kidney, black, Azufrado, white, pinto, navy). However, the nutraceutical properties and health benefits of less commercial beans are still unknown.

The scope of this chapter is to review the role of some bioactive compounds present in common beans related with beneficial effects in human, not including phytochemicals, which will be reviewed in the following chapter.

HEALTH BENEFITS OF COMMON BEAN INTAKE

Epidemiological and clinical studies support that consumption of common bean is important to reduce risk for chronic diseases such as coronary heart disease, diabetes mellitus, obesity and cancer [10-13]. Despite the recommendation to increase the intake of beans by health organizations, their consumption has decreased with urbanization [13].

The cholesterol-lowering property of whole beans has been demonstrated in rats [14], in hypercholesterolemic men and in normo-cholesterolemic adults [15]. Generally small (6-10%) but statistically significant reductions in total and low density lipoprotein-cholesterol occur when beans are added to the diet [13, 16]. Also, bean consumption can improve lipid profiles associated with cardiovascular disease, and lowered serum high density lipoprotein-cholesterol and low density lipoprotein-cholesterol without affecting serum triglycerides, very low density lipoprotein- cholesterol or glucose [17]. Consumption of kidney beans for 14 days gave significant decreases for both total (6 %) and low density lipoprotein- cholesterol (9%) and no significant increase in high density lipoprotein-cholesterol was observed [18]. A 1% decline in serum total cholesterol reduces the risk of coronary heart disease by 2% while each 1% decline in serum LDL cholesterol decreases the risk to about 1% [19].

The contribution of common beans on the reduction of coronary heart disease risk has also been reported by a few studies [19, 20]. The cardioprotective effects of consumption of cooked beans was investigated in Sprague–Dawley rats and a diet-induced obesity model in C57Bl/6 mice [21]. Also, individuals consuming dry beans at least 4 times per week were found to have a 22% lower risk of heart disease than individuals consuming dry beans less than once per week [12].These benefits could be due to the combination of several biological activities (antioxidant, antihypertensive and hypocholesterolemic) of compounds contained in beans.

The intake of common beans is also protective against numerous types of cancer. The effect of bean consumption has been studied in colon cancer in mice/rats [22, 23], mammary cancer in rats [24-27] and breast cancer in women [28]. The anticarcinogenic activity of beans is related to the presence of resistant starch, soluble and insoluble dietary fiber, phenolic compounds as well as other microconstituents such as phytic acid, protease inhibitors and saponins [29].

Several animal studies have been published about the hypoglycemic effect of the common bean and its possible role against diabetes mellitus [11, 20]. Most of the animal studies report the antihyperglycemic effects of common bean extracts [30, 31] or a purified component, such as alpha amylase inhibitor [32, 33]. Also, the results showed that the hypoglycemic effect is more powerful in diabetic than in healthy animals [34]. However, there is a lack of human studies examining the hypoglycemic effects of whole beans. Recently, Olmedilla-Alonso *et al,* reported that common bean intake rendered a significant reduction in the triglycemic response in diabetic patients [35].

Very few studies have examined the relationship between pulse consumption and lower body weight in humans. Bean consumers had better overall nutrient intake levels, better body weights and waist circumferences, and lower systolic blood pressure in comparison to non-consumers [36]. It is necessary to conduct longer term clinical studies involving a large amount of subjects to establish a solid basis for the utilization of beans and/or its extracts in weight loss programs [11]. A study with a *P. vulgaris* dry extract resulted in substantial reductions in food intake and body weight in genetically obese Zucker *fa/fa* rats. Additionally, acute treatment with *P. vulgaris* dry extract suppressed postprandial glycemia [37].

All these benefits associated with common bean consumption have been attributed to the presence of multiple functional components in the bean, such as dietary fiber, oligosaccharides, lectins, alpha-amylase inhibitors and peptides, which will be described below.

CARBOHYDRATES AND FIBER

Carbohydrates are the major components of legumes (24–68%); mainly in form of starch (22–45%) and the rest is comprised of non-starch polysaccharides (dietary fiber) and oligosaccharides [38]. Crude fiber is a measure of the cellulose and lignin content mainly, constitutes from 20% to 50% of dietary fiber and from 3% to 8% of whole seed [39].

Dietary fiber plays an important physiological role and has the potential to improve human health and lower the risk of many diet related non-infectious diseases [40]. The main bioactive functions that have been attributed to dietary fibers are cholesterol reduction, modulation of blood glucose levels, prevention of certain cancers, prevention of constipation, and pre-biotic effects [41-43]. Dietary fiber represents 15-48% in different types of common beans [11] and more new data has been published recently for Bayo Madero, Azufrado Higuera, Negro 8025 and bean varieties grown in Canada [43-45]. Cooking significantly increase insoluble dietary fiber and total dietary fiber content in seven varieties of beans (on a dry weight basis) [45]. Feregrino- Pérez *et al.*, found a chemopreventive effect of a polysaccharide extract (rich in insoluble fiber) from black beans on azoxymethane-induced colon cancer [46] and chemoprotective effect on HT-29 colon adenocarcinoma cells [47]. However, fiber intake from beans was not related with a lower risk of developing breast cancer in post-menopausal women in a long-term study [42].

OLIGOSACCHARIDES

The oligosaccharides are often considered negative compounds associated with flatulence and intestinal discomfort in humans. Technically, they are known as α-galactosides (raffinose family of oligosaccharides, RFO), and derived from sucrose and have 1–3 α-1,6-linked galactosyl units attached [48]. They are commonly known as raffinose (1 galactosyl unit), stachyose (2 galactosyl units), and verbascose (3 galactosyl units), and also have been shown to be able to stimulate the growth and activity of beneficial bacteria or prebiotic agent [38]. Also, oligosaccharides affect different gastrointestinal activities and lipid metabolism, to enhance immunity, and to reduce diabetes, obesity and cardiovascular risk [49]. α-Galactosides are not digested in the upper part of the gastrointestinal tract, due to the absence of α -galactosidase among human endogenous enzymes, and are therefore available for bacterial fermentation in the colon to produce a mixture of short-chains fatty acids which may have several beneficial implications for health, such as protective effects against colorectal cancer[50]. Although prebiotic effects have been demonstrated in lupin and soybean [48], there is a lack of studies of such effects with α -galactosides from common bean.

The highest content of these oligosaccharides obtained in our laboratory with different bean varieties cultivated in Mexico was found in the Flor de Mayo bean (30 mg/g), and the lowest was in the Morado bean (9 mg/g). Raffinose content ranged from 3.2 to 16.1 mg/g. Stachyose ranged from 4 to 15 mg/g and verbascose content ranged from 0.24 to 1.68 mg/g (unpublished data). These values are comparable to those observed by Doria *et al.* [1], except for the raffinose and verbascose content, 50.9 to 63.8 mg/g and 2.2 to 5.1 mg/g, respectively.

PROTEIN

Beans are a rich and inexpensive source of proteins (20–25%) in the diet, and is considered as the 'poor man's meat' [19].The major components of common bean protein are the globulins (65 % w/w), followed by albumins (15% w/w). Enzyme inhibitors and lectins that are present in beans reduce protein digestibility and nutrient absorption respectively, and lectins may also exhibit certain toxic effects at the intestinal level. Nevertheless, these undesirable effects are mostly eliminated by cooking [10].

Seed protein contents (N x 5.7) in studied beans in our laboratory ranged from 11 g/100g in Era 270308 bean to 25 g/100g in Flor de Mayo bean (unpublished data). The values observed in Mexican varieties are lower than those observed in Egyptian kidney beans (28g /100g) and Jamapa beans (27g/100g) [51, 52]. However, most of the studies use a factor of 6.25 for the conversion of nitrogen to protein in *Phaseolus vulgaris* L. seeds, which means a slightly increase in protein content.

In addition, similarities can be observed in the SDS-PAGE profiles of bean varieties (**Figure 2**). The main proteins were phaseolin with a molecular weight of 40-46 kDa being the most abundant, followed by the phytohemagglutinin group, which includes proteins of 23, 25 and 28 kDa. The number of subunits among phaseolins ranged from 2 to 6 bands with specific molecular weight for each subunit [53], which can be noted in the SDS-PAGE. Bands at 10 kDa corresponding to protease inhibitor, and 12.4, 15, 30 and 45 kDa corresponding to α-amylase inhibitor and its β subunits [9, 53, 54]. Also, some minor differences were found between the varieties. Blanco, Rojo and Peruano beans showed unique bands with molecular weight of 34-36 kDa and 13 kDa, respectively ("a" and "b" in Figure 2). Therefore, bean varieties have dissimilar protein content and patterns.

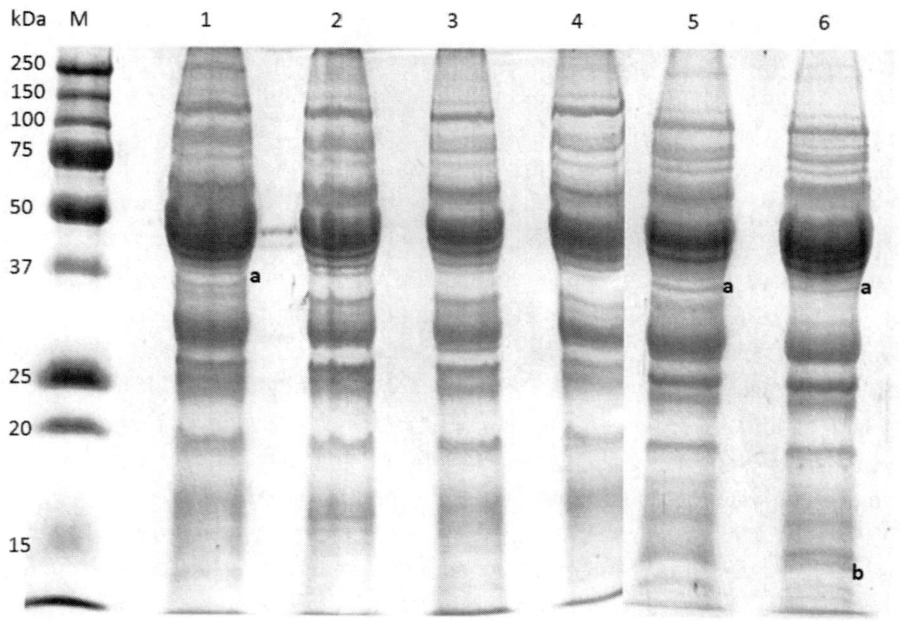

Figure 2. Electrophoretic (SDS–PAGE) pattern of *Phaseolus vulgaris* L. flours. M= Molecular weight standards (kDa), 1=Blanco, 2= Pinto Saltillo, 3= Negro Bola, 4=Negro Americano, 5=Rojo, 6= Peruano.

Some of the proteins contained in bean seeds are also implicated in the prevention of some chronic disorders such as cancer, heart disease and diabetes, such as lectins, alpha amylase inhibitor and trypsin inhibitors.

LECTINS

Lectins are defined as plant proteins possessing at least one non-catalytic domain that enables them to selectively recognize and binds reversibly to a specific mono- or oligosaccharide present on glycoproteins and glycolipids [55]. Lectins of common beans are called phytohemagglutinin. It consists of five different tetramers formed by two polypeptide chains: leukoagglutinating phytohemagglutinin (L-PHA) and erythroagglutinating phytohemagglutinin (E-PHA). They consists of five tetrameric lectins designated L_4, L_3E_1, L_2E_2, L_1E_3, and E_4 based on the number of L and E polypeptide subunits/ molecule [56, 57].

Lectin contents in beans can vary greatly (from 5-20%, w/w) [10]. Lectins are glycoproteins that selectively and reversibly bind carbohydrates. This property gives them a diversity of biological functions, many of which are still unknown, but generally have been considered to be toxic and antinutritional [57]. Recent studies showed that lectins may play a key role in preventing certain cancers [55, 58], in the activation of certain innate defense mechanisms and be used as a therapeutic agent for preventing or controlling obesity [59]. In addition, lectins exhibit, immunomodulatory, antifungal, antiviral, and HIV-1 reverse transcriptase inhibitory activities [57].

Within the context of cancer, bean lectin domains are reported to possess significant anti-proliferative and apoptosis-inducing activities [55]. Hemagglutinin from *P. vulgaris* cultivar "Legumi secchi" induced apoptosis in human breast cancer MCF-7 cells [60]. The lectin from *P. vulgaris* var. extralong autumn purple inhibited in a dose- and time-dependent way the proliferation of human breast cancer MCF-7cells, human liver cancer Hep G2 cells and including nasopharyngeal carcinoma cells (CNE-1, CNE-2, HNE-2) [61]. Also, PHA induced mitotic index in blood lymphocytes, which was significantly associated with an increased risk of breast cancer. Therefore, PHA may serve as an independent biomarker for breast cancer risk [62]. Hokkaido red bean hemagglutinin possesses a potent anti-proliferative activity on HepG2 cells, but it did not have anti-fungal effect against *Mycophaerella arachidicola, Fusarium oxysporum* and *Copriuns comatus*, in contrast to red kidney bean and French bean lectin [63, 64]. A dimeric lectin from *P. vulgaris* cv. Blue tiger king exhibited selective toxicity to human hepatoma Hep G2 cells and lacks significant toxicity on normal liver WRL 68 cells, so this lectin can be considered as a good candidate for treatment of liver tumor [65].

On the other hand, inclusion of raw kidney beans in the diet of obese rats reduced lipid accumulation that was related to a decrease of insulin levels caused by lectins [57, 59]. Lectin from kidney bean seeds inhibited HIV-1 reverse transcriptase, an enzyme crucial for HIV replication, as well as β-glucosidase, which has a role in HIV-1 envelope protein gp120 processing. This inhibitory effect of PHA on HIV-1 reverse transcriptase represents an important element of the antiretroviral chemotherapy [57, 61, 63, 64].

ALPHA AMYLASE INHIBITOR

α-amylase inhibitors (α-AIs) are kinds of glycosidase inhibitors such as both insect and mammalian α-amylases (α-1, 4, glucan-4-glucanogydrolases, EC3. 2. 1) [66]. α- AIs show sequence homology to the lectins but do not bind carbohydrates. In the common bean several of these lectin-like proteins have been identified, and have been shown to function as defensive proteins [56].

Common beans have 3 isoforms of α-amylase inhibitor (α-AI1, α-AI2, α-AIL) [67].The selective inhibition of human α-amylase also plays an important role in reducing digestive-starch degradation in patients suffering from diabetes and controlling human weight [33]. The α-AI1prevents starch digestion by completely blocking access to the active site of the α-amylase enzyme. The health benefits of α-AIs have been characterized and tested in numerous clinical studies described in several recent reviews [32, 33, 67]. Recently, the *P. vulgaris* (rich in α–amylase inhibitor) dry extract used by Carai *et al.*, inhibited pancreatic α-amylase, resulting in the deceleration of carbohydrate metabolism and absorption [37]. Also, the effect of an extract rich in α–amylase inhibitor on food (regular food pellets, starch-enriched diet, and chocolate-flavored beverage) reduced intake, body weight, and glycemia in rats [31]. These studies show that α-AIs from beans have potential as active ingredients against obesity and diabetes.

TRYPSIN INHIBITORS

Trypsin inhibitors (60 to 85 amino acids, MW 8-10 kDa) belong to the Bowman-Birk inhibitor family. They have seven disulfide bridges and contain two active sites which inhibit the proteolytic enzymes trypsin and chymotrypsin [59].

Trypsin inhibitors may inhibit carcinogenesis and therefore be used as chemopreventive, antiproliferative agents in transformed cells and anti-inflammatory agents [68]. However, most of these biological activities are reported for other legumes such as soybean, lentil, and tepary beans (*Phaseolus acutifolius*) [69-72], but there are few studies about health benefits of trypsin inhibitors from *P. vulgaris.* Recently, nine novel polypeptides with trypsin inhibitor activity were identified in common bean seeds. Three of these polypeptides showed additional high homology with lectins, likely indicating that they have lectin properties, while the other five showed high homology with α-amylase inhibitors, indicating that they probably have a dual inhibitory effect against trypsin and the α-amylase enzyme. These polypeptide inhibitors could have bifunctional unique properties [73]. White cloud bean trypsin inhibitors demonstrate antiproliferative activity against L1210 cells (human leukemia) but do no inhibit mycelial growth (MBL2, murine lymphoma) or HIV-1 reverse transcriptase [74]. Also, the trypsin inhibitor from *P. vulgaris* cv. brown kidney bean inhibited the proliferation of human breast cancer MCF7 cells. On the other hand, it only slightly inhibited proliferation of hepatoma HepG2 and embryonic liver WRL68 cells, but it is remarkable its high heat (up to 90 °C) and pH stability (1-14) [75]

PROTEIN HYDROLYSATES

In recent years, much research has been focused on the generation of bioactive peptides from food sources, and common bean proteins are not the exception. Bioactive peptides are short amino acid chains (2–20 amino acids, in some cases up to 40 amino acids) that are inactive within the sequence of the parent protein [76]. These peptides can be released by gastrointestinal digestion or during food processing (ripening, fermentation, cooking) and also by storage or in vitro hydrolysis by proteolytic enzymes and exert one or more biological activities [77]

Antihypertensive, antioxidant, antiinflamatory and metal chelating activities have been previously reported in bean protein hydrolysates. Torruco-Uco *et al.* evaluated the angiotensin converting enzyme (ACE)-I-inhibitory activity of Jamapa bean protein hydrolysates produced by enzymatic hydrolysis with Alcalase and Flavourzyme individually, obtaining IC_{50} values of 0.061 and 0.127 mg protein/mL, respectively [52]. Akillioglu and Karakaya assessed the ACE-I-inhibitory activity of common dry bean and pinto bean protein extracts. The effects of different heat treatment periods were investigated and the stability of the ACE-I-inhibitory activity was determined. Their results revealed that heat treatment increased the release of ACE-I-inhibitory peptides with IC_{50} values between 0.78 and 0.83 mg protein/mL, which were resistant to in vitro gastrointestinal digestion [78]. ACE inhibitory properties from navy, black and small red bean hydrolysates produced by Alcalase /Flavourzyme and Alcalase/papain sequential digestions demonstrated that a pre-heat treatment of isolates improved the ACE inhibitory activities. Also, *in vitro* digestibility simulation showed that the hydrolysates could be resistant or liberate new ACE inhibitory peptides after gastrointestinal hydrolysis [79]. Recently, the ACE-I inhibitory activity was evaluated in protein hydrolysates from normal and anthracnose disease-damaged Jamapa bean seeds. Damage by anthracnose disease to Jamapa beans does not affect the ACE-I-inhibitory activity that can be released by treatment of protein concentrates with Alcalase, resulting in IC_{50} values as low as 0.0191 mg protein /mL [80]

Protein isolates from hard-to-cook Jamapa bean were hydrolysed using pepsin-pancreatin and Alcalase™-Flavourzyme™ systems and antioxidant activity was measured using the ABTS decolorisation and DPPH assay [81, 82]. Also, some of the hydrolysates resulting from digestion of protein concentrates from three cultivars of Azufrado (sulphur yellow) beans using Thermolysin™, Alcalase™ or Pancreatin™ exhibited a high DPPH and ABTS scavenging activity [83]. Antioxidant and metal chelating activities in the hydrolysates resulting from hydrolysis of a protein isolate, purified phaseolin, and a lectin fraction using pepsin and pancreatin have also been described. The highest antioxidant activity was found in the hydrolysates resulting from hydrolytic treatment of whole protein isolates, as compared to a lectin extract and a phaseolin preparation. Nevertheless, the lectin and phaseolin hydrolysates, especially the latter, had higher copper and iron chelating activity, which constitutes a powerful antioxidant mechanism [9]. Also, these authors studied the antioxidant and metal chelating activity of peptide fractions coming from the isolate and phaseolin. Results suggest that the peptide fractions have antioxidant activity higher than the original whole hydrolysates, as determined using three different in vitro methods (ABTS, β-carotene bleaching method and reducing power) and have similar activities except for iron chelation. It seems that Phaseolin is the major contributor to the antioxidant and copper chelating activities of the hydrolyzed protein isolate [84].

Protein hydrolysates of the common bean varieties Negro 8025 and Pinto Durango inhibited inflammation in lipopolisaccharide-induced RAW 264 macrophages through suppression of NF-κB pathways. The authors suggest that hydrolysates from the common bean can be used to combat inflammatory and oxidative-associated diseases [85]. Nevertheless, further purification should be done in order to identify the sequence of the peptides that are responsible for high activity, as well as *in vivo* assays (in animal models and human clinical trials) that confirm the physiological effects of bioactive peptides liberated from common bean proteins.

CONCLUSION

Data from several *in vitro* and *in vivo* studies indicates that the common bean contains several bioactive compounds in addition to their nutritional value. They may be useful in decreasing coronary heart disease, diabetes mellitus, obesity and cancer risk, observed in several epidemiological studies. Dietary fiber, lectins, α-amylase inhibitors and bioactive peptides are the main contributors to the mentioned benefits. Although, there is high variability in the concentrations of particular components among bean varieties, it is likely that the nutraceutical properties of beans remain constant with slightly variations. In addition, there is still more to learn about the mechanism of those bioactive compounds on health and the possible relationship between two or more bioactive components contained in the common bean needs to be analyzed.

REFERENCES

[1] Doria, E., B. Campion, F. Sparvoli, A. Tava, and E. Nielsen, Anti-nutrient components and metabolites with health implications in seeds of 10 common bean (*Phaseolus vulgaris* L. and *Phaseolus lunatus* L.) landraces cultivated in southern Italy. *Journal of Food Composition and Analysis*, 2012. 26(1–2): p. 72-80.

[2] Bitocchi, E., L. Nanni, E. Bellucci, M. Rossi, A. Giardini, P.S. Zeuli, G. Logozzo, J. Stougaard, P. McClean, G. Attene, and R. Papa, Mesoamerican origin of the common bean (*Phaseolus vulgaris* L.) is revealed by sequence data. *Proceedings of the National Academy of Sciences*, 2012.

[3] Mensack, M.M., V.K. Fitzgerald, E.P. Ryan, M.R. Lewis, H.J. Thompson, and M.A. Brick, Evaluation of diversity among common beans (*Phaseolus vulgaris* L.) from two centers of domestication using 'omics' technologies. BMC Genomics, 2010. 11: p. 686.

[4] Gill-Langarica, H.R., J.S. Muruaga-Martínez, M.L.P. Vargas-Vázquez, R. Rosales-Serna, and N. Mayek-Pérez, Genetic diversity analysis of common beans based on molecular markers. *Genetics and Molecular Biology*, 2011. 34: p. 595-605.

[5] Blair, M.W., L.M. Díaz, H.R. Gill-Langarica, R. Rosales-Serna, N. Mayek-Perez, and J.A. Acosta-Gallegos, Genetic Relatedness of Mexican Common Bean Cultivars Revealed by Microsatellite Markers. *Crop Science*, 2011. 51(6): p. 2655-2667.

[6] Blair, M., L. González, P. Kimani, and L. Butare, Genetic diversity, inter-gene pool introgression and nutritional quality of common beans (*Phaseolus vulgaris* L.) from Central Africa. *Theoretical and Applied Genetics*, 2010. 121(2): p. 237-248.

[7] FAOSTAT, Food and Agricultural Organization of the United Nations. Available from: *http://faostat.fao.org/, 2013.*

[8] Análisis de la Cadena de Valor del Frijol. Secretaría de Economía. Dirección General de Industrias Básicas. Available from: *www.economía.gob.mx*, 2012.

[9] Carrasco-Castilla, J., A.J. Hernández-Álvarez, C. Jiménez-Martínez, C. Jacinto-Hernández, M. Alaiz, J. Girón-Calle, J. Vioque, and G. Dávila-Ortiz, Antioxidant and metal chelating activities of *Phaseolus vulgaris* L. var. Jamapa protein isolates, phaseolin and lectin hydrolysates. *Food Chemistry*, 2012. 131(4): p. 1157-1164.

[10] Campos-Vega, R., G. Loarca-Piña, and B.D. Oomah, Minor components of pulses and their potential impact on human health. *Food Research International,* 2010. 43(2): p. 461-482.

[11] Kadouh, H. and K. Zhou, *Nutraceutical and Health Properties of Common Beans (Phaseolus vulgaris), in Cereals and Pulses*, Y.R. Tsao and F. Shahidi, Editors. 2012, Wiley-Blackwell: Oxford, UK. p. 187-198.

[12] Iqbal, A.M., F.A. Nhevi, A. Dar Sher, and G. Ali, Health benefits associated with consumption of dry beans (*Phaseolus vulgaris* L.). *Trends in Biosciences*, 2012. 5(2): p. 97-100.

[13] Bennink, M.R., Health benefits associated with consumption of beans. *Annual Report of the Bean Improvement Cooperative* (BIC), 2010. 53: p. 2-3.

[14] Pereira, L.L.S., C.A. Pereira, R.V.d. Sousa, C.D.d. Santos, C.F.d. Moraes, and L.C. Sátiro, White Bean Flour (*Phaseolus vulgaris*). Therapeutic and Toxicological Research in Wistar Rats. *Journal of Applied Pharmaceutical Science*, 2012. 2(3): p. 1-7.

[15] Abeysekara, S., P.D. Chilibeck, H. Vatanparast, and G.A. Zello, A pulse-based diet is effective for reducing total and LDL-cholesterol in older adults. *British Journal of Nutrition*, 2012. 108(SupplementS1): p. S103-S110.

[16] Bazzano, L.A., A.M. Thompson, M.T. Tees, C.H. Nguyen, and D.M. Winham, Non-soy legume consumption lowers cholesterol levels: A meta-analysis of randomized controlled trials. *Nutrition, Metabolism and Cardiovascular Diseases*, 2011. 21(2): p. 94-103.

[17] Finley, J.W., J.B. Burrell, and P.G. Reeves, Pinto bean consumption changes SCFA profiles in fecal fermentations, bacterial populations of the lower bowel, and lipid profiles in blood of humans. *The Journal of Nutrition*, 2007. 137(11): p. 2391-8.

[18] Trinidad, T.P., A.C. Mallillin, A.S. Loyola, R.S. Sagum, and R.R. Encabo, The potential health benefits of legumes as a good source of dietary fibre. *British Journal of Nutrition*, 2010. 103(04): p. 569-574.

[19] Hayat, I., A. Ahmad, T. Masud, A. Ahmed, and S. Bashir, Nutritional and Health Perspectives of Beans (Phaseolus vulgaris L.): An Overview. *Critical Reviews in Food Science and Nutrition,* 2013.

[20] Hutchins, A.M., D.M. Winham, and S.V. Thompson, Phaseolus beans: impact on glycaemic response and chronic disease risk in human subjects. *British Journal of Nutrition*, 2012. 108(SupplementS1): p. S52-S65.

[21] Zhu, Z., W. Jiang, and H.J. Thompson, Edible dry bean consumption (*Phaseolus vulgaris* L.) modulates cardiovascular risk factors and diet-induced obesity in rats and mice. *British Journal of Nutrition*, 2012. 108(SupplementS1): p. S66-S73.

[22] Bobe, G., K.G. Barrett, R.A. Mentor-Marcel, U. Saffiotti, M.R. Young, N.H. Colburn, P.S. Albert, M.R. Bennink, and E. Lanza, Dietary Cooked Navy Beans and Their Fractions Attenuate Colon Carcinogenesis in Azoxymethane-Induced Ob/Ob Mice. *Nutrition and Cancer*, 2008. 60(3): p. 373-381.

[23] Mentor-Marcel, R.A., G. Bobe, K.G. Barrett, M.R. Young, P.S. Albert, M.R. Bennink, E. Lanza, and N.H. Colburn, Inflammation-Associated Serum and Colon Markers as Indicators of Dietary Attenuation of Colon Carcinogenesis in ob/ob Mice. *Cancer Prevention Research*, 2009. 2(1): p. 60-69.

[24] Thompson, M.D., M.A. Brick, J.N. McGinley, and H.J. Thompson, *Chemical Composition and Mammary Cancer Inhibitory Activity of Dry Bean Crop Science,* 2009. 49(1): p. 179-186.

[25] Thompson, M.D., M.M. Mensack, W. Jiang, Z. Zhu, M.R. Lewis, J.N. McGinley, M.A. Brick, and H.J. Thompson, Cell signaling pathways associated with a reduction in mammary cancer burden by dietary common bean (*Phaseolus vulgaris L.*). *Carcinogenesis*, 2012. 33(1): p. 226-232.

[26] Thompson, M.D., H.J. Thompson, M.A. Brick, J.N. McGinley, W. Jiang, Z. Zhu, and P. Wolfe, Mechanisms Associated with Dose-Dependent Inhibition of Rat Mammary Carcinogenesis by Dry Bean (Phaseolus vulgaris, L.). *The Journal of Nutrition*, 2008. 138(11): p. 2091-2097.

[27] Mensack, M.M., J.N. McGinley, and H.J. Thompson, Metabolomic analysis of the effects of edible dry beans (*Phaseolus vulgaris* L.) on tissue lipid metabolism and carcinogenesis in rats. *British Journal of Nutrition*, 2012. 108(SupplementS1): p. S155-S165.

[28] Murtaugh, M.A., C. Sweeney, A.R. Giuliano, J.S. Herrick, L. Hines, T. Byers, K.B. Baumgartner, and M.L. Slattery, Diet patterns and breast cancer risk in Hispanic and non-Hispanic white women: the Four-Corners Breast Cancer Study. *The American Journal of Clinical Nutrition*, 2008. 87(4): p. 978-984.

[29] Patterson, C.A., H. Maskus, and C. Dupasquier, Pulse Crops for health. *Cereals Foods Worlds,* 2009. 54(108-113).

[30] Helmstädter, A., Beans and diabetes: Phaseolus vulgaris preparations as antihyperglycemic agents. *Journal of Medicinal Food*, 2010. 13: p. 251-254.

[31] Fantini, N., C. Cabras, C. Lobina, G. Colombo, G.L. Gessa, A. Riva, F. Donzelli, P. Morazzoni, E. Bombardelli, and M.A.M. Carai, Reducing Effect of a *Phaseolus vulgaris* Dry Extract on Food Intake, Body Weight, and Glycemia in Rats. *Journal of Agricultural and Food Chemistry*, 2009. 57(19): p. 9316-9323.

[32] Preuss, H.G., Bean Amylase Inhibitor and Other Carbohydrate Absorption Blockers: Effects on Diabesity and General Health. *Journal of the American College of Nutrition*, 2009. 28(3): p. 266-276.

[33] Obiro, W.C., T. Zhang, and B. Jiang, The nutraceutical role of the Phaseolus vulgaris α-amylase inhibitor. *British Journal of Nutrition*, 2008. 100(01): p. 1-12.

[34] Yu, L., R. Tsao, and F. Shahidi, Cereals and Pulses: Nutraceutical properties and health benefits. *Functional Food Science and Technology*, ed. F. Shahidi. 2012, Oxford, UK: Wiley-Blackwell. 400.

[35] Olmedilla-Alonso, B., M.M. Pedrosa, C. Cuadrado, M. Brito, C. Asensio-S-Manzanera, and C. Asensio-Vegas, Composition of two Spanish common dry beans (*Phaseolus vulgaris*), 'Almonga' and 'Curruquilla', and their postprandial effect in type 2 diabetics. *Journal of the Science of Food and Agriculture*, 2012: p. n/a-n/a.

[36] Papanikolaou, Y. and V.L. Fulgoni, Bean Consumption Is Associated with Greater Nutrient Intake, Reduced Systolic Blood Pressure, Lower Body Weight, and a Smaller Waist Circumference in Adults: Results from the National Health and Nutrition Examination Survey 1999-2002. *Journal of the American College of Nutrition*, 2008. 27(5): p. 569-576.

[37] Carai, M.A.M., N. Fantini, B. Loi, G. Colombo, G.L. Gessa, A. Riva, E. Bombardelli, and P. Morazzoni, Multiple cycles of repeated treatments with a Phaseolus vulgaris dry extract reduce food intake and body weight in obese rats. *British Journal of Nutrition*, 2011. 106(05): p. 762-768.

[38] McCrory, M.A., B.R. Hamaker, J.C. Lovejoym, and P.E. Eischelsdoerfer, Pulse consumption, satiety, and weight management. *Advances in Nutrition*, 2010. 1: p. 17-30.

[39] Tiwari, B.K., A. Gowen, and B. McKenna, Pulse Foods. Processing, quality and nutraceutial applications. Food Science and Technology, International Series, ed. S.L. Taylor. 2011, San Diego, CA, USA: Elsevier Inc. 483.

[40] Eyaru, R., A.K. Shrestha, and J. Arcot, Effect of various processing techniques on digestibility of starch in Red kidney bean (*Phaseolus vulgaris*) and two varieties of peas (*Pisum sativum*). *Food Research International*, 2009. 42(8): p. 956-962.

[41] McClements, D.J., E.A. Decker, Y. Park, and J. Weiss, Structural Design Principles for Delivery of Bioactive Components in Nutraceuticals and Functional Foods. *Critical Reviews in Food Science and Nutrition*, 2009. 49(6): p. 577-606.

[42] Park, Y., L.A. Brinton, A.F. Subar, A. Hollenbeck, and A. Schatzkin, Dietary fiber intake and risk of breast cancer in postmenopausal women: the National Institutes of Health–AARP Diet and Health Study. *The American Journal of Clinical Nutrition*, 2009. 90(3): p. 664-671.

[43] Campos-Vega, R., R. Reynoso-Camacho, G. Pedraza-Aboytes, J.A. Acosta-Gallegos, S.H. Guzman-Maldonado, O. Paredes-Lopez, B.D. Oomah, and G. Loarca-Piña, Chemical Composition and In Vitro Polysaccharide Fermentation of Different Beans (*Phaseolus vulgaris* L.). *Journal of Food Science*, 2009. 74(7): p. T59-T65.

[44] Chung, H.-J., Q. Liu, K. Peter Pauls, M.Z. Fan, and R. Yada, In vitro starch digestibility, expected glycemic index and some physicochemical properties of starch and flour from common bean (*Phaseolus vulgaris* L.) varieties grown in Canada. *Food Research International*, 2008. 41(9): p. 869-875.

[45] Wang, N., D.W. Hatcher, R.T. Tyler, R. Toews, and E.J. Gawalko, Effect of cooking on the composition of beans (*Phaseolus vulgaris* L.) and chickpeas (*Cicer arietinum* L.). *Food Research International*, 2010. 43(2): p. 589-594.

[46] Feregrino-Perez, A.A., L.C. Berumen, G, Garcia-Alcocer, R,n,G. Gievara-Gonzalez, M Ramos-Gomez, R.a. Reynoso-Camacho, J.A. Acosta-gallegos, and G. Loarca-Pina. Composition and Chemopreventive Effect of Polysaccharides from Common Beans (*Phaseolus vulgaris* L.) on Azoxymethane-Induced Colon Cancer. *Journal of Agricultural and Food Chemistry*, 2008. 56(18): p. 8737-8744.

[47] Cruz-Bravo, R.K., R. Guevara-Gonzalez, M. Ramos-Gomez, T. Garcia-Gasca, R. Campos-Vega, B.D. Oomah, and G. Loarca-Piña, Fermented Nondigestible Fraction from Common Bean (*Phaseolus vulgaris* L.) Cultivar Negro 8025 Modulates HT-29 Cell Behavior. *Journal of Food Science*, 2011. 76(2): p. T41-T47.

[48] Martínez-Villaluenga, C., J. Frias, and C. Vidal-Valverde, Alpha-Galactosides: Antinutritional Factors or Functional Ingredients? *Critical Reviews in Food Science and Nutrition*, 2008. 48(4): p. 301-316.

[49] Qiang, X., C. YongLie, and W. QianBing, Health benefit application of functional oligosaccharides. *Carbohydrate Polymers*, 2009. 77(3): p. 435-441.

[50] Tang, Y., Y. Chen, H. Jiang, and D. Nie, The role of short-chain fatty acids in orchestrating two types of programmed cell death in colon cancer. *Autophagy*, 2011. 7(2): p. 235-237.

[51] Khattab, R.Y., S.D. Arntfield, and C.M. Nyachoti, Nutritional quality of legume seeds as affected by some physical treatments, Part 1: Protein quality evaluation. LWT - *Food Science and Technology*, 2009. 42(6): p. 1107-1112.

[52] Torruco-Uco, J., L. Chel-Guerrero, A. Martínez-Ayala, G. Dávila-Ortíz, and D. Betancur-Ancona, Angiotensin-I converting enzyme inhibitory and antioxidant activities of protein hydrolysates from Phaseolus lunatus and *Phaseolus vulgaris* seeds. *LWT - Food Science and Technology*, 2009. 42(10): p. 1597-1604.

[53] Montoya, C.A., P. Leterme, N.F. Victoria, O. Toro, W.B. Souffrant, S. Beebe, and J.-P. Lallès, Susceptibility of Phaseolin to in Vitro Proteolysis Is Highly Variable across Common Bean Varieties (*Phaseolus vulgaris*). *Journal of Agricultural and Food Chemistry*, 2008. 56(6): p. 2183-2191.

[54] Rui, X., J.I. Boye, S. Ribereau, B.K. Simpson, and S.O. Prasher, Comparative study of the composition and thermal properties of protein isolates prepared from nine Phaseolus vulgaris legume varieties. *Food Research International*, 2011. 44(8): p. 2497-2504.

[55] Fu, L.-l., C.-c. Zhou, S. Yao, J.-y. Yu, B. Liu, and J.-k. Bao, Plant lectins: Targeting programmed cell death pathways as antitumor agents. *The International Journal of Biochemistry & Cell Biology*, 2011. 43(10): p. 1442-1449.

[56] Nasi, A., G. Picariello, and P. Ferranti, Proteomic approaches to study structure, functions and toxicity of legume seeds lectins. Perspectives for the assessment of food quality and safety. *Journal of Proteomics*, 2009. 72(3): p. 527-538.

[57] Zhang, J., J. Shi, S. Ilic, S. Xue, and Y. Kakuda, Biological Properties and Characterization of Lectin from Red Kidney Bean (*Phaseolus Vulgaris*). *Food reviews international*, 2009. 25(1): p. 12-27.

[58] Shi, Z., N. An, S. Zhao, X. Li, J.K. Bao, and B.S. Yue, In silico analysis of molecular mechanisms of legume lectin-induced apoptosis in cancer cells. *Cell Proliferation*, 2013. 46(1): p. 86-96.

[59] Roy, F., J.I. Boye, and B.K. Simpson, Bioactive proteins and peptides in pulse crops: Pea, chickpea and lentil. *Food Research International*, 2010. 43(2): p. 432-442.

[60] Lam, S.K. and T.B. Ng, Apoptosis of human breast cancer cells induced by hemagglutinin from Phaseolus vulgaris cv. Legumi secchi. *Food Chemistry*, 2011. 126(2): p. 595-602.

[61] Fang, E.F., P. Lin, J.H. Wong, S.W. Tsao, and T.B. Ng, A Lectin with Anti-HIV-1 Reverse Transcriptase, Antitumor, and Nitric Oxide Inducing Activities from Seeds of

Phaseolus vulgaris cv. Extralong Autumn Purple Bean. *Journal of Agricultural and Food Chemistry*, 2010. 58(4): p. 2221-2229.

[62] Kosti, O., C. Byrne, C. Cocilovo, S.C. Willey, and Y.-L. Zheng, Phytohemagglutinin-Induced Mitotic Index in Blood Lymphocytes: A Potential Biomarker for Breast Cancer Risk. Breast Cancer: *Basic and Clinical Research*, 2010. 4: p. 73-83.

[63] Wong, J.H., C.T. Wan, and T.B. Ng, Characterisation of a haemagglutinin from Hokkaido red bean (*Phaseolus vulgaris* cv. Hokkaido red bean). J*ournal of the Science of Food and Agriculture*, 2010. 90(1): p. 70-77.

[64] Lam, S.K. and T.B. Ng, Isolation and characterization of a French bean hemagglutinin with antitumor, antifungal, and anti-HIV-1 reverse transcriptase activities and an exceptionally high yield. *Phytomedicine*, 2010. 17(6): p. 457-462.

[65] Fang, E., W. Pan, J. Wong, Y. Chan, X. Ye, and T. Ng, A new *Phaseolus vulgaris* lectin induces selective toxicity on human liver carcinoma Hep G2 cells. *Archives of Toxicology*, 2011. 85(12): p. 1551-1563.

[66] Yang, M.Y., X.Q. Zhang, Y. Ma, J. Shen, J.R. Song, and H.L. Zhu, Purification and partial characterization of a glycoprotein alpha-amylase inhibitor from white kidney bean (*Phaseolus vulgaris* L.) *Journal of Food Biochemistry*, 2008. 32(1): p. 72-84.

[67] Barrett, M.L. and J.K. Udani, A proprietary alpha-amylase inhibitor from white bean (*Phaseolus vulgaris*): a review of clinical studies on weight loss and glycemic control. *Nutrition Journal*, 2011. 10: p. 24.

[68] Clemente, A., G. Sonnante, and C. Domoney, Bowman-Birk Inhibitors from Legumes and Human Gastrointestinal Health: Current Status and Perspectives. *Current Protein and Peptide Science*, 2011. 12(5): p. 358-373.

[69] Caccialupi, P., L.R. Ceci, R.A. Siciliano, D. Pignone, A. Clemente, and G. Sonnante, Bowman-Birk inhibitors in lentil: Heterologous expression, functional characterisation and anti-proliferative properties in human colon cancer cells. *Food Chemistry*, 2010. 120(4): p. 1058-1066.

[70] Clemente, A., F.J. Moreno, M.d.C. Marín-Manzano, E. Jiménez, and C. Domoney, The cytotoxic effect of Bowman–Birk isoinhibitors, IBB1 and IBBD2, from soybean (Glycine max) on HT29 human colorectal cancer cells is related to their intrinsic ability to inhibit serine proteases. *Molecular Nutrition & Food Research*, 2010. 54(3): p. 396-405.

[71] García-Gasca, T., M. García-Cruz, E. Hernandez-Rivera, J. López-Matínez, A.L. Castañeda-Cuevas, L. Yllescas-Gasca, A.J. Rodríguez-Méndez, E. Mendiola-Olaya, J.L. Castro-Guillén, and A. Blanco-Labra, Effects of Tepary Bean (*Phaseolus acutifolius*) Protease Inhibitor and Semipure Lectin Fractions on Cancer Cells. *Nutrition and Cancer,* 2012. 64(8): p. 1269-1278.

[72] Jieliang, L., Y. Li, R.C. Denise, W. Xu, L. Jinping, L.K. Dennis, P. Yuri, and H. Wen-Zhe, Soybean-derived Bowman-Birk inhibitor inhibits neurotoxicity of LPS-activated macrophages. *Journal of Neuroinflammation*, 2011. 8(1): p. 15-15.

[73] Alves, M., I. Chaves, D. Carrilho, M. Veloso, and C.P. Ricardo, Detection of novel trypsin inhibitors in the cotyledons of Phaseolus vulgaris seeds. *Journal of Plant Physiology*, 2010. 167(10): p. 848-854.

[74] Sun, J., H. Wang, and T.B. Ng, Trypsin Isoinhibitors with Antiproliferative Activity toward Leukemia Cells from Phaseolus vulgaris cv White Cloud Bean. *Journal of Biomedicine and Biotechnology*, 2010. 2010.

[75] Chan, Y., Y. Zhang, and T. Ng, Brown Kidney Bean Bowman–Birk Trypsin Inhibitor is Heat and pH Stable and Exhibits Anti-proliferative Activity. *Applied Biochemistry and Biotechnology*, 2013. 169(4): p. 1306-1314.

[76] Carrasco-Castilla, J., A. Hernández-Álvarez, C. Jiménez-Martínez, G. Gutiérrez-López, and G. Dávila-Ortiz, Use of Proteomics and Peptidomics Methods in Food Bioactive Peptide Science and Engineering. *Food Engineering Reviews*, 2012. 4(4): p. 224-243.

[77] Di Bernardini, R., P. Harnedy, D. Bolton, J. Kerry, E. O'Neill, A.M. Mullen, and M. Hayes, Antioxidant and antimicrobial peptidic hydrolysates from muscle protein sources and by-products. *Food Chemistry*, 2011. 124(4): p. 1296-1307.

[78] Akıllıoğlu, H. and S. Karakaya, Effects of heat treatment and in vitro digestion on the Angiotensin converting enzyme inhibitory activity of some legume species. *European Food Research and Technology*, 2009. 229(6): p. 915-921.

[79] Rui, X., J.I. Boye, B.K. Simpson, and S.O. Prasher, Angiotensin I-converting enzyme inhibitory properties of Phaseolus vulgaris bean hydrolysates: Effects of different thermal and enzymatic digestion treatments. *Food Research International*, 2012. 49(2): p. 739-746.

[80] Hernández-Álvarez, A.J., J. Carrasco-Castilla, G. Dávila-Ortiz, M. Alaiz, J. Girón-Calle, J. Vioque-Peña, C. Jacinto-Hernández, and C. Jiménez-Martínez, Angiotensin-converting enzyme-inhibitory activity in protein hydrolysates from normal and anthracnose disease-damaged *Phaseolus vulgaris* seeds. *Journal of the Science of Food and Agriculture*, 2013. 93(4): p. 961-966.

[81] Ruiz-Ruiz, J., G. Dávila-Ortíz, L. Chel-Guerrero, and D. Betancur-Ancona, Angiotensin I-Converting Enzyme inhibitory and antioxidant peptide fractions from hard-to-cook bean enzymatic hydrolysates *Journal of Food Biochemistry*, 2011.

[82] Segura-Campos, M., J. Ruiz-Ruiz, L. Chel-Guerrero, and D. Betancur-Ancona, Antioxidant activity of Vigna unguiculata L. walp and hard-to-cook Phaseolus vulgaris L. protein hydrolysates. CyTA - *Journal of Food*, 2012: p. 1-8.

[83] Valdez-Ortiz, A., C.I. Fuentes-Gutiérrez, L.J. Germán-Báez, R. Gutiérrez-Dorado, and S. Medina-Godoy, Protein hydrolysates obtained from Azufrado (sulphur yellow) beans (*Phaseolus vulgaris*): Nutritional, ACE-inhibitory and antioxidative characterization. LWT - *Food Science and Technology*, 2012. 46(1): p. 91-96.

[84] Carrasco-Castilla, J., A.J. Hernández-Álvarez, C. Jiménez-Martínez, C. Jacinto-Hernández, M. Alaiz, J. Girón-Calle, J. Vioque, and G. Dávila-Ortiz, Antioxidant and metal chelating activities of peptide fractions from phaseolin and bean protein hydrolysates. *Food Chemistry*, 2012. 135(3): p. 1789-1795.

[85] Oseguera-Toledo, M.E., E.G. de Mejia, V.P. Dia, and S.L. Amaya-Llano, Common bean (*Phaseolus vulgaris L.*) hydrolysates inhibit inflammation in LPS-induced macrophages through suppression of NF-κB pathways. *Food Chemistry*, 2011. 127(3): p. 1175-1185.

In: Seeds as Functional Foods and Nutraceuticals
Editors: R. Mora-Escobedo, J. De J. Berrios et al.
ISBN: 978-1-62808-489-4

Chapter 8

CHANGES OF PHENOLIC COMPOUNDS AND ANTIOXIDANT ACTIVITY INDUCED BY GERMINATION OF *VICIA FABA*

***Dulce María Valdez-Anguiano*[1], *Gloria Dávila-Ortiz*[1], *B. Dave Oomah*[2], *Anaberta Cardador-Martínez*[3] *and Cristian Jiménez-Martínez*[*1]**

[1] Graduados e Investigación en Alimentos, Depto Ingenieria Bioquimica, Escuela Nacional de Ciencias Biológicas, Instituto Politécnico Nacional, México D.F., México
[2]National Bioproducts and Bioprocesses Program, Pacific Agri-Food Research Centre, Agriculture and Agri-Food Canada, Summerland, British Columbia, Canada
[3]Tecnológico de Monterrey, Campus Querétaro, Querétaro, México

ABSTRACT

In Mexico, the seeds of faba beans (*V. faba*) are a food preferably consumed in dry form. This seed has good nutritional value (proteins and carbohydrates) that contributes to the proper functioning of the human body. Moreover, it has some constituents which are considered as non-nutritive compounds and have demonstrated important biological activity, such as phenolic compounds. During seed germination changes occur (caused by the hydrolysis of reserve storage material and synthesis of components in order to produce a new plant) in the concentration, distribution and structure of phenolic compounds, thereby modifying their potential as antioxidant agents which may participate in the scavenging of various free radicals, chelation of metals or inhibition of some enzymes. Therefore, the objective of this research was to evaluate the change in the nutritional and non-nutritional composition, as well as in the biological activity of phenolic compounds during germination in the different structures (cotyledon, hull and embryonic axis) of three varieties (Col-160, Col-181 and Col-281) of *V. faba*. Changes in

* Corresponding author; E-mail:crisjm_99@yahoo.com]. Graduados e Investigación en Alimentos. Escuela Nacional de Ciencias Biológicas. Instituto Politécnico Nacional. Prolongación de Carpio y Plan de Ayala. Col. Sto. Tomás. Del. Miguel Hidalgo. C.P.11340. México, Distrito Federal. Telephone +52 5557296000x62462.

the composition of phenolic compounds were evaluated for different germination periods (0, 8, 12, 16 days), besides their antioxidant capability by different methods (ABTS, DPPH, FRAP, ORAC, and chelation of Fe and Cu). During germination, it was observed that while the total concentration of phenolic compounds remained constant in the cotyledon and decreased in the hull, this value significantly increased in the embryonic axis with time; the antioxidant activity significantly increased as well in the latter structure, unlike in the other ones, throughout the germination process. Therefore, the germination is a low cost process that can modify the composition and nutritional quality of seeds, so that germinated faba bean seeds might be used as food with potential benefit for health.

INTRODUCTION

The faba bean is a seed that belongs to the Leguminosae family, genus Vicia (name given to it due to botanical kinship with the vetch) and species faba [1]. Bean seed (*Vicia faba*) is known since ancient times for its various applications, whether therapeutic, ornamental or food [2]. Internationally, the faba bean is of great economic importance especially in East and Mediterranean countries. In Mexico, it is cultured in the central states of the country, being the state of Puebla the leading producer of dry faba beans [3-5].

Dry legumes have high protein content (15 to 45%), which is primarily found in cotyledons and embryonic axes (0.213 and 0.334 g/100 g, respectively). The protein content in *V. faba* represents about 26% of its total composition; however, this fluctuates according to genetic variability, cultivation method and ripeness of the seed [6, 7]. Carbohydrates are *V. faba* major components, mainly starch and oligosaccharides (stachyose, raffinose and verbascose).

Among the plants used for human consumption, legumes are important because of their nutritive potential. However, they have different compounds that may cause concentration-dependent changes in the bioavailability of nutrients, by interference with the uptake, utilization and digestibility of the latter, and adverse effects in humans and animals (growth inhibition, damage in some tissues or in organism function) when consumed in large amounts [8]. For that reason, they were considered 'anti-nutritional factors', but the negative effects of these compounds can be reversed, depending on their concentration and frequency of consumption, and their interaction with other components of the diet; so that these factors are considered as pro-nutrients as well. These compounds without any defined nutritional function and which are not considered as essential have a significant impact on the course of diseases and, thus, they are also identified as phytochemicals. Legumes containing these compounds are considered as functional foods since they provide health benefits [9, 10]. Many of these substances have been identified by their adverse effects, so that their removal was assumed as necessary for a better use of legumes. Nonetheless, recent studies have shown that among the beneficial effects these compounds may act as prebiotics, protectors of circulatory system, antihypertensives, blood glucose and cholesterol regulators, anticarcinogenics, immune response enhancers, among others; they are currently called 'active nutritional factors' or 'bioactive compounds' [11]. Such compounds include: protease inhibitors, phytic acid, saponins, lectins, polyphenolic compounds, among others [12-15]. Some of these compounds are involved in the secondary metabolism of legumes by acting as reserve compounds (inositol phosphate, α-galactosides) that are synthesized and accumulated during the germination process or as a defense mechanism of the plant against attack by bacteria, viruses, fungi, insects and animals, including humans [16]. These compounds have varied nature: proteins (protease inhibitors, α-amylase inhibitors, lectins), carbohydrates (α-galactosides, vicine, convicine, saponins), non-protein amino acids (L-DOPA acid, β-oxalyl-diamino-propionic acid (β-ODAP), polyphenols (condensed tannins), alkaloids, etc., so that their extraction and quantification are very specific; they are not equally present in all legumes, and their physiological effects in humans and other animals may differ.

PHENOLIC COMPOUNDS

Polyphenols are a large group of compounds including potential chemopreventive ones. They are present in all plants and, therefore, in the diet, but their concentration varies depending on the type of food [17, 18]. Chemically, they can be defined as substances with an aromatic ring to which one or more hydroxyl groups are attached, including functional derivatives. The main polyphenols in legumes and cereals are flavonoids and tannins [18]. Phenolic compounds are derivatives of phenylalanine and, in smaller proportion, tyrosine [19-21]. This group includes a variety of phenolic derivatives including acid phenols, phenylpropanoids, benzoic acid and its derivatives, flavonoids, stilbenes, tannins, lignins and lignans; some of them are soluble in water, and the others in organic solvents [22]. Polyphenols are a heterogeneous group, from a metabolic point of view, and are biosynthesized via the shikimic acid pathway, which is responsible of most of the phenolic compounds biosynthesis in plants [23].

Some of the sensory properties of the vegetable-origin products such as colour, flavour and astringency are due to the presence of phenolic compounds. There are more than 8000 structures and these range from simple molecules (phenolic acids with a 6 carbon atoms ring) to highly polymerized compounds (includes tannins) [18]. These compounds are essential for growing plants, since they act against predators and insect pathogens [22], and also contribute to the pigmentation of edible plants. Besides, they function as antibiotics, natural pesticides, signalling substances in order to establish symbiosis with Rhizobium, UV light protective agents, insulation materials to form a cell impermeable wall to gas and water, and as structural materials to provide stability to the plants [22].

The reducing capability of phenolic compounds allows polyphenols to prevent oxidative damage and diseases. The tetramers and pentamers show the maximum capability to scavenge free radicals, since larger compounds exhibit lesser accessibility to the hydroxyl groups required to capture these radicals which cause oxidative degradation of proteins, unsaturated lipids, carbohydrates and nucleic acids, besides stimulating macrophages and monocytes which release cytokines (inducers of inflammation) and participating in the pathogenesis of ulcers, diabetes, atherosclerosis, carcinogenesis and aging [24, 25].

GERMINATION PROCESS

Germination is a simple, low cost food processing method which naturally increases the digestibility of proteins and carbohydrates, as well as the degradation of these biomolecules for use them in cellular respiration, energy production, and synthesis of new cellular constituents of the developing embryo; hence, this process can produce significant changes in the biochemical, nutritional and sensory characteristics of legumes. In addition, the germination process improves the nutritional quality of legumes, not only by diminishing non-nutritional compounds, but also by the increase in the level of free amino acids which can be used in the synthesis of new proteins in the developing seedling [26-29].

Due to the importance of the availability of functional foods, it is necessary to acquire knowledge about the changes occurring in the concentration of phenolic compounds and their antioxidant activity during germination; even if some data have been reported previously, the evolution of the optimal period when the germination process is extending beyond eight days remains unknown.

PHENOLIC COMPOUNDS IN FABA BEANS

In this chapter, three varieties of faba bean (Vicia faba) Col 160, Col 181 and Col 281 were studied. The three varieties had a seed coat of different color and flowering cycles (early, intermediate and late). Faba beans were obtained from the Collection of Mature Faba bean from the Germplasm of the Central Valley of Mexico, Colegio de Postgraduados University campus Puebla.

Table 1. Total phenolic compounds in ethanol extracts from seed structures of *faba bean* (V. faba) during different germination time (mg eq (+) catechin/g dry sample)

Variety/ flowering	Structure	Time (Days)			
		0	8	12	16
Col-160	Cotyledon	24.17 ± 0.43^{d}	26.43 ± 0.37^{b}	24.38 ± 0.56^{cd}	29.76 ± 0.39^{a}
Early	Hull	15.18 ± 0.19^{a}	6.91 ± 0.37^{b}	4.51 ± 0.13^{cd}	4.91 ± 0.05^{a}
	Embryonic axis	---	138.21 ± 1.91^{c}	153.59 ± 1.74^{a}	148.20 ± 0.97^{b}
Col-181	Cotyledon	21.07 ± 0.40^{d}	24.09 ± 0.52^{a}	22.66 ± 0.25^{b}	22.45 ± 0.22^{bc}
Intermediate	Hull	11.44 ± 0.21^{a}	4.89 ± 0.17^{b}	4.21 ± 0.10^{c}	3.70 ± 0.34^{d}
	Embryonic axis	---	139.48 ± 2.25^{b}	164.23 ± 1.68^{a}	138.69 ± 1.61^{bc}
Col-281	Cotyledon	23.29 ± 0.41^{d}	$23.60 \pm 0.31d^{c}$	25.05 ± 0.59^{b}	26.24 ± 0.56^{a}
Late	Hull	14.90 ± 0.79^{a}	7.41 ± 0.20^{b}	4.98 ± 0.05^{c}	4.72 ± 0.12^{cd}
	Embryonic axis	---	141.51 ± 2.44^{c}	145.49 ± 4.47^{a}	144.82 ± 1.51^{ab}

* Results correspond to mean values after triplicate-measurements ± standard deviation, expressed as mg equivalent to (+) catechin/g dry sample. ** Different letters in the same row indicate significant differences by Duncan´s method ($p < 0.05$).

The concentration of total phenolic compounds in ungerminated faba bean seeds varied between 32.51-39.35 mg eq of (+)-catechin corresponding to early and late flowering varieties. Total phenolic content for whole seed was obtained by the addition of phenolic content of cotyledon and hull (Table 1). Although the intermediate flowering variety had the maximum colour intensity in the hull, its phenolic content was the lowest, which suggests that the colour is not necessarily associated with the content of phenolic compounds (Table 1). When the seed structures (cotyledon and hull) were separated from each other, it was observed that the highest fraction of ethanol-soluble phenolic compounds was found in the cotyledon, being almost twice of the extracted from the hull.

In studies performed by Oomah et al., [30] and Pastor-Cavada et al., [31], a concentration of phenolic compounds ranging from 4.6 to 40.63 mg/g in ungerminated faba bean seeds of different varieties, was reported, so that the values observed in this study are within this range; these studies reported also the influence of genotype and place of production as determinant factors for variation in phenolic content [30, 32].

Beans (*Phaseolus vulgaris*) are an extensively studied legume, in which phenolic compounds range between 4.15 and 13.26 mg/g (+)-catechin [33] in ketone extracts, and 11.61 to 55.30 mg/g in methanol extracts [34]; on the other hand, in faba bean, values from 2.19 to 5.55 mg/g have been observed, which are close to the lower limits referenced herein [35]. In other legumes such as *Vicia sativa*, soy bean and chickpea, phenolic concentrations are around 11.3, 3.6 and 15.2 mg / g, respectively [36, 37].

During the *Vicia faba* seeds germination process, the concentration of phenolic compounds in the cotyledon of the three varieties was maintained nearly constant throughout the germination period, showing only a maximum increase in the order of 14.33% (16 days). While in the hull, a significant reduction was observed, being higher at eight days (57.26%) in the variety Col-181. After analysis to the ketone, aqueous and methanol extracts from the hull of germinated *Phaseolus vulgaris* seeds, phenolic content values ranging from 2.07-158.2 mg/g, 4.21-66.02 mg/g, and 56.31-57-04 mg/g, [33, 34] were respectively obtained. While the hull of chickpea and horse gram seeds, showed a concentration of 75.9 and 484.6 mg/g, respectively [37].

As germination progressed (8 days) and the embryonic axis appeared, it was observed that the concentration of phenolic compounds in this structure was almost five times higher than that found in the cotyledon, being noticeably the variety Col-281 (141.51 mg/g (+)-catechin). As can be seen, although the concentration increases at 12 days, it is slightly reduced at 16 days. Previous studies have reported the increase that occurs in the phenolic composition in embryonic axes of mung bean from 9.76 mg/g to 20.53 mg/g after 7 days of germination [38]; while in chickpea the phenolic concentration is 26.4 mg/g after germination [37]. The results suggest that the germination process has a significant effect on the increase of this type of compounds in the embryo axis.

Ferric-Reducing Antioxidant Power (FRAP) in Faba Bean Ethanolic Extracts

The FRAP (Ferric-Reducing Antioxidant Power) method is based on the reductive capability (transfer of an electron from the antioxidant molecule to the oxidant one) [39] of one sample, i.e. it determines the ferric reducing capability of the sample. At low pH and in the presence of a reducer (antioxidant) compound, the complex tripyridyl triazine (TPTZ)-Fe (III) is reduced to the ferrous form, producing an intense blue colour with maximum absorbance at 595 nm. The FRAP activity showed by the ethanolic extracts of phenolic compounds from seed structures of faba bean germinated for various periods is shown in Table 2. As observed, the FRAP activity was lower for the cotyledons of the three varieties than in hull, being nearly constant during germination in both structures. The FRAP activity has been evaluated in cotyledon of ungerminated faba bean seeds, resulting in 2.67 mg eq Fe (II) SO4/g. In other legumes, like chickpeas, values close to those reported in this chapter were obtained [40]; while other foods such as corn, peas, wheat, lentils, soybeans, pine nuts and peanuts have shown FRAP activities between 0.12 - 3.87 mg eq Fe (II) SO4/g [41].

In the hull, the highest FRAP antioxidant activity was found in the ungerminated Col-160 (33.19 ± 0.48 mg eq Fe (II) SO4/g); however, in the three varieties a decrease (28.52 %) was observed with germination time (**Table 2**). Although there are no reports about the determination of FRAP antioxidant activity in the hull, it is important to mention that its activity is similar to other foods such as raisins and apricot [40].

The FRAP antioxidant activity in the embryonic axis was greater than in the other two structures (hull and cotyledon), observing that the FRAP value is proportional to germination time. Varieties Col-160 and Col-181 showed the greatest increase in FRAP activity (≈ 158%), at 16 days. Similarly to the hull, the FRAP activity has not been reported for the embryonic axis; nonetheless, it is possible to do comparisons with some herbs and spices such as paprika, whose antioxidant activity is similar to that found in the embryonic axis at the sixteenth day of germination process [40]. On the other hand, this activity has been reported in Phak Miang (*Gnetum gnemon*) (a Thai plant), with lower values than those determined in this study (323.72-354.34 mg eq Fe (II) SO4/g) (41). It is also suggested that the reaction to determine FRAP activity involves an electron transfer mechanism, in which the main functional groups of the phenolic compounds present in embryonic axis increase the FRAP activity.

Table 2. Ferric reducing antioxidant power (FRAP) of ethanolic extracts from seed structures of *V. faba* after different periods of germination (mg eq Fe (II) SO4/g dry sample)

Variety/ flowering	Structure	Time (Days) 0	8	12	16
Col-160	Cotyledon	3.30±0.08^{d}	5.97±0.12^{a}	4.41±0.10^{c}	5.43±0.09^{b}
Early	Seed coat	33.19±0.48^{a}	10.76±0.19^{b}	6.57±0.10^{c}	5.54±0.11^{d}
	Embryonic axis	---	223.21±6.73^{c}	577.22±16.02^{b}	664.05±10.69^{a}
Col-181	Cotyledon	2.17±0 03^{c}	3.67±0.07^{b}	4.09±0.11^{a}	4.11±0.03^{a}
Intermediate	Seed coat	21.71±0.48^{a}	6.13±0.29^{b}	5.26±0.10^{c}	3.76±0.10^{d}
	Embryonic axis	---	231.69±2.85^{c}	599.98±11.71^{b}	645.90±19.34^{a}
Col-281	Cotyledon	3.38±0.06^{d}	5.43±0.16^{c}	5.56±0.09^{b}	6.41±0.05^{a}
Late	Seed coat	29.22±1.17^{a}	10.89±0.25^{b}	6.70±0.19^{c}	4.88±0.07^{d}
	Embryonic axis	---	318.47±7.40^{c}	536.90±10.38^{b}	635.44±8.37^{a}

* Results correspond to mean values after in triplicate-measurements ± standard deviation, expressed as mg equivalent to Fe (II) SO4/g dry sample). ** Different letters in the same row indicate significant differences by Duncan´s method ($p < 0.05$).

Oxygen Radical Absorbance Capacity (ORAC) in Faba Bean Ethanolic Extracts

ORAC antioxidant activity (Oxygen Radical Absorbance Capacity) is the only method that combines time and degree of inhibition of a compound. This assay depends on the inhibition of free-radical-caused damage by a fluorescent probe, which produces a decrease in fluorescence intensity. The presence of the antioxidant inhibits free radical damage and the effect can be determined by fluorescent compounds. This assay is used in food products [42].

As can be seen, the ORAC activity in ungerminated seeds was greater in the hull than in the cotyledon, the latter being similar in all three varieties. However, in the case of the hull, ORAC activity was lower in variety Col-181 (37%). By evaluating changes in ORAC antioxidant activity during different periods of germination, it was found that the antioxidant activity in cotyledon has a constant increase as germination proceeded. Among the three varieties, the Col-181 (intermediate) had the lowest increase in activity. In the Col-160 and Col-281 increases were observed (165 and 117%., respectively) at 16 day in relation to the initial antioxidant activity (**Table 3**). Studies in faba bean seeds have reported values ranging 37.88-62.79 mg Trolox eq/g [30], which are above the ones obtained in this study. While in P. vulgaris and lentil ungerminated seeds, values have been reported between 13.30 and 94.90 mg eq Trolox/g, similar to those obtained for different germination times; in *P. vulgaris* and soya bean, reported values are higher (72.0 to 157.9 mg eq Trolox / g) [33, 35, 41].

Table 3. Oxygen Radical Absorbance Capacity (ORAC) of ethanolic extracts from seed structures of V. faba after different periods of germination (mg eq Trolox/g dry sample)

Variety/ flowering	Structure	Time (Days) 0	8	12	16
Col-160	Cotyledon	23.50±2.18^{c}	37.67±4.30^{b}	42.74±4.59^{b}	62.30±5.17^{a}
Early	Seed coat	36.55±1.76^{a}	29.40±2.25^{b}	18.29±1.84^{c}	18.63±1.58^{c}
	Embryonic axis	---	229.61±9.76^{b}	453.67±10.29^{a}	463.94±7.72^{a}
Col-181	Cotyledon	20.56±3.06^{c}	28.36±3.10^{b}	32.90±4.67^{a}	34.57±3.00^{a}
Intermediate	Seed coat	24.07±2.19^{a}	15.76±0.80^{b}	13.06±1.14^{c}	11.40±1.59^{c}
	Embryonic axis	---	197.76±11.26^{b}	413.36±34.66^{a}	407.78±12.42^{a}
Col 281	Cotyledon	23.66±1.66^{d}	32.88±3.36^{c}	39.54±3.10^{a}	51.49±4.40^{a}
Late	Seed coat	37.93±2.76^{a}	27.84±1.54^{b}	17.92±0.71^{c}	17.37±0.79^{c}
	Embryonic axis	---	271.89±12.04^{c}	342.89±4.44^{b}	436.17±17.28^{a}

* Results correspond to mean values after in triplicate-measurements ± standard deviation, expressed as mg equivalent to Trolox/g dry sample). ** Different letters in the same row indicate significant differences by Duncan´s method ($p < 0.05$).

Regarding the hull, in the three varieties, as germination proceeds the activity decreases about 50% at 16 days. In the first eight days, the variety Col-181 had the highest reduction (34.52%) and at 12 days the variety Col-160 decreases 50%. In previous studies, the hull of different varieties of *P. vulgaris* which were not subjected to germination showed antioxidant activity ranging from 4.92 to 342.57mg Trolox eq / g [33].

As can be seen, among the three structures, the embryonic axis had the highest antioxidant activity and it was similar in all three varieties. At 12 days of germination process, increases were observed in the activity of the different varieties, being the highest the embryonic axis of Col-181 (109.02%); also, this variety showed a decrease in the activity at 16 days, while other varieties followed a pattern of increases of this property during germination (Table 3). Previously, the antioxidant activity of aqueous extracts from wheat germ at different stages of germination has been evaluated, in which an increase in activity at 8 and 10 days was observed, while in the ethanol extracts the activity was kept constant. However, the reported values for both types of extracts are lower than those determined in the present study [43].

ABTS Scavenging Capacity in Faba Bean Ethanolic Extracts

The capacity to scavenge of radical ABTS (2,2 '-azino-bis (3-ethylbenzothiazoline-6-sulphonic acid) has been frequently used for the evaluation of some individual compounds and complex mixtures of various origins. In this method, radical ABTS is directly generated in stable form by using potassium persulfate; the generation of the radical before antioxidants are added to the mixture prevents interferences in the radical formation [42].

The ungerminated cotyledon's inhibitory activity of radical ABTS presented similar values close to 16%. As germination proceeded, increases in ABTS scavenging were observed. After 8 days of germination, ABTS discoloration was greater (48%) in the cotyledon of variety Col-160, while this value decreased in the variety Col-281. After this period, the ABTS discoloration activity was different, observing that the variety Col -160 decreased, with the greatest reduction at 12 days (19.81%); in the Col-181, increments up to 21% occurred at 16 days of germination, while the variety Col-281 showed a reduction of 2.79% at 12 days and an increase of 126% at 16 days (Table 4). It has been identified the presence of antioxidant activity in different varieties of amaranth (*Amaranthus cruentus*), around 11.42 ± 1.2 and 12.71 ± 1.1, while quinoa (*Chenopodium quinoa*) had higher antioxidant capability compared to seeds of faba bean and amaranth [44].

Discoloration of ABTS radical in the hull was greater in ungerminated variety Col-160 (32.31 ± 0.90%). It is worth to mention that ABTS discoloration capacity decrease during germination times in the three varieties tested up to 76.63, 67.22 and 48-24% by Col-281, Col-181 and Col-160, respectively (**Table 4**).

Table 4. ABTS scavenging capacity of ethanolic extracts from seed structures of *V. faba* after different periods of germination

Variety/ flowering	Structure	Time (Days) 0	8	12	16
Col-160	Cotyledon	16.99±0.72^{c}	25.29±0.31^{a}	20.28±0.57^{b}	17.67±0.88^{c}
Early	Seed coat	32.31±0.90^{a}	8.04±0.51^{b}	5.12±0.18^{c}	2.65±0.28^{d}
	Embryonic axis	---	75.08±0.90^{a}	76.40±1.02^{a}	75.35±0.88^{a}
Col-181	Cotyledon	16.95±0.88^{cd}	18.00±0.68^{c}	18.39±0.65^{cb}	22.39±0.74^{a}
Intermediate	Seed coat	12.78±0.52^{a}	4.18±0.26^{b}	1.37±0.07^{c}	1.17±0.08^{c}
	Embryonic axis	---	77.40±0.73^{a}	72.99±0.95^{c}	74.92±0.21^{b}
Col-281	Cotyledon	16.46±0.70^{b}	11.13±0.27^{c}	10.82±0.39^{c}	24.49±0.73^{a}
Late	Seed coat	20.45±0.21^{a}	4.78±0.26^{b}	3.64±0.19^{c}	0.91±0.07^{d}
	Embryonic axis	---	78.33±1.85^{b}	77.42±0.98^{b}	86.88±0.90^{a}

* Results correspond to mean values after in triplicate-measurements ± standard deviation. ABTS antioxidant activity expressed as percentage ABTS discoloration. ** Different letters in the same row indicate significant differences by Duncan´s method ($p < 0.05$).

The maximum capacity to scavenge ABTS in the embryonic axis was observed in variety Col-281 (78.33 ± 1.85%) at 8 days of germination. This capacity showed different behaviours according to the variety analyzed, Col-181 tend to decrease their capacity to scavenge ABTS free radical during germination on the contrary, Col -281 increased from 78.33 to 86.66 after 16 days of germination. In the Col-160 the capacity of discoloration ABTS solution was maintained during the formation of embryonic axes (**Table 4**). In previous studies in lentil and wheat sprouts a tendency to increase the antioxidant activity of radical ABTS was observed during periods of 0, 8 and 10 days of germination [43, 45].

DPPH Free Radical Scavening Capacity in Faba Bean Ethanolic Extracts

The organic radical DPPH (2,2-diphenyl-1-picrylhydrazyl), which is relatively stable, has been widely used to determine the antioxidant activity of some individual phenolic compounds, as well as in extracts from different types of food [42].

From the two structures obtained from ungerminated faba seeds, hull had the greatest DPPH free radical scavenging capacity, an average of seven-times greater than cotyledon (**Table 5**). Ungerminated cotyledon capacity ranged from 10.44 to 14.52 % DPPH discoloration, while ungerminated hull showed values of 75.32 and 88.60, corresponding to Col-181 and Col-281 respectively.

Table 5. DPPH free radical scavenging capacity of ethanolic extracts from seed structures of *V. faba* after different periods of germination

Variety/ flowering	Structure	Time (Days) 0	8	12	16
Col-160 Early	Cotyledon	10.44±0.76^{d}	20.15±0.25^{b}	13.42±0.60^{c}	21.54±2.18^{ab}
	Seed coat	87.60±0.69^{a}	47.40±1.61^{b}	26.20±1.20^{c}	23.77±0.74^{d}
	Embryonic axis	---	84.27±0.51^{b}	85.75±0.22^{a}	84.74±0.33^{b}
Col-181 Intermediate	Cotyledon	14.52±1.14^{b}	24.02±0.76^{a}	5.35±0.26^{d}	7.45±0.51^{c}
	Seed coat	75.32±1.84^{a}	25.87±0.26^{b}	18.68±0.96^{c}	3.29±0.19^{d}
	Embryonic axis	---	81.16±0.96^{b}	84.53±0.62^{a}	84.61±0.13^{a}
Col-281 Late	Cotyledon	12.62±0.65^{d}	24.02±0.62^{c}	25.87±0.19^{b}	33.65±0.33^{a}
	Seed coat	88.60±0.07^{a}	52.61±0.32^{b}	36.76±0.48^{c}	28.73±0.83^{d}
	Embryonic axis	---	82.21±0.44^{b}	82.72±0.70^{b}	85.37±0.63^{a}

* Results correspond to mean values after in triplicate-measurements ± standard deviation. DPPH antioxidant activity expressed as percentage DPPH discoloration. ** Different letters in the same row indicate significant differences by Duncan´s method ($p \leq 0.05$).

Col-160 cotyledon DPPH discoloration started with 10.44 %, and after 16 days of germination this value was increased up to 21.54 %. On the other hand, Col-181 and Col-281 cotyledons showed a general tendency to decrease their DPPH discoloration capacity.

Previous works reported for ungerminated *P. vulgaris* seeds inhibitory activity around 27.5 to 30.7% [34]; as well, in chickpea and horse gram values of 34.2 and 26.3% have been respectively observed, which are above those reported in this study. High inhibitory activity in lentils and soybeans [37, 41] has been reported also.

Regarding hull, DPPH discoloration capacity decreases along the germination time. The greatest diminution was observed in Col 181 variety, followed by Col-160 and finally Col-281 (95.6, 73, 62 % reduction, respectively). These results suggest that compounds present in seed coat could be used by the growing plant. Besides, the reduction in DPPH discoloration activity corresponds with the reduction in phenolic compounds during germination (Table 1). The highest DPPH radical discoloration occurred at 8 days of germination in the variety Col-181 (65.65%), while at 12 days the variety Col-160 (44.73%) was the highest. At 16 days the greatest reduction occurred in the variety Col-181 (82.39%); besides, this variety showed the greatest reduction compared to the initial discoloration (Table 5). The inhibitory activity in seed coat in *P. vulgaris*, chickpea and horse gram has been reported previously in the range of 13.1 to 30.7%, which is lower than that reported in this chapter; nevertheless, the percentage of inhibition observed for the different periods germination are in this range [34, 37].

The embryonic axis had an average of 83 .9 % DPPH discoloration, no matter variety, structure or germination time, values were very similar. The highest DPPH radical scavenging capacity in the embryonic axis was observed in the variety Col-160 (84.27 ± 0.51%) at 8 days of germination, though at 12 days variety Col-181 (84.51%) was the highest. At 16 days, the inhibition decreased 1.18% in the variety Col-160, but increased in Col-281 and maintained in Col-181 (**Table 5**).

In previous works on the embryonic axis of chickpea seeds the DPPH radical scavenging activity values was reported 26.4%, while in horse gram it was 15.4% [37]; also, in a previous study on leaves and stems of Phak Miang (*Gnetum gnemon*) an inhibitory activity of 51.67 to 52.23% was reported; these values were lower than those quantified in the present work for different germination stages [46].

Conclusion

From the results obtained in this study, from the three structures evaluated in faba bean seeds at different germination stages, it is worth mentioning the important phenolic content mainly located in the embryonic axis. The determination of the antioxidant activity of these compounds, by using four techniques with different mechanisms of action as free radical scavengers (DPPH, ABTS), hydrogen donors (FRAP), or singlet oxygen attenuators (inhibiting the reaction of Fenton), as well as their potential effect as metal chelators (Fe +, Cu +), allowed to verify the correlation between the concentration of phenolic compounds and the potential functional activity of the faba bean sprouts. Although the faba bean is not widely consumed in Mexico, germination may reduce some undesirable factors, adding value to the germinated product, since this would improve the digestibility and uptake of proteins and starch, while natural antioxidant compounds are simultaneously obtained along with the former. All this might be helpful to produce a nutritive, low cost functional food with good sensory properties.

ACKNOWLEDGMENTS

We wish thank the Instituto de Ciencia y Tecnologia del Distrito Federal (ICyT-DF), proyect PY-ICYT-D.F.-258/2011 and Instituto Politecnico Nacional for financial support for this research.

REFERENCES

[1] Fuccillo D., Sears L., and Stapleton P., *Biodiversity in Trust: Conservation and Use of Plant Genetic Resources in CGIAR Centres2007: Cambridge University Press.* 388.

[2] Giambanco de Ena H. Historia de las habas. 2008; Available from: *http://www.horticom.com/pd/imagenes/70/682/70682.pdf.*

[3] SIAP. Cierre de la producción agrícola por estado. 2007; Available from: *http://www.siap.gob.mx/index.php?option=com_wrapper&view=wrapper&Itemid=351.*

[4] FAO. Agricultural Production. 2010; Available from: *http://faostat.fao.org/site/default.aspx.*

[5] SDR. Cultivos del Estado de Puebla. 2010; Available from: *http://puebla.gob.mx.*

[6] FAO. Food composition table for use in East Asia. 1972; Available from: *http://www.fao.org/docrep/003/x6878e/X6878E03.htm.*

[7] INCAP. Tabla de composición de alimentos de américa latina. 2000; Available from: *http://www.tabladealimentos.net/tca.*

[8] Champ M.M.J., Non-nutrient bioactive substances of pulses. *British J Nutr*, 2002. 88(SUPPL. 3): p. S307-S319.

[9] Silvera M.B., Monereo S., and Molina B., *Alimentos funcionales y nutrición óptima. ¿Cerca o lejos? Rev Esp Salud Pública, 2003.* 77(3): p. 317-331.

[10] Bello J., *Alimentos saludables. Calidad de vida, alimentos y salud humana 2005, Madrid, España: Ediciones Díaz de Santos.*

[11] Muzquiz M., Welham T., Altares P., Goyoaga C., Cuadrado C., Romero C., Guillamon E., and Domoney C., The effect of germination on seed trypsin inhibitors in Vicia faba and *Cicer arietinum. J Sci Food Agric,* 2004. 84(6): p. 558-562.

[12] Liener I.E., *Toxic constituents of plant foodstuffs1980: Academic Press.*

[13] Janardhanan K., Gurumoorthi P., and Pugalenthi M., Nutritional potential od five accessions of a south Indiand tribal pulse, *Mucuna pruriens* var utilis I. The effect of processing methods on the content of L-Dopa, phytic acid, and oligosaccharides. *Tropical and Subtropical Agroecosystems,* 2003. 1: p. 141-152.

[14] Serrano J. and Goñi I., Papel del frijol negro *Phaseolus vulgaris* en el estado nutricional de la población guatemalteca. *Archivos Latinoamericanos deNutricion,* 2004. 54: p. 36-44.

[15] Bartholomai G.B., Tosi E., and González R., Caracterización de compuestos nutritivos, no nutritivos y calidad proteica., in *Programa Iberoamericano de Ciencia y Tecnología para el Desarrollo 2000, Eudeba*: Buenos Aires, Argentina.

[16] Chung K.T., Wong T.Y., Wei C.I., Huang Y.W., and Lin Y., Tannins and human health: A review. *Critical Reviews in Food Science and Nutrition*, 1998. 38(6): p. 421-464.

[17] Bravo L., Polyphenols: Chemistry, dietary sources, metabolism, and nutritional significance. *Nutrition Reviews,* 1998. 56(11): p. 317-333.

[18] Ucero A., Haza A.I., Arranz N., García A., and Morales P., Agentes quimio preventivoos de la dieta(II): Indoles y polifenoles. *Alimentaria*, 2007. 7(388): p. 72-78.

[19] Van Sumere C.F., *Phenols and phenolic acids, in Methods in plant biochemistry.* Press A., Editor 1989. p. 29-73.

[20] Shahidi F., Antioxidants in food and food antioxidants. *Nahrung - Food,* 2000. 44: p. 158-163.

[21] Shahidi F., *Antioxidant in plants and oleaginous seeds, in Free radicals in food: chemistry, nutrition and health effects*. Morello M.J., Shahidi F., and Ho C.-T.-. Editors. 2002, ACS Symposium series Washington, D.D. p. 162-165.

[22] Butler L.G., Protein-polyphenol interactions: nutritional aspects, in *Proc 16th Int Conf Groupe Polyphenols1992*. p. 125-138.

[23] Taiz L. and Zeiger E., *Fisiología Vegetal2006: Universitat Jaume I. Servei de Comunicació i Publicacions.*

[24] Rice-Evans C.A., Miller N.J., and Paganga G., Antioxidant properties of phenolic compounds. *Trends in Plant Science*, 1997. 2(4): p. 152-159.

[25] Valko M., Leibfritz D., Moncol J., Cronin M.T., Mazur M., and Telser J., Free radicals and antioxidants in normal physiological functions and human disease. *Int J Biochem Cell Biol,* 2007. 39(1): p. 44-84.

[26] Kollöffel C. and van Dijke H.D., Mitochondrial Arginase Activity from Cotyledons of Developing and Germinating Seeds of Vicia faba L. *Plant Physiology,* 1975. 55(3): p. 507-510.

[27] Muzquiz M., Cuadrado C., Guillamon E., Goyoaga C., Altares P., Varela A., and Pedrosa M.M., *Calidad nutritiva de judía grano, in III Seminario de Judía de la Península Ibérica2003.*

[28] Cruz-Huerta E. and Verdalet-Guzmán I. Tortillas de maíz: una tradición muy nutritiva 2007; Available from: *http://www.uv.mx/cienciahombre/revistae/vol20num3/articulos/tradicion/index.html.*

[29] López-Amorós M.L., Hernández T., and Estrella I., Effect of germination on legume phenolic compounds and their antioxidant activity. *Journal of Food Composition and Analysis,* 2006. 19(4): p. 277-283.

[30] Oomah B.D., Luc G., Leprelle C., Drover J.C.G., Harrison J.E., and Olson M., Phenolics, phytic acid, and phytase in canadian-grown low-tannin faba bean (Vicia faba L.) genotypes. *Journal of Agricultural and Food Chemistry,* 2011. 59(8): p. 3763-3771.

[31] Pastor-Cavada E., Juan R., Pastor J.E., Alaiz M., and Vioque J., Nutritional Characteristics of Seed Proteins in 28 Vicia Species (Fabaceae) from Southern Spain. *Journal of Food Science*, 2011. 76(8): p. C1118-C1124.

[32] Pastor-Cavada E., Juan R., Pastor J.E., Alaiz M., Girón-Calle J., and Vioque J., Antioxidative activity in the seeds of 28 vicia species from southern spain. *Journal of Food Biochemistry,* 2011. 35(5): p. 1373-1380.

[33] Oomah B.D., Corbe A., and Balasubramanian P., Antioxidant and Anti-inflammatory Activities of Bean (*Phaseolus vulgaris* L.) Hulls. *Journal of Agricultural and Food Chemistry,* 2010. 58: p. 8225-8230.

[34] Cardador-Martínez A., Loarca-Piña G., and Oomah B.D., Antioxidant Activity in Common Beans (Phaseolus vulgaris L.). *Journal of Agricultural and Food Chemistry,* 2002. 50: p. 6975-6980.

[35] Oomah B.D., Blanchard C., and Balasubramanian P., Phytic acid, phytase, minerals, and antioxidant activity in Canadian dry bean (*Phaseolus vulgaris* L.) cultivars. *Journal of Agricultural and Food Chemistry,* 2008. 56(23): p. 11312-11319.

[36] Megías C., Pastor-Cavada E., Torres-Fuentes C., Girón-Calle J., Alaiz M., Juan R., Pastor J., and Vioque J., Chelating, antioxidant and antiproliferative activity of *Vicia sativa* polyphenol extracts. *European Food Research and Technology,* 2009. 230(2): p. 353-359.

[37] Sreerama Y.N., Sashikala V.D., and Pratape V.M., Variability in the distribution of phenolic compounds in milled fractions of chickpea and horse gram: evaluation of their antioxidant properties. *Journal of Agricultural and Food Chemistry,* 2010. 58: p. 8322-8330.

[38] Troszyńska A., Wotejszo A., and Narolewska O., Effect of germination time on the content of phenolic compound ans sensory quality of mung bean (Vigna radiata L.) sprouts. *Polish Journal of Food and Nutrition Sciences*, 2006. 15(56): p. 453-459.

[39] Prior R.L. and Cao G., In vivo antioxidant capacity: comparison of differebt analytical methods. *Free Radic Biol Med,* 1999. 27: p. 1173-1181.

[40] Pellegrini N., Serafini M., Salvatore S., Del-Rio D., Bianchi M., and Brighenti F., Total antioxidant capacity of spices, dried fruits, nuts, pulses, cereals and sweets consumed in Italy assessed by thre different in vitro assays. *Mol Nutr Food Res,* 2006. 50: p. 1030-1038.

[41] Xu B.J., Yuan S.H., and Chang S.K.C., Comparative analyses of phenolic composition, antioxidant capacity, and color of cool season legumes and other selected food legumes. *Journal of Food Science*, 2007. 72(2): p. S167-S177.

[42] Dudonne S.p., Vitrac X, Coutiere P., Woillez M. and Mreillon J.-M,. Comparative Study of Antioxidant Properties and Total Phenolic Content of 30 Plant Extracts of Industrial Interest Using DPPH, ABTS, FRAP, SOD, and ORAC Assays. *Journal of Agricultural and Food Chemistry,* 2009. 57(5): p. 1768-1774.

[43] Kulkarni S.D., Tilak J.C., Acharya R., Rajukar N.S., Devasagayam T.P.A., and Reddy V.R., Evaluation of the antioxidant activity of wheatgrass (*Triticum aestivum* L.) as a function of growth under different conditions. *Phytotherapy Research*, 2006. 20: p. 218-227.

[44] Paśko P., Bartoń H., Zagrodzki P., Gorinstein S., Fołta M., and Zachwieja Z., Anthocyanins, total polyphenols and antioxidant activity in amaranth and quinoa seeds and sprouts during their growth. *Food Chemistry*, 2009. 115(3): p. 994-998.

[45] Świeca M., Gawlik-Dziki U., Kowalczyk D., and Złotek U., Impact of germination time and type of illumination on the antioxidant compounds and antioxidant capacity of *Lens culinaris* sprouts. *Scientia Horticulturae,* 2012. 140(0): p. 87-95.

[46] Rungruang Y. and Chanthachum S., *Antioxidant Activity of Chlorophyll Ethanolic Extracts From Phak Miang* (*Gnetum gnemon* Linn.).

In: Seeds as Functional Foods and Nutraceuticals
Editors: R. Mora-Escobedo, J. De J. Berrios et al.
ISBN: 978-1-62808-489-4

Chapter 9

PHYTOCHEMICALS AND ANTIOXIDANT ACTIVITY IN CRIOLLO CORN VARIETIES FROM MEXICO AFTER NIXTAMALIZATION AND TORTILLA PREPARATION

Cristian Jiménez-Martínez[1,*], Anaberta Cardador-Martínez[2], Abel Gil-Muñoz[3] and Gloria Dávila-Ortiz[1]

[1]Graduados e Investigación en Alimentos, Depto Ingenieria Bioquimica
Escuela Nacional de Ciencias Biológicas, Instituto Politécnico Nacional, México
[2]Instituto Tecnológico y de Estudios Superiores de Monterrey,
Campus Querétaro, Avenida Epigmenio González No. 500
Fraccionamiento San Pablo, México
[3]Colegio de Posgraduados Campus Puebla, México-Puebla,
Santiago Momoxpan, San Pedro Cholula, Puebla, CP, Mexico

ABSTRACT

Corn *Zea mays* is a major grain worldwide and a significant source carbohydrate and phytochemical compounds. An evaluation of total phenolic compounds (TPC), anthocyanin and carotenoid contents in 25 criollo corn varieties, including white, yellow, blue, red and spotted varieties was made, subjected to no treatment (whole grain), to the thermal/alkaline process of nixtamalization and to tortilla preparation. The untreated blue varieties had the highest overall content of TPC, anthocyanins and carotenoids, followed by the spotted, red, yellow and white varieties. After thermal alkaline processing, TPC increased in the yellow and white varieties, less so that in the red and spotted varieties and decreased in the blue varieties. Anthocyanin content in the colored varieties decreased notably with processing, while carotenoid content generally increased in yellow and white varieties. Antioxidant activity was highest in the untreated yellow and white varieties, followed by the spotted, red and blue varieties. Thermal alkaline processing diminished this activity in the yellow and white varieties but increased it in the spotted, red and blue varieties. The tested criollo corn varieties are good sources of phenolic compounds, anthocyanins and carotenoids, and exhibit antioxidant activity,

* E-mail: crisjm_99@yahoo.com. Tel. 52 5557296000 ext 62462 fax 62463.

although these vary between varieties after thermal alkaline processing with nixtamalization.

INTRODUCTION

Originally domesticated in Mesoamerica, corn is the oldest cultivated grain in the Americas, and was probably used as food as early as 7000 BCE [1]. Its cultivation spread across the Americas, and, after European contact, to the rest of the world. For ancient complex societies of the Western Hemisphere, corn was a pillar of the food systems, and continues to be a fundamental foodstuff around the world wide, as well as one of the best available forage grains [2].

Well before European contact, the cultures of what today is Mexico cultivated corn in diverse agroclimatic conditions. Over centuries, the inhabitants of different regions developed technology and appropriate production systems for corn cultivation based on human labor and variety classification. These varieties were adapted for use in production of processed foods such as tortillas, a Mesoamerican dietary staple for millennia. Selected, regionally-adapted varieties are known as criollo varieties [3].

Corn generally provides 74 g kg^{-1} carbohydrates, 9 g kg^{-1} protein, 3.4 g kg^{-1} fat and 1 g kg^{-1} fiber [4], and is also a significant source of phytochemicals, including phenolic acids, flavonoids, anthocyanins and carotenoids. Carotenoids are a family of pigments with color intervals ranging from yellow to orange, found in plants, over 600 have been identified from natural contexts to date. Xanthophylls (lutein, zeaxanthin and cryptoxanthin) and carotenoids (α- and β-carotenoids) are the dominant phytochemicals in white and yellow corn [5-8]. Carotenoid intake has been associated with reduced risk of chronic diseases such as cancer, and cardiovascular disorders, as well as improved vision [9, 10]. Blue, black and purple corns contain mainly anthocyanins, which are water-soluble anthocyanin glucosides which form part of phenolic compounds known as flavonoids. These pigments are reported to improve health due to their antioxidant, anti-inflammatory and anticarcinogenic activities and hypoglycemic effects [11, 12].

Corn is an integral ingredient in Mexican culinary culture, usually in the form of the thin, soft rounds originally called *tlaxcalli* by the ancient Aztec-Mexica people and tortillas by the Spanish. To prepare corn grain for tortilla manufacture, it is processed by nixtamalization, an ancient process still in common use. The grain is first cooked in a water/calcium hydroxide ($Ca(OH)_2$) solution, soaked overnight in the same water, washed to remove the pericarp and then milled to produce a dough. Small portions of dough are then flattened into discs about 20 cm in diameter and cooked on a hot surface [13]. Nixtamalization and subsequent heating causes changes in the processed corn's structure, chemical composition and nutritional value. The alkaline/thermal treatment (i.e. nixtamalization), washing and handling for tortilla preparation lead to partial pericarp elimination, the most obvious physical change in the grain. Nixtamalization increases calcium content, improves niacin bioavailability and significantly reduces mycotoxin levels. It also affects grain nutritional properties by increasing resistant starch content, improving protein biological quality, and decreasing anthocyanin, vitamin, mineral and phytic acid contents [5].

The present study objective was to evaluate the effect of nixtamalization and tortilla manufacture on original total phenolic compounds, anthocyanin and carotenoid contents and antioxidant capacity in 25 criollo corn varieties (including white, yellow, red, purple blue and spotted colors) from Puebla state, Mexico.

Material and Methods

Biological Material

Grain samples from 25 criollo corn varieties were collected in Puebla State, Mexico, were donated from germoplasm bank, by Dr. Abel Gil Muñoz, research-professor, Colegio de Posgraduados, Puebla. The seeds were collected in the 2006 year .All varieties have high yield per hectare. Description of the origin and physical characteristics of corn varieties are given in Table 1.

Nixtamalization and Tortilla Preparation

The corn samples were processed with a traditional nixtamalization method [14]. For each variety, 50 g of grain were weighed out and added to 200 mL water containing 1% calcium hydroxide. This mixture was heated to boiling temperature (93 °C), the time needed to separate the pericarp from the endosperm and the grain was soft, was taken as a cooking time.

The cooked corn was allowed to rest overnight in the same water which was separated to analyze its TPC (Total phenolic content) and total solids contents. The resulting nixtamalized grain was washed with two volumes of water and divided into two equal samples, the first of which was lyophilized and the second of which was milled in a traditional stone handcrafted mill adding water until forming dough (56% moisture content). Small balls of dough (20g) were flattened into discs (tortillas), which were cooked on a hot plate at 185 °C for 15 s on one side, 30 s on the other, and then returned to the original side to repeat the cooking until the steam produced within the tortilla is sufficient to inflate [15], allowed to cool to room temperature, frozen and lyophilized (-45°C, 0.32 mBar). Once dried, both batches were ground to meal for use in the analyses.

Total Phenolics Compounds Quantification

Total phenolic content was estimated by using the Folin-Ciocalteu colorimetric method [16]. Briefly, the 0.02 mL of the extracts was oxidized with 0.1 mL of 0.5 N Folin-Ciocalteu reagent, and then the reaction was neutralized with 0.3 mL sodium carbonate solution (20%). The absorbance values were obtained by the resulting blue color measured at 760 nm with a Genesys 10UV spectrophotometer (Rochester, NY, USA) after incubation for 2 h at 25°C.

Table 1. Origin and morphological characteristics of criollo corn varieties from Puebla state, Mexico

Sample	Region[1]	Location	Grain color	Length (cm)	Wide (cm)	Height (cm)	Weight (kg/hL)
B-14	VP	Santa Ma. Zacatepec	Blue	$1.51^{g,h,i,j}$	0.93^{f}	$0.49^{a,b,c,d,e}$	68
B-19	LM	Ntra. Sra. Del Monte	Blue	$1.41^{c,d,e,f,g,h}$	$0.75^{a,b}$	$0.48^{a,b,c,d}$	70
B-23	IP	S. Miguel Tianguistengo	Blue	$1.38^{b,c,d,e,f}$	$0.75^{a,b}$	$0.50^{a,b,c,d,e}$	65
B-27	VP	Huejotzingo	Blue	$1.31^{a,b,c,d}$	$0.78^{a,b,c,d}$	$0.48^{a,b,c,d}$	67
B-34	IP	Sta. Ma. Nepopualco	Blue	$1.50^{g,h,i,j}$	$0.87^{e,f}$	$0.49^{a,b,c,d,e}$	64
B-41	LM	S. Miguel Canoa	Blue	$1.48^{f,g,h,i,j}$	$0.79^{a,b,c,d}$	$0.50^{g,h,i}$	71
P-33	IP	Sta. Ma. Nepopualco	Purple	$1.56^{j,k}$	$0.81^{a,b,c,d,e}$	$0.52^{c,d,e,f,g,h}$	68
P-48	IP	S. Juan Tlale	Multicolored (Purple)	$1.41^{c,d,e,f,g,h}$	0.74^{a}	$0.51^{a,b,c,d,e,f}$	66
P-49	IP	S. Matías Atzala	Multicolored (Purple)	1.24^{a}	$0.80^{a,b,c,d,e}$	$0.57^{f,g,h,i}$	68
R-32	IP	Sta. Ma. Nepopualco	Red	$1.53^{h,i,j}$	$0.76^{a,b}$	$0.51^{b,c,d,e,f}$	69
R-40	LM	Tepatlaxco	Red	$1.29^{a,b,c}$	0.94^{f}	$0.52^{c,d,e,f,g}$	69
S-47	LM	Acajete	Spotted	$1.30.05^{a,b,c,d}$	$0.85^{d,e}$	$0.46^{a,b,c}$	69
W-20	VP	Huejotzingo	White	$1.28^{a,b}$	$0.77^{a,b,c}$	$0.54^{d,e,f,g,h,i}$	68
W-28	VP	Santa Ma. Zacatepec	White	$1.43^{e,f,g,h,i}$	$0.77^{a,b,c}$	0.44^{a}	69
W-80	IP	Iztaccíhuatl-Popocatépetl	White	$1.32^{a,b,c,d,e}$	$0.81^{a,b,c,d,e}$	$0.56^{e,f,g,h,i}$	69
W-89	LM	Ntra. Sra. Del Monte	White	$1.43^{d,e,f,g,h}$	$0.76^{a,b}$	$0.47^{a,b,c,d}$	64
W-103	LM	La Malinche	White	1.68^{k}	$0.75 \pm 0.13^{a,b}$	$0.49^{a,b,c,d,e}$	66
W-106	IP	La Preciosita	White	$1.55^{i,j}$	$0.82^{b,c,d,e}$	$0.53^{c,d,e,f,g,h,i}$	63
W-107	IP	Tlahuapan	White	$1.57^{j,k}$	$0.84^{c,d,e}$	$0.47^{a,b,c,d}$	68
W-116	LM	Acajete	White	$1.53^{h,i,j}$	$0.81^{a,b,c,d,e}$	$0.50^{a,b,c,d,e}$	69
W-122	LLS	Los Llanos de Serdán	Creamy White	$1.40^{c,d,e,f,g}$	$0.81^{b,c,d,e}$	$0.48^{a,b,c,d}$	66
WC-36	LM	Tepatlaxco	White Cacahuacintle	$1.31^{a,b,c,d}$	$0.79^{a,b,c,d}$	0.59^{i}	59
WC-42	LM	S. Miguel Canoa	White Cacahuacintle	$1.33^{a,b,c,d,e}$	1.02^{g}	$0.60^{h,i}$	68
Y-24	IP	S. Rafael Ixtapaluca	Yellow	$1.29^{a,b,c}$	$0.79^{a,b,c,d}$	$0.51^{c,d,e,f}$	69
Y-46	LM	Acajete	Yellow	$1.29^{a,b,c}$	$0.79^{a,b,c,d}$	$0.51^{c,d,e,f}$	69

[1]VP = Valley of Puebla; LM = La Malinche; IP = Iztaccíhuatl-Popocatépetl; LLS = Llanos de Serdán.
Means in the same column followed by different letter are statically different.

Quantification was done on the basis of a standard curve of gallic acid [16]. Results were expressed as mg of gallic acid equivalent per 100 g of dry weight.

Anthocyanin Quantification

Anthocyanin extraction and quantification was done following Abdel-Aal and Hucl [17]. 1 g of each blue purple or spotted corn sample was added to 20 mL acidified methanol (1% HCl in methanol), and agitated for 2 h at room temperature in darkness. An acid pH was used to take the anthocyanins to the flavylium ion form, which exhibits coloration, and thus be able to quantify them by spectrophotometry. The resulting solution was analyzed between 400 and 600 nm to identify the maximum absorption wavelength. Total anthocyanins concentration was calculated as mg cyanidin 3- glucoside equivalent per 100 g of dry sample.

Carotenoids Quantification

Carotenoids were extracted according to the method of Fish *et al.*, [18] using low hexane volumes. Briefly, 0.5 g of ground sample, 5 mL acetone, 5 mL ethanol and 10 mL hexane were placed in a 50 mL Erlenmeyer flask and agitated for 15 min in darkness. Distilled water (3 mL) was added; the mixture agitated again for 5 min and set aside to allow the two phases were separated. The hexane phase was recovered using Pasteur pipettes and analyzed at 503 nm in a Genesys 10UV spectrophotometer (Rochester, NY, USA). Carotenoid content was expressed as μg equivalent β-carotene per 100 g of sample.

Antioxidant Capacity (DPPH Method)

Measurement of antioxidant capacity was done using 2,2-Diphenyl-1-picrylhydrazyl (DPPH) (also known as 1,1-diphenyl-2-picrylhydrazyl) as a free radical. Reduction of DPPH by an antioxidant or a free radical produces decreased absorbance at 515 nm. Measurement was done following Brand- Williams et al. [19] method as adapted for use with microplates. A solution of 20 μL extract or BHT with 200 μL DPPH was prepared at a 125 μM concentration, kept at room temperature in darkness and gauged with 80% methanol. Extracts and standards were prepared in triplicate. After 60 min, the plate was read at 520 nm in a spectrophotometer, and antioxidant capacity (AOC) calculated as a percentage of DPPH discoloration [20].

Results and Discussion

Physical Characteristics and Alkaline Treatment of Grain Seeds

Description of grains is presented in Table 1. From the 25 varieties studied, 13 have light colored seed coat while the others have dark colored seed coat. There were no differences in length, wide and height of dark and light colored seeds. The average weight in kg/hL was similar among all the grains, in light colored seeds ranged between 59 and 69 kg/hL; while in dark colored seeds the variation was from 64 and 71 kg/hL. The general average weight was 68 kg/hL (Table 2). These results indicated that criollo corn varieties weighed less than white hybrid varieties, as reported by Plugfelder *et al.*, [21] Serna-Saldivar *et al.*, [22]. Despite the criollo origin of seeds, only 5 varieties of light color seed coat and 4 dark colored seed coat, did not reach the weight recommended as a criteria for grain grading (70 kg) [22].

Corn varieties usually required varying periods of time to achieve the same degree of cooking (loosening of the pericarp) as shown in Table 2. The time needed to cook the dark colored seeds varied from 42 to 95 min, with an average of 63 min.

Table 2. Physical and chemical analysis of steep liquor obtained from the thermal alkaline process of criollo corn varieties from Puebla State

Sample	Cooking time (min)	Total solids (%)	TPC (mg/100 g corn)	% TPC[1]
Dark colored corn				
B-14	82	1.23	2.42 ± 0.26 [a]	1.95 ± 0.21
B-19	61	1.85	1.75 ± 0.06 [e,f,g]	1.53 ± 0.06
B-23	78	1.67	1.29 ±0.06 [g]	1.21 ± 0.06
B-27	69	1.10	1.68 ± 0.03 [f,g]	1.32 ± 0.03
B-34	59	1.28	2.96 ± 0.14 [a,b]	3.01 ± 0.14
B-41	42	1.42	2.72 ± 0.41 [d,e,f]	2.22 ± 0.34
P33	47	1.52	1.94 ± 0.16 [d,e,f]	2.40 ± 0.20
P-48	50	1.50	2.28 ± 0.15 [c,d,e]	2.34 ± 0.15
P-49	69	1.25	1.97 ±0.51 [d,e,f]	2.33 ± 0.61
R-32	47	1.52	2.34 ± 0.09 [c,d,e]	3.51 ± 0.13
R-40	95	1.47	2.50 ± 0.13 [b,c,d]	3.41 ± 0.18
S-47	56	1.35	3.45 ± 0.34 [a]	7.82 ± 0.76
Light colored corn				
W-103	67	1.54	1.11 ± 0.07 [g]	1.63 ± 0.10
W-106	60	1.93	2.01 ± 0.05 [d,e,f]	3.57 ± 0.09
W-107	56	0.99	1.80 ± 0.05 [f]	3.68 ± 0.68
W-116	78	1.78	4.17 ± 0.04 [a]	6.64 ± 0.06
W-122	78	2.73	2.74 ± 0.07 [b]	5.21 ± 0.14
W-20	70	1.35	2.20 ± 0.07 [c,d,e]	3.69 ± 0.12
W-28	50	1.38	2.86 ± 0.09 [b,c]	4.39 ± 0.60
W-80	65	1.05	2.12 ± 0.34 [d,e,f]	4.16 ± 0.66
W-89	67	1.43	3.85 ± 0.08 [a]	5.69 ± 0.12
WC-36	85	1.26	4.15 ± 0.19 [a]	7.05 ± 0.32
WC-42	63	1.64	2.42 ± 0.15 [b,c,d]	3.83 ± 0.24
Y-24	73	1.82	1.85 ± 0.07 [e,f]	2.74 ± 0.10
Y-46	56	1.38	2.32 ± 0.30 [b,c,d,e]	3.42 ± 0.44

[1]% TPC= Percentage of total phenolics determined in steep liquor compared gainst TPC in NT corn.
Means followed by different letter in the same column are significantly different (Tuckey, alfa= 0.05).

On the other hand, light colored seeds ranged from 56 to 85 min, averaging 67 min (Table 2). These results indicated that dark colored seeds are softer than the light ones. Trejo-González *et al.*, [15] reported cooking times as short as 13 min and up to 70 min for 18 varieties of hybrid and native varieties from several regions of México. Serna Saldivar *et al.*, [22] reported cooking times up to 100 min for hybrid corn. The time required for cooking the corn with lime water apparently depends on characteristics other than endosperm type—most probably composition and thickness of the pericarp, distinct lime treatment conditions are required for different corn varieties [15].

The nixtamalization process induces changes in corn seeds. These changes include the solubilization and dissolution of pericarp tissues, especially at the endocarp layer, which facilitates the physical removal of the pericarp during subsequent agitation and rinsing processes. After the steeping time, the cooked grains are separated from the steep liquor (known as nejayote), which has become rich in lime and loosened pericarp tissues.

Dry matter losses ranged from 0.99 to 2.73 %, which are in the lower end of the values reported for several corn varieties (0.5 -14.5%) [21-23]. According to several authors, the lower dry matter loss translates into higher yields of dry masa flour or tortillas [2, 22, 23]. These authors established that extent of cooking; lime concentration, pericarp removal and grain hardness are correlated to dry matter losses. Pericarp removal during nixtamalization is critically important because its presence affects product color, texture and tortilla yields In the tortilla industry, the most outstanding kernels are those that loose most of the pericarp during lime cooking, steeping and nixtamal washing and yet do not have high dry matter loses [22].

Total Phenolics Compounds

Non treated dark colored corn varieties had higher TPC values than the light ones, with some having values almost twice those of the white varieties (Table 3). The spotted and red varieties had levels below those of the blue varieties, but generally higher than the yellow and white varieties. TPC in corn previous to treatment had ranges of 44.96 to 68.23 mg / 100 g for white corn, 49.77 to 67.66 mg / 100 g for yellow corn (Table 3), and 66.51 to 127.11 mg/ 100 g for dark colored corn varieties (Table 3). These results contrast with De la Parra *et al.*, [5], who reported no significant differences in TPC concentrations among white, yellow, red and blue varieties.

Thermal-alkaline treatment in the form of nixtamalization increased TPC content in the yellow, purple, spotted and white varieties while lowering it in the blue and varieties. In corn, phenolic compounds include flavonoids and phenolic acids [24] which are concentrated in the pericarp grain or found esterified to the heteroxylans that form the hemicelluloses [25]. The most important phenolic acids in corn grain are ferulic, p-coumaric and synapic acids [26]. The increases in TPC in the yellow and white varieties were probably due to the hydrolysis of bound phenolics. Thermal processing may release more bound phenolic acids from the breakdown of cellular constituents. Although disruption of cell walls also releases the oxidative and hydrolytic enzymes that can destroy these phenolics, thermal processing will deactivate these enzymes to avoid the loss of phenolic compounds [27]. In contrast, the decreases in TPC in the blue and spotted varieties may be attributed to formation of phenolic compound-protein complexes that decrease the proportion of phenolic compounds available for the color reaction.

Table 3. Total phenolic, anthocyanins and carotenoids contents in light and dark colored seeds of criollo corn varieties from Puebla state before and after thermal alkaline process (mg/100g sample)

	Total phenolic content			Anthocyanins		
Sample	NT	Nixtamal	Tortilla	NT	Nixtamal	Tortilla
B-14	123.88 ± 8.0 [a,b]	54.62 ± 8.0 [g]	79.00 ± 14.0 [d]	88.28 ± 12 [a]	8.26 ± 0.024 [e,f]	11.39 ± 0.27 [c,d]
B-19	114.10 ± 6.0 [a,b,c]	99.76 ± 3.0 [d,e]	113.83 ± 10.0 [c]	60.79 ± 1.0 [c]	14.18 ± 0.18 [b]	18.65 ±0.96 [a]
B-23	106.56 ± 4.0 [b,c]	97.74 ± 4.0 [d,e]	104.56 ± 6.0 [c]	68.80 ± 1.2 [b,c]	9.01 ± 0.10 [d,e]	11.75 ±0.45 [c]
B-27	127.11 ± 4.0 [a]	118.11 ± 10.0 [c,d]	105.66 ± 7.0 [c]	85.43 ± 3.0 [a]	8.37 ± 0.25 [e,f]	8.41 ± 0.23 [g,h]
B-34	98.37 ± 4.0 [c,d]	73.86 ± 5.0 [f]	105.53 ± 5.0 [c]	50.32 ± 1.3 [d]	8.65 ± 0.60 [e]	10.57 ±0.27 [d,e]
B-41	122.30 ± 14.0 [a,b]	145.49 ± 14.0 [b]	153.79 ± 6.1 [b]	70.83 ± 1.2 [b]	9.00 ± 0.07 [d,e]	11.01 ± 0.19 [c,c]
P-33	81.03 ± 5.0 [d,e,f]	174.32 ± 9.0 [a]	124.24 ± 1.0 [c]	24.42 ± 1.2 [f,g]	6.42 ± 0.51 [g]	5.32 ± 0.15 [i]
P-48	97.56 ± 2.0 [c,d]	169.58 ± 6.0 [a]	198.53 ± 8.2 [a]	72.65 ± 1.4 [b]	15.79 ± 0.77 [a]	14.71 ± 0.27 [b]
P-49	84.31 ± 3.0 [d,e]	40.16 ± 4.0 [g]	46.48 ± 2.2 [e]	32.76 ± 1.3 [e,f]	12.64 ± 0.97 [e]	11.54 ± 0.28 [c]
R-32	66.51 ± 6.0 [f]	113.55 ± 2.0 [e]	151.21 ± 5.3 [b]	38.66 ± 1.3 [e]	8.21 ± 0.24 [e,f]	9.83 ± 0.24 [e,f]
R-40	73.41 ± 9.0 [e,f]	136.40 ± 7.0 [b,c]	118.73 ± 7.4 [c]	13.33 ± 1.3 [h]	10.09 ± 0.68 [d]	8.12 ± 0.29 [h]
S-47	44.10 ± 10.0 [g]	175.81 ± 9.0 [a]	200.38 ± 13.1 [a]	18.57 ± 1.2 [g,h]	7.31 ± 0.42 [f,g]	9.26 ± 0.16 [f,g]
			Tota phenolic content Carotenoids			
W-103	68.23 ± 1.0 [a]	131.75 ± 5.0 [d]	107.11 ± 5.0 [e,f]	38.34 ± 3.0 [d,e,f]	64.91 ± 3 [b,c]	55.93 ± 5 [b,c,d]
W-106	56.35 ± 11 [a,b,c]	144.53 ± 14 [b,c,d]	111.21 ± 5.0 [d,e]	42.65 ± 1.0 [d,e]	44.52 ± 3 [e,f]	34.26 ± 2 [f]
W-107	44.96 ± 2.0 [c]	149.56 ± 10 [b,c,d]	118.07 ± 11 [d,e]	47.63 ± 8.0 [c,d]	40.22 ± 2 [e,f]	40.15 ± 3 [e,f]
W-116	62.88 ± 8.0 [a,b]	140.73 ± 13 [c,d]	87.12 ± 6.0 [f]	60.75 ±1.0 [b]	45.48 ± 2 [e,f]	52.50 ± 1 [c,d,e]
W-122	52.63 ± 2.0 [a,b,c]	142.52 ± 17 [c,d]	118.35 ± 3.0 [d,e]	28.06 ± 4.0 [f]	34.86 ± 3 [e,f]	49.12 ± 4 [c,d,e]
W-20	59.68 ± 9.0 [a,b,c]	140.17 ± 17 [c,d]	100.79 ± 11 [e,f]	55.77 ± 1.0 [b,c]	61.88 ± 5 [c,d]	42.93 ± 3 [d,e,f]
W-28	61.49 ± 2.0 [a,b]	144.61 ± 5.0 [b,c,d]	167.84 ± 13 [a,b,c]	29.93 ± 3.0 [e,f]	42.70 ± 1 [e,f]	51.77 ± 1 [c,d,e]
W-80	51.12 ± 6.0 [b,c]	159.57 ± 5.0 [a,b,c]	128.83 ± 5.0 [d]	63.94 ±10 [a,b]	51.93 ± 6 [d,e]	57.22 ± 14 [b,c]
W-89	67.63 ± 7.0 [a]	144.17 ± 3.0 [b,c,d]	180.02 ± 4.0 [a,b]	75.41 ± 7.0[a]	67.59 ±9 [a,b,c]	34.84 ± 3 [f]
WC-36	58.89 ± 4.0 [a,b,c]	136.33 ± 13 [c,d]	183.77 ± 9.0 [a]	64.08 ± 3.0 [a,b]	44.70 ± 2 [e,f]	68.57 ± 2 [a,b]
WC-42	63.23 ± 4.0 [a,b]	141.09 ± 10 [c,d]	100.44 ± 8.0 [e,f]	76.81 ± 7.0 [a]	80.11 ± 9 [a]	78.82 ± 7 [a]
Y-24	67.66 ± 11 [a]	177.81 ± 5.0 [a]	161.91 ± 12 [b,c]	33.93 ± 1.0 [e,f]	39.79 ± 2 [e,f]	50.44 ± 4 [c,d,e]
Y-46	49.77 ± 4.0 [b,c]	170.17 ± 6.0 [a,b]	150.81 ± 4.0 [c]	40.32 ±3.0 [d,e,f]	75.09 ± 6 [a,b]	78.15 ± 2 [a]

NT= Not Treated, B= blue, P= purple, R= red, S= spotted W= white, WC= Creamy-White and Y= yellow color of seeds.

Means indicated by different letter in the same column are significantly different (Tuckey, alfa= 0.05) and they were determined between light and dark colored seeds.

After nixtamalization and tortilla preparation, TPC increased significantly ($p<0.0001$) in the yellow and white varieties to double or triple the initial TPC, and to a lower degree in the spotted and red varieties (Table 3). This increase is probably related to changes in grain structure due to thermal/alkaline hydrolysis with $Ca(OH)_2$ during nixtamalization; such as breaking of fiber components located in the pericarp, aleurone cell wall, and endosperm cell walls, which leads to release of most phenolic compounds [5, 25, 28]. The presence of pericarp residues after nixtamalization and washing also probably had some impact on TPC since the pericarp has the highest phenolics content of any grain element. In contrast, TPC decreased in the blue varieties after processing, due largely to removal of anthocyanins, which were reduced up to 24.28 % in R-40, and 90.65 % in B-14.

The chemical hydrolysis releases and solubilizes phenolics that leach into the cooking liquor (also known as nejayote). In our work, the maximum concentration of TPC in steep liquor was found in WC-36 and W-116, 4 mg/100 g of corn, this value represented 7.05 % and 6.64 %, respectively, of the total concentration of phenolics in no treated corn. Light colored seeds (white and yellow) released more phenolics into steep liquor than dark colored seeds (blue, purple and red).

When phenolics in steep liquor were expressed as percentage of initial total phenolics in corn (NT), light colored seeds showed values between 1.63% and 7.05%, most of them were near to the high value. Only S-47 among the dark colored seeds had a high percentage loss of phenolics into the steep liquor (7.82 %); blue varieties showed minor losses about 1.5 %, and purple corn loss around 2.5% of phenolics into steep liquor (Table 2). Baking of masa to obtain tortillas had a slight effect on the phenolic concentration of TPC, this result agree with the data obtained by Del Pozo-Insfran [29] who processed white and blue corn into tortillas and chips. Previous studies [5, 27] showed that there were significant increases in the content of total free phenolics following thermal treatment with time and alkaline treatment, which is consistent with the data of this work.

Anthocyanins

Anthocyanins are present in most of the 41 colored corn varieties described in Mexico [30]. The black, purple and red colors in many of these varieties are caused by the presence of anthocyanins in the pericarp, the aleurone layer or in both of these grain structures [31, 32]. In the present study, the blue, red and spotted varieties had anthocyanin values ranging from 18.57 to 88.28 mg cyanidin-3-glucoside equiv. / 100 g sample (Table 3), which is higher than reported for commercial colored varieties [5]. Overall, the blue varieties had the highest levels, with particularly high values for the B-14 (88.28 mg cyanidin-3-glucoside equiv./100 g sample) and B-27 (85.43 μg cyanidin-3-glucoside equiv/ 100 g sample) varieties from the Valley of Puebla. The red and spotted varieties had values (13.33-38.66 mg cyanidin-3-glucoside equiv/ 100 g sample) approximately 75% lower than those of the blue varieties. Purple corn showed values between those of red and blue corn (Table 3). De la Parra et al., [5] observed that anthocyanin content depends on corn variety, type and color, with higher values in blue varieties than in red varieties.

Nixtamalization and tortilla preparation reduced anthocyanin levels in the blue varieties by 76-90 %, with particularly notable decreases in the B-14 and B-27 varieties (Table 3). Purple corn decreased its anthocyanin concentration approximately in 73 %. Reductions in spotted varieties were less drastic (24 %) although their initial levels were also lower than in the blue varieties. The effects of nixtamalization and tortilla preparation on anthocyanin content were similar (Table 3), a phenomenon also reported by De la Parra *et al.*, [5].

When processed in alkaline conditions the structures of anthocyanins are transformed into a colorless carbinol, or pseudobase, with a higher proportion of open form of the colorless chalcone.[33]. The calcium hydroxide increases the pH effect, which is the most important factor influencing the structural transformation of anthocyanins during the nixtamalization process, at an alkaline pH the flavyl cation begins to hydrate, giving rise to a colorless carbinol or pseudobase, in equilibrium with the open form of chalcon, which is also colorless [34].

Anthocyanins are mainly located in the pericarp and aleurone layer [32], which are both largely removed during nixtamalization, washing and subsequent discard of the waste water [5]. There are previous reports of anthocyanin loss after nixtamalization of blue corn varieties. When soaked in an alkaline solution and cooked at >90 °C, these compounds also experience a synergistic effect that leads to significant structural modifications [34, 35]. Salinas *et al.* [32] reported that the color of nixtamalized maize flour and total anthocyanin losses are influenced by their location in the maize. Higher losses occurred in samples where the anthocyanins were located in the pericarp and in nixtamalized flours from these samples.

Carotenoids

Carotenoid content among the untreated white and yellow corn varieties ranged from 33.93-76.81 μg β-carotene equivalent/100 g sample, with the highest values in the white cacahuacintle WC-42 and white W-89 varieties (Table 3). These values are higher to values reported for some hybrid genotypes of different colors by Kurilich and Juvik [6]. In contrast, the carotenoids contents observed in the present study are lower than reported by De la Parra *et al.*, [5] for white and yellow corn. Higher concentrations of carotenoids have been observed for several varieties of yellow corn [5, 6, 8], however in our study, some white varieties such as W-89 and WC-42 had higher concentration than the yellow ones. Nixtamalization and tortilla preparation increased slightly, but not significantly the carotenoid content in some of the tested white and yellow varieties (Table 3) but a decrease was observed for. W-116, W-80, W-89 and WC- 36.

Increases in thermal processed corn had been reported previously for white and yellow corn [8]. Carotenoid content varies widely in different vegetables, within a given food and in products processed similarly. This variation may be partially due to the quantification method and retention calculation; for example, increases in β-carotene and other provitamins after thermal processing cannot be associated with enzymes since these are denaturalized at high temperatures. Heat also induces chemical changes in provitamins such as isomerization and epoxidation. Increases in β-carotene levels can also be associated with the fact that carotenoids are more easily extracted from processed samples than from fresh foods, in which these molecules are physically protected or bonded to other compounds in the food that do not allow solvent penetration and compound extraction. Higher carotenoid content per food weight unit may be due to omission of moisture in calculations and loss of soluble solids while lower contents per weight unit can result from water and/or oil absorption [36-38]. On the other hand, also a decrease in carotenoid content has been assessed in corn after alkaline cooking, attributing these losses to alkaline treatment [5].

Antioxidant Capacity

Antioxidant capacity in the tested varieties ranged from 24.6 to 39.0 in the white varieties, 26.9 to 34.4 in the yellow varieties, 29.1to 42.0 in the spotted and red varieties, and 20.3 to 27.3 in the blue varieties (Figure 1A and B).

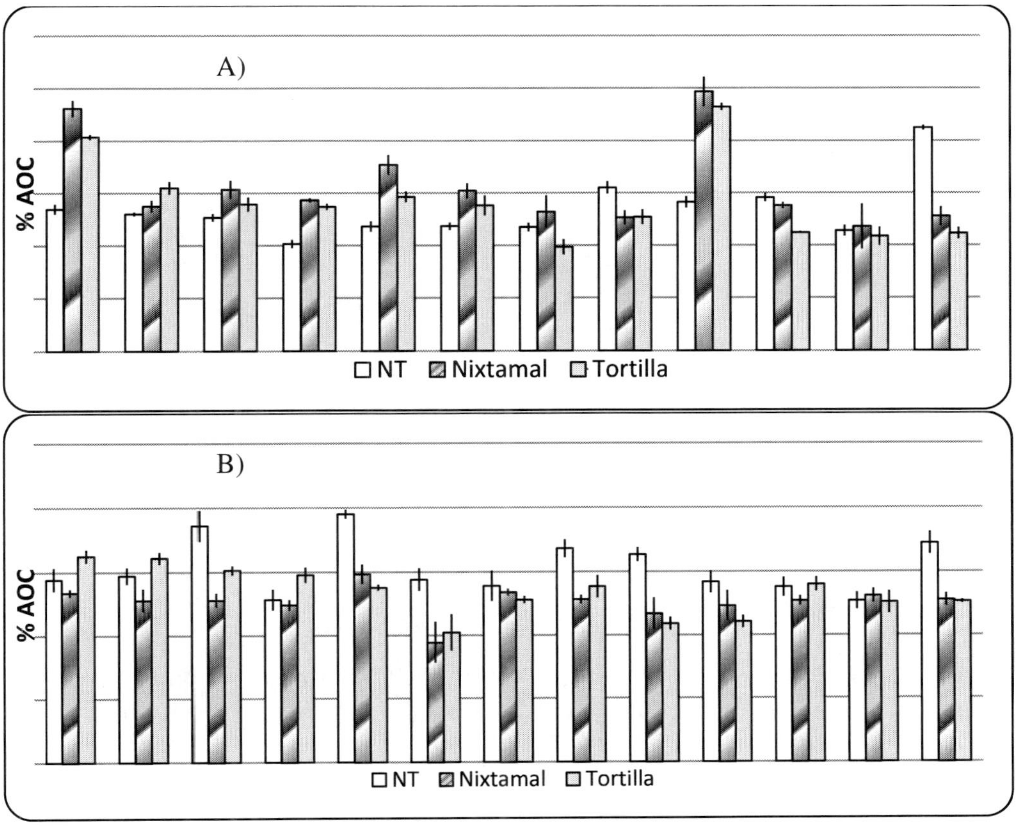

Figure 1. Antioxidant capacity in A) dark colored seeds and B) light colored seeds of criollo corn varieties from Puebla state, Mexico, after no treatment (NT), nixtamalization and preparation as tortillas. B= blue, P= purple, R= red, S= spotted, W= white, WC= Creamy-White and Y= yellow.

In the white and yellow varieties, the heat and alkaline pH of nixtamalization diminished AOC, while in the red, blue and spotted varieties it generally increased AOC. In these latter varieties, this increase was highest (50%) in the red R-49 and blue B-14 varieties from the Iztaccíhuatl-Popocatépetl region. This effect is probably due to changes in pigment compound structure caused by the alkaline medium.

USDA data on 100 foods show that cooking can increase or decrease antioxidant activity depending on antioxidant compound molecular structure [39]. In these data, phenolic compound antioxidant capacity in corn dough and tortillas is shown to decrease by an average of 25% after thermal alkaline treatment versus unprocessed grain. This coincides with the decreased antioxidant capacity behavior observed in most of the corn varieties studied here. In the present case, this behavior is probably due to the loss of some antioxidant phenolic compounds in the soaking and washing stages of nixtamalization, and detachment of the pericarp during cooking [40].

Table 4 shows the effect of seed color on the content of total phenolics, antocyanins, carotenoids and antioxidant capacity of corn seeds. Without considering treatment, color did not affect the TPC, the average concentration was of 120 mg gallic acid equivalent per 100 of sample either light colored seeds or dark colored seeds. Although dark colored seeds shown higher TPC values than light colored ones, this difference is compensated by the increase in TPC observed after alkaline treatment of light colored corn.

Table 4. Effect of seed color on the total phenolics content, antocyanins, carotenoids and antioxidant capacity of corn seeds

Seed Color	TPC[1]	Anthocyanins[2]	Carotenoids[3]	AOC[4]
Blue	108.01[a]	30.76[a]	-	29.36[a]
Purple	112.91[a]	21.80[a,b]	-	30.58[a]
Red	109.97[a]	14.71[b]	-	24.51[b]
Spotted	140.09[a]	11.71[a,b]	-	30.07[a,b]
White	109.52 [a]	-	50.03[b]	27.57[a,b]
White-Cacahuacintle	113.96 [a]	-	68.84[a]	26.02[a,b]
Yellow	129.69 [a]	-	52.95[b]	27.02[a,b]
Creamy-White	108.01 [a]	-	37.34[c]	32.09[a]

[1]Total phenolic content expressed as mg gallic acid equivalent per 100 g of dry sample.
[2]Anthocyanins content expressed as mg cyaniding-3-glucoside equivalent per 100 g of dry sample.
[3]Carotenoid content expressed as μg β-carotene equivalent per 100 g of sample.
[4]AOC= antioxidant capacity. Means followed by different letter in the same column are significantly different (Tuckey, alfa= 0.05).

Table 5. Effect of treatment (Nixtamalization and Tortilla preparation) on total phenolics, anthocyanins, carotenoids and antioxidant capacity of criollo corn seeds

Treatment	Dark colored seeds			Light colored seeds		
	TPC[1]	Anthocyanins[2]	AOC[3]	TPC[1]	Carotenoids[4]	AOC[3]
NT	94.94[b]	52.07[a]	27.12[b]	58.80[c]	50.39[a]	30.68[a]
Nixtamal	116.61[a]	9.83[b]	28.48[a,b]	147.92[a]	53.37[a]	25.31[b]
Tortilla	125.16[a]	10.88[b]	31.36[a]	132.02[b]	53.44[a]	26.79[b]

[1]Total phenolic content expressed as mg gallic acid equivalent per 100 g of dry sample.
[2]Anthocyanins content expressed as mg cyaniding-3-glucoside equivalent per 100 g of dry sample.
[3]AOC= antioxidant capacity.
[4]Carotenoid content expressed as mg b-carotene equivalent per 100 g of sample.
Means followed by different letter in the same column are significantly different (Tuckey, alfa= 0.05).

Anthocyanin concentration is modified by the color of grain seed, being blue and purple corn the varieties that showed significantly ($p<0.0001$) higher concentration than red and spotted corn. In light colored corn, there were significant differences in carotenoid content due to the variations in color, being white Cacahuacintle corn which had the highest carotenoid concentration, followed by white, yellow and creamy-white varieties. Antioxidant capacity show little differences due to color grain, either light or dark colored seeds showed similar values of AOC (Table 4).

The effect of treatment on TFC, anthocyanins, carotenoids and antioxidant capacity of criollo corn seeds is shown in Table 5. Nixtamalization and tortilla preparation increased significantly ($p<0.0001$) the TPC as compared to not treated corn, but there was no difference between nixtamalization and tortilla preparation in dark colored corn. A similar effect was observed for light colored corn. Thermal processing of corn resulted in an average decrease of four times in anthocyanin concentration. Caroteoinds were not affected by nixtamalization, as observed in Table 5. Finally, AOC increased slightly in dark colored corn, but decreased significantly in light colored seeds.

The present results clearly demonstrate that temperature and alkalinity are critical elements in the modification of corn's natural properties for human consumption. Corn, and other cereals, have significant antioxidant capacity, which may depend on the amount of ingested TPC, as shown in studies of whole-grain wheat, barley, oats, buckwheat and rye [41, 42]; corn [43]; corn byproducts such as corn steep liquor [44]; and seed oil extraction residue byproducts [45].

CONCLUSION

The different criollo corn varieties tested here can contribute with phenolic compounds and antioxidant capacity to products incorporating them, both improving product quality and providing added value to these grains. Consumption of antioxidants via whole grains helps reduce the risk of chronic degenerative diseases such as colon-rectal cancer, breast cancer, coronary diseases and diabetes.

REFERENCES

[1] MacNeish R. S., Investigaciones arqueológicas en el Valle de Tehuacán. *Revista de Arqueología Mexicana*, 1995. 13: p. 18-23.

[2] Serna Saldivar S. O., Gómez H. M., and Rooney L. W., Technology, chemistry and technology, alkaline cooked corn products, in Advances in Cereal Science and Technology. Pomeranz Y., Editor 1990, American Association of cereal Chemistry: St. Paul, M. N. p. 243–307.

[3] Rojas G. and Rovalo M., Fisiología vegetal aplicada. 3 ed1985, Madrid: Interamericana Mc-Graw Hill.

[4] Paliwal L. P., Granados G., and Marathée J. P., El Maíz en los Trópicos: mejoramiento y producción. 2001, Roma Italia: FAO.

[5] de la Parra C., Serna Saldivar S. O., and Liu R. H., Effect of Processing on the Phytochemical Profiles and Antioxidant Activity of Corn for Production of Masa, Tortillas, and Tortilla Chips. *Journal of Agricultural and Food Chemistry*, 2007. 55(10): p. 4177-4183.

[6] Kurilich A. C. and Juvik J. A., Quantification of Carotenoid and Tocopherol Antioxidants in Zea mays. *Journal of Agricultural and Food Chemistry*, 1999. 47(5): p. 1948-1955.

[7] Moros E. E., Darnoko D., Cheryan M., Perkins E. G., and Jerrell J., Analysis of Xanthophylls in Corn by HPLC. *Journal of Agricultural and Food Chemistry*, 2002. 50(21): p. 5787-5790.

[8] Scott C. E. and Eldridge A. L., Comparison of carotenoid content in fresh, frozen and canned corn. *Journal of Food Composition and Analysis*, 2005. 18(6): p. 551-559.

[9] Mayne S. T., Beta-carotene, carotenoids, and disease prevention in humans. *FASEB. J.*, 1996. 10(7): p. 690-701.

[10] Bendich A. and Olson J. A., Biological actions of carotenoids. *FASEB. J.*, 1989. 3(8): p. 1927-1932.

[11] Kong J.-M., Chia L.-S., Goh N.-K., Chia T.-F., and Brouillard R., Analysis and biological activities of anthocyanins. *Phytochemistry*, 2003. 64(5): p. 923-933.

[12] Lila M. A., Anthocyanins and Human Health: An In Vitro Investigative Approach. *Journal of Biomedicine and Biotechnology*, 2004. 5: p. 306-313.

[13] Paredes-López O., Guevara-Lara F., and Bello-Pérez L. A., Los alimentos mágicos de las culturas indígenas mesoamericanas 2006, Mexico DF: Fondo de cultura económica.

[14] Serna-Saldivar S. O., Gomez M. H., Almeida-Dominguez H. D., Islas-Rubio A., and Rooney. L. W., A Method to Evaluate the Lime-Cooking Properties of Corn (Zea mays). *Cereal Chemistry*, 1993. 70(6): p. 762-764.

[15] Trejo-Gonzalez A., Feria-Morales A., and Wild-Altamirano C., The Role of Lime in the Alkaline Treatment of Corn for Tortilla Preparation, in Modification of Proteins 1982, American Chemical Society: Washington, D. C. p. 245-263.

[16] Singleton V. L., Orthofer R., Lamuela-Raventós R. M., and Lester P., Analysis of total phenols and other oxidation substrates and antioxidants by means of Folin-Ciocalteu reagent, in Methods in Enzymology 1999, Academic Press. p. 152-178.

[17] Abdel-Aal E. S. M. and Hucl P., A Rapid Method for Quantifying Total Anthocyanins in Blue Aleurone and Purple Pericarp Wheats. *Cereal Chemistry*, 1999. 76(3): p. 350-354.

[18] Fish W. W., Perkins-Veazie P., and Collins J. K., A Quantitative Assay for Lycopene That Utilizes Reduced Volumes of Organic Solvents. *Journal of Food Composition and Analysis*, 2002. 15(3): p. 309-317.

[19] Brand-Williams W., Cuvelier M. E., and Berset C., Use of a free radical method to evaluate antioxidant activity. *LWT - Food Science and Technology*, 1995. 28(1): p. 25-30.

[20] Burda S. and Oleszek W., Antioxidant and Antiradical Activities of Flavonoids. *Journal of Agricultural and Food Chemistry*, 2001. 49(6): p. 2774-2779.

[21] Pflugfelder R. L., Rooney L. W., and Waniska R. D., Dry matter losses in commertial corn masa production. *Cereal Chemistry*, 1988. 65(2): p. 127-132.

[22] Serna-Saldivar S., Amaya Guerra C., Herrera Macias P., Melesio Cuellar J., Preciado Ortiz R., Terron Ibarra A., and Vazquez Carrillo G., Evaluation of the Lime-Cooking and Tortilla Making Properties of Quality Protein Maize Hybrids Grown in Mexico. *Plant Foods for Human Nutrition (Formerly Qualitas Plantarum)*, 2008. 63(3): p. 119-125.

[23] Rosentrater K. A., A review of corn masa processing residues: Generation, properties, and potential utilization. *Waste Management*, 2006. 26(3): p. 284-292.

[24] Arnason J. T., Gale J., Conilh de Beyssac B., Sen A., Miller S. S., Philogene B. J. R., Lambert J. D. H., Fulcher R. G., Serratos A., and Mihm J., Role of phenolics in resistance of maize grain to the stored grain insects, Prostephanus truncatus (Horn) and Sitophilus zeamais (Motsch.). *Journal of Stored Products Research*, 1992. 28(2): p. 119-126.

[25] Duodu K. G., Taylor J. R. N., Belton P. S., and Hamaker B. R., Factors affecting sorghum protein digestibility. *Journal of Cereal Science*, 2003. 38(2): p. 117-131.

[26] Classen D., Arnason J. T., Serratos J. A., Lambert J. D. H., Nozzolillo C., and Philogéne B. J. R., Correlation of phenolic acid content of maize to resistance to

Sitophilus zeamais, the maize weevil, in CIMMYT'S collections. *Journal of Chemical Ecology*, 1990. 16(2): p. 301-315.

[27] Dewanto V., Wu X., and Liu R. H., Processed Sweet Corn Has Higher Antioxidant Activity. *Journal of Agricultural and Food Chemistry*, 2002. 50(17): p. 4959-4964.

[28] Doner L. W. and Hicks K. B., Isolation of Hemicellulose from Corn Fiber by Alkaline Hydrogen Peroxide Extraction. *Cereal Chemistry*, 1997. 74(2): p. 176-181.

[29] Del Pozo-Insfran D., Brenes C. H., Serna Saldivar S. O., and Talcott S. T., Polyphenolic and antioxidant content of white and blue corn (Zea mays L.) products. *Food Research International*, 2006. 39(6): p. 696-703.

[30] Ortega P. R. A., Sánchez J., Castillo G. F., and Hernández J. M., Estado actual de los estudios sobre maíces nativos de México, in *Avances en el estudio de los recursos fitogenéticos de México*. R. Ortega P. G.P.H., F. Castillo G., V. A. González H., y M Livera M., Editor 1991, SOMEFI, A. C.: México. p. 161-196.

[31] Salinas M., Soto H., Martínez B., González H., and Ortega P., Análisis de antocianinas en maíces de grano azul y rojo provenientes de cuatro razas. *Revista Fitotecnia Méxicana*, 1999. 22: p. 161-174.

[32] Salinas-Moreno Y., Martínez-Bustos F., Soto-Hernández M., Ortega-Paczka R., and Arellano-Vázquez J. L., Efecto de la nixtamalización sobre las antocianinas del grano de maíces pigmentados. *Agrociencia*, 2003. 37(6): p. 617-628.

[33] Cevallos-Casals B. A. and Cisneros-Zevallos L., Stability of anthocyanin-based aqueous extracts of Andean purple corn and red-fleshed sweet potato compared to synthetic and natural colorants. *Food Chemistry*, 2004. 86(1): p. 69-77.

[34] Cortés G. A., Salinas M. Y., Martín-Martinez E. S., and Martínez-Bustos F., Stability of anthocyanins of blue maize (Zea mays L.) after nixtamalization of seperated pericarp-germ tip cap and endosperm fractions. *Journal of Cereal Science*, 2006. 43(1): p. 57-62.

[35] Markakis P., Stability of anthocyanins in foods, in Anthocyanins as food colors. Markakis P., Editor 1982, Academic Press: New York. p. 163-180.

[36] Dietz J. M., Kantha S. S., and Erdman J. W., Reversed phase HPLC analysis of α- and β-carotene from selected raw and cooked vegetables. *Plant Foods for Human Nutrition (Formerly Qualitas Plantarum)*, 1988. 38(4): p. 333-341.

[37] Nagra S. A. and Khan S., Vitamin A (beta-carotene) losses in Pakistani cooking. *Journal of the Science of Food and Agriculture*, 1989. 46(2): p. 249-251.

[38] Speek A. J., Speek-Saichua S., and Schreurs W. H. P., Total carotenoid and [beta]-carotene contents of Thai vegetables and the effect of processing. *Food Chemistry*, 1988. 27(4): p. 245-257.

[39] Wu X., Beecher G. R., Holden J. M., Haytowitz D. B., Gebhardt S. E., and Prior R. L., Lipophilic and Hydrophilic Antioxidant Capacities of Common Foods in the United States. *Journal of Agricultural and Food Chemistry*, 2004. 52(12): p. 4026-4037.

[40] Velasco-Martinez M., Angulo O., Vazquez-Couturier D. L., Arroyo-Lara A., and Monroy-Rivera J. A., Effect of dried solids of nejayote on broiler growth. *Poult. Sci.*, 1997. 76(11): p. 1531-1534.

[41] Adom K. K. and Liu R. H., Antioxidant Activity of Grains. *Journal of Agricultural and Food Chemistry*, 2002. 50(21): p. 6182-6187.

[42] Zielinski H. and Kozlowska H., Antioxidant Activity and Total Phenolics in Selected Cereal Grains and Their Different Morphological Fractions. *Journal of Agricultural and Food Chemistry*, 2000. 48(6): p. 2008-2016.

[43] Cevallos-Casals B. A. and Cisneros-Zevallos L., Stoichiometric and Kinetic Studies of Phenolic Antioxidants from Andean Purple Corn and Red-Fleshed Sweetpotato. *Journal of Agricultural and Food Chemistry*, 2003. 51(11): p. 3313-3319.

[44] Niwa T., Doi U., Kato Y., and Osawa T., Antioxidative Properties of Phenolic Antioxidants Isolated from Corn Steep Liquor. *Journal of Agricultural and Food Chemistry*, 2000. 49(1): p. 177-182.

[45] Matthaus B., Antioxidant Activity of Extracts Obtained from Residues of Different Oilseeds. *Journal of Agricultural and Food Chemistry*, 2002. 50(12): p. 3444-3452.

In: Seeds as Functional Foods and Nutraceuticals
Editors: R. Mora-Escobedo, J. De J. Berrios et al.
ISBN: 978-1-62808-489-4

Chapter 10

CHIA SEEDS: A NEW NUTRACEUTICAL OPTION

Ma. de la Paz Salgado-Cruz [1], ***Georgina Calderón-Domínguez*** [1,*], ***J. Jorge Chanona-Pérez*** [1], ***Alan J. Hernández-Álvarez*** [1], ***Mayra Díaz-Ramírez*** [1] ***and Reynold R. Farrera-Rebollo*** [1]

[1]Escuela Nacional de Ciencias Biológicas, Instituto Politécnico Nacional, Departamento de Ingeniería Bioquímica, Plan de Ayala y Carpio s/n. Col Santo Tomas, D.F., Mexico

ABSTRACT

Chia is an annual plant which belongs to the Lameacea family. Worldwide, the family Lameacea is composed of approximately 224 genera and 5600 species, while in Mexico, where this seed is native to, 26 genera and approximately 512 species have been reported. Nonetheless a marginally cultivated crop today, chia has recovered interest as a result of the many investigations that have aimed awareness to the nutritional features of this seed. Chia seed has been reported to be an excellent source of protein and vitamin "B". However lipids and antioxidants can be considered the most studied components. This has been related to their neutraceutic activity, resulted from their ω-3 fatty acids content. Regarding chia antioxidants, such as quercetin, myricetin, kaemperol, and caffeic acid, potent antioxidant activity has been reported. On the other hand, chia seed possesses no toxic compounds as compared to flax seed. Another chia component is fibre, which has been researched for its outstanding soluble and insoluble fibre ratio that could be applied in the food and pharmaceutical industries due to its physical properties.

ORIGIN AND TAXONOMY

Recent research has focused on the search and appreciation of ancient cultures that used food components and/or derivatives which are applicable to the development of food, and contribute to nutrition by providing simultaneous beneficial health effects. In this context the

[*] E-mail: gcalderon@ipn.mx.

cultivation of Chia seeds (*Salvia hispanica* L.), native to Mexico, and commonly known as “salvia española”, “chía mejicana”, “chia negra” or ‘‘chia” (commercial chia) is important.

The word "chia" is a Spanish adaptation from the Nahuatl term "chian" or "chien" (plural), which means "seed from which oil is obtained" (Di Sapio et al., 2012; Watson, 1938; Ayerza and Coates, 2009, 2005; INIFAP, 1995; León, 2005). Its origin has been ascribed to different parts of Mexico (mountainous areas of western and central) and Guatemala (northern) (Beltrán-Orozco and Romero, 2003; Cornejo-Tenorio and Ibarra-Manriquez, 2011; Cahill, 2004), with the largest genetic diversity occurring in the Pacific Ocean side (Miranda, 1978; Cahill, 2004; Hernandez-Gomez et al., 2008). This crop, prior to the arrival of the Spanish in Mexican territory, was considered as a staple food to Mesoamerican cultures. Archeological studies have suggested that this crop was cultivated in the Valley of Mexico since 2600 (BC), initially by Teotihuacan and Toltec civilizations, and continuing its cultivation by others ethnic groups up to the occupation of the Aztec Empire. Tenochtitlan, the capital of the Aztec Empire, received between 5,000 and 15,000 tons of chia as an annual tribute from conquered nations. Chia seed was not just used as food, but also it was offered as a tribute to the Aztec gods; hence this crop was one of the main Aztecs diet components along with quinoa, amaranth, corn and some varieties of beans (Rodríguez Vallejo, 1992, Monroy Torres et al, 2008; Rovati, Escobar and Prado, 2012). The importance of these four crops in the Aztec diet is well depicted in the Florentine Codex (Ayerza and Coates, 2009; Ayerza and Coates, 2005). The use of chia in pagan religious ceremonies was one reason the Spanish conquistadors tried to eliminate it, and replace it by species brought from the old world (Ayerza, 2008).

Salvia hispanica L. (Chia seed) is an annual summer crop (September to November) that belongs to the Lameacea or Labiateae families. Its flowers are purple or white and are produced in numerous clusters in a spike at the end of each stem. This family comprises 7 subfamilies, which constitutes about 300 genera and 7500 species approximately (Brummitt, 1992; Stevens, 2001) widely distributed in warm and temperate regions in both hemispheres (Di Sapio, et al., 2012), although the species distribution is influenced by vegetation, weather, altitude and biogeographic affinity. Salvia (Mentheae Tribe) is the most diverse genus of this family (Cornejo-Tenorio and Ibarra-Manríquez, 2011) and consists of about 900 species in the family Lamiaceae (Reyes-Caudilo et al., 2008). The name Salvia comes from the Latin word "Salvare", the healer. Since ancient periods species of Salvia have been used as a traditional medicine treatment for stomach illnesses and common cold. The volatile oils of several species are used as antiseptic while tannins as local anti-inflammatory agents, and its bitter taste produces a pleasant sensory feeling in the mouth and throat (Baytop, 1999; Ulubelen, 2000, Gören et al., 2005, Lu and Foo, 2002). Nowadays this genus has been the object of numerous chemical studies which describes the isolation of diterpenoids, tanshinones and polyphenols from different parts of the plants (Reyes-Caudilo et al., 2008).

According to the taxonomy proposed by Linnaeus, the systematic position of chia (*Salvia hispanica*L.) is (Walker et al., 2004; González, 2010):

Kingdom: Plantae
  Subkingdom: Tracheobionta (vascular plants)
    Superdivision: Spermatophyta (seed plants)
      Division: Magnoliophyta or Angiospermae (flowering plants)
        Class: Magnoliopsida or Dicotyledoneae
          Subclass: Asteridae

Order: Lamiales
Family: Lamiaceae
Subfamily: Nepetoideae
Tribe: Mentheae
Genus: Salvia
Species: hispánica

This plant grows to 0.3-1.5 m (Figure 1 a -b), with blue, purple or white (Figure 1b) hermaphrodites flowers, producing a small seed (1.5 to 2 mm) with three unequal semiaxes, an oblong-elliptical outline, and a slightly navicular form, which resembles the sesame seed ovoid form.

Seeds are indehiscent nucules of greyish brown colour with dark brown spots, and some coloured whitish (Di Sappio et al., 2011, Rovati, Escobar and Prado, 2012). Structurally they are formed by three layers of cells (Figure 2), where the first one corresponds to the epidermal layer (EpL), also called Exotesta (17 -18μm).

Figure 1. Chia seed planting, a) flowering, b) harvest. Photos provided by Nopala Villagran, Hidalgo growers community. Chia seed is a seasonal and irrigated crop.

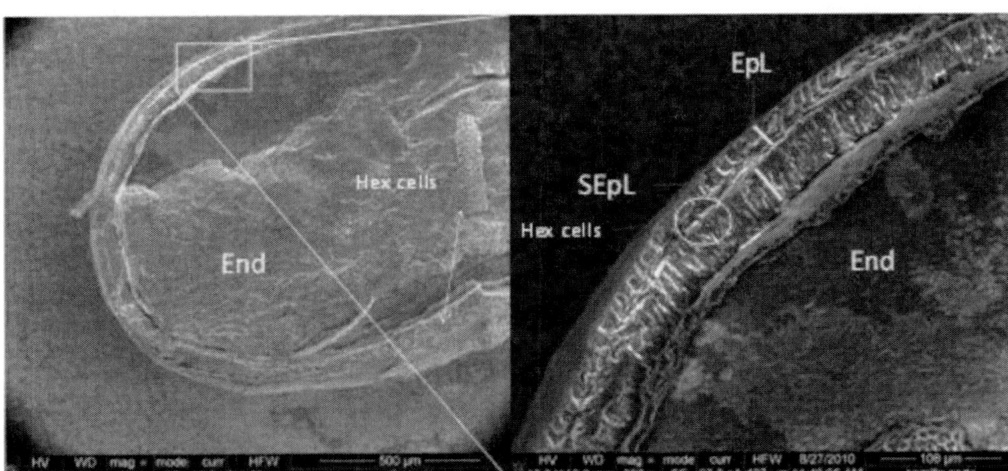

Figure 2. Chia seed microstructure. EpL: epidermal layer or exotesta cells; SEpL: second layer or subepidermal cells; End: third layer or core endosperm; Hex cells: polygonal isodiametric cells.

The second layer or subepidermal (SEpL), thicker than the first one (40 to 42μm), shows a more compact cell structure, while the third one or core endosperm (End), constitutes approximately 80% of the seed. The epidermis of the chia seed is formed by a reticulated and slightly rough surface, with a layer of cells with an irregular geometric shape (hexagonal, Hex cells), and arranged in a grid (Salgado-Cruz 2013).

Global Seeds Markets: Situation and Outlook

Currently this crop remains negligible with few applications. It is still used for the preparation of a beverage that resembles "flax seed tea" (Baughman and Jamieson 1929, León, 2005). The roasted and grounded seeds (chiapinolli) are eaten as gruel and the oil is used as a body emollient and pottery lacquer (Reyes-Caudilo et al., 2008). Furthermore is also used like ChiaPets (www.chiapet.com), which consists of an unglazed ceramic animal or tree that can be filled with water and covered with chia seeds which sprout to produce the effect of fur, hair, skin or leaves (Jamboonsri, 2010). The curative properties of the seeds are also appreciated, for example, for treating eye obstructions, infections and respiratory malaises. The seeds soaked in water or fruit juice were and are still consumed in some regions as a refreshing drink (Cahill, 2003; Reyes-Caudilo et al., 2008). Although nowadays its cultivation is not only limited to the Americas but also extended to other areas such as Australia and Southeast Asia (Jamboonsri et al., 2012), chia seed is not a well-known food, but its global production is increasing due to its healthy properties and popularity (Muñoz et al., 2012). Chia seed has been approved for its consumption as a food in N. America, Australasia, Argentina, and Asia. Mintel GNPD (Global New Products Database) reported over 277 products released in the last 12 months globally containing Chia seed; 100% whole packaged Chia seeds are being sold in 340 chemists and natural food stores throughout Australia as well as an ingredient in various products and breads in over 700 nationwide retail sites ((EC) No 258/97, 2011). In 2009, it was approved as a novel food by the European Parliament and the European Council (Commission, E.U., 2009) including chia seeds and grounded Chia for its inclusion in bread at a maximum level of five percent (5%) (European Commission, 2009), although these has been modified in 2013 as shown in Table 1 (EC No 258/97, 2013).

Regarding its agronomic characteristics, chia is grown in tropical and subtropical areas where latitudes range from 20°55'N to 25°05'S, however in higher latitudes it cannot produce seeds since it is weakened by frost before maturation. The crop cycle varies from 90 to 150 days depending on the latitude where it is planted. Chia is sown at a rate of 6-8 kilograms per hectare, with row spacing of 0.70 or 0.80 m in Argentina and Bolivia and of 0.75 m in Mexico.

Table 1. Uses of chia (Salvia hispanica) seed (EC No 258/97, 2013)

Product	Percent
Baked products	not more than 10 %
Breakfast cereals	not more than 10 %
Fruit, nut and seed mixes	not more than 10 %
Pre-packaged Chia seed as such	not more than 15 g per day

Chia seeds require wet soil to germinate, but once the seedlings are established, chia grows well with limited water (Jamboonsri, 2010), this feature gives the mucilage prevalence among those plant species adapted to surviving on arid sandy soils (Deng et al., 2009). The first 45 growth days are critical since weeds can compete for light, nutrients and water.

Nowadays, a commercial chia is grown in Mexico, Australia, Argentina, Bolivia, Guatemala, Ecuador and Australia. In 2008, Bolivia exported over 2,300 million dollars from Chia seed trade to markets in the United States, Chile, Argentina, UAE, Canada, the Netherlands, Spain and Colombia in order of importance. In addition in 2008, "The Ord Valley", in the far northwest of Australia, was the largest producer of chia seeds in the world, representing two-thirds of world production (Matt, 2008; Webster et al., 2009, Ixtaina, 2010; IBCE, 2009; Jamboonsri et al., 2012).

Agricultural production in Mexico (2011) is reported in the states of Jalisco and Puebla, with harvested areas of 2720 and 30 (Ha) and 3427.60 and 21 (Ton) production in irrigated and seasonal methods, respectively (SAGARPA-SIAP, 2011). Although harvest in other states such as Hidalgo, Colima, Jalisco and Michoacán, and these data does not appear in the statistics (Cahill, 2004; Cahill and Ehdaie, 2005), its production has increased since 2000. Due to mentioned above, any new industrial crop for which there is a market, and appears adapted to the region, would be attractive.

Composition and Nutritional Profile: Physicochemical Properties

Table 2, shows the values of chia seed chemical analysis where the composition of this seed can be observed as an excellent source of protein, fiber, fat, vitamins and minerals, which places it as a very good nutritive alternative. These data shows slight differences due to both genetic and environmental conditions (Ali et al., 2010). It has been reported that chemical composition of chia is influenced by the soil quality, environmental temperature, light intensity, soil type and nutrients availability (Ayerza and Coates, 2004; Ayerza and Coates, 2009; Ayerza, 2009; Ayerza, 2011).

Protein and Amino Acid Profile

Chia seed has a protein content of 17-22%. This concentration relates favorably with other grains and seeds for human consumption such as wheat (14%), corn (14%), rice (8.5%), oats (15.3%), barley (9.2%), amaranth (14.8%) and sunflower seed (16.32%) (Mera, 2009; INNSZ, 2000; Ayerza and Coates, 2005; Sandoval-Oliveros and Octavio Paredes-López, 2013). Protein obtained from defatted and without mucilage chia seed shows that this fraction is constituted by 17.3% of crude albumins, 52% of globulins, 12.7% of prolamins, and 14.5% of glutelins, while 3.4% protein was not recovered; this protein pattern shows some similarities with other important seeds such as peas, lupins, and cotton, varying its content according to the botanical source, type of variety, meal preparation, extraction method, among other factors (Sandoval-Oliveros and Paredes-López, 2013). However, unlike other seeds and cereals, chia protein (Table 3) has no limiting amino acids for adults, besides glutamic acid content is found in a large proportion and nowadays this amino acid has been the subject of several studies in humans and animals for its participation in relevant metabolic functions, such as carbohydrate metabolism in cerebral tissue, related to the hyperinsulinism/ hypermmonemia syndrome (Kelly and Stanley, 2001).

Defatted chia flour presents *in vitro* digestibility values (77.5%) similar to those reported for *Phaseolus vulgaris* (77.5%) and higher than those for some cereals such as corn (66.6%), rice (59.4%), sorghum (59.1%) and wheat (52.7%) (Sandoval-Oliveros and Paredes-López, 2013). Apparent protein digestibility of previously hydrolyzed chia seed (Salgado-Cruz, 2006, Hsu et al., 1997) showed a higher value (77.5) than corn (73). It is also important to mention that there are no reports of antinutritional factors in Chia, excluding the presence of protease inhibitors that could retard the *in vitro* digestibility (Sandoval-Oliveros and Paredes-López, 2013).

The chemical coefficient shows that the limiting amino acids for chia seeds are the sulfur containing amino acids (Met+Cys) (Table 4),however it covers the FAO requirements for adults; tryptophan was not determined.

Table 2. Nutritional Profile of Chia seed

Nutrients	Percentage %	Minerals	(mg / 100g)	Vitamins	(mg / 100g)
Protein	17.4 - 22.4[a] 20.0 - 22.0[b] 18.6 -20.03[c]	Potassium[a]	510 - 710	Thiamin (Vit. B1)[a]	0.79 - 0,81
Carbohydrate	37.1 - 42.6[a] 25.0 - 41.0[b] 37.19-43.49[c]	Iron[a]	5,7 – 15	Niacin (Vit. B3)[a]	7.8 – 9.4
Ash	4.5 - 5.6[a] 4.0 -6.0[b] 5.17-5.31[c]	Phosphorus[a]	600 – 870	Ascorbic acid (Vit. C)[a]	<1 – 6
Dry matter	95 - 96.8[a] 91.0 -96.0[b] --------[c]	Sodium[a]	<0.1 – 6	Retinol (Vit. A)[a]	16 IU
Fat	28.5 - 34.7[a] 30.0-35.0[b] 26.87-32.77[c]	Calcium[a]	500 – 640	Riboflavin (Vit. B2)[a]	0,05
Fiber	34.45[a] 18.0 – 30.0[b*] 19.4-20.3[C]	Magnesium[a]	310 - 430	Cobalamin (Vit B12)[a]	90 – 110 ng
				Alpha-tocopherol (Vit. E)[a]	<0.1 – 0.3

[a]Source: Australia Government National Measurement Institute, Certificate of Analysis, Job Number: CHIA01/090219, Date Sampled 18 February 2009.

[b](EC) No 258/97 (2013).

[C] Alvarado, (2011).

Crude Fibre: fibre made mainly of indigestible cellulose, pentosans and lignin

Table 3. Amino acid analysis of Chia seed protein hydrolyzate compared with FAO requirements

Amino acid protein	Chia -s[1] (g/16 gN)	Chia- p[2] (g/16 gN)	Amino acid requirements FAO Adult[2] (g/16 gN)	Preschool[2] (g/16 gN)	School[2] (g/16 gN)
Histidine (HIS)	2.57	2.45	1.6	1.9	1.9
Isoleucine (ILE)	3.21	3.35	1.3	2.8	2.8
Leucine (LEU)	5.89	5.99	1.9	6.6	4.4
Lysine (LYS)	4.44	3.6	1.6	5.8	4.4
Methionine+cysteine (MET + CYS)	1.836	1.4	1.7	2.5	2.2
Phenylalanine+tyrosine (PHE + TYR)	7.48	7.52	1.9	6.3	2.2
Treonina (THR)	3.43	3.23	0.9	3.4	2.8
Triptofano (TRY)	-	-	0.5	1.1	0.5
Valine (VAL)	5.1	5.32	0.3	3.5	2.5
Serine (SER)	4.86	4.43	-	-	-
Glutamic acid (GLU)	12.4	13.65	-	-	-
Glycine (GLY)	4.22	4.03	-	-	-
Alanine (ALA)	4.31	4.41	-	-	-
Arginine (ARG)	8.9	8.63	-	-	-
Proline (PRO)	4.4	3.92	-	-	-
Asparagine (ASP)	7.64	7.36	-	-	-
Total	80.64	79.52	-	-	-

Note: S:solvent extraction[1], P:press extraction;[2] (Ting et. al., 1990).; [2](Bethesda, 1991). FAO/WHO (1989).

Table 4. Chia seed Chemical Protein Coefficient (C.C)*

	Preschool		**School**		**Adult**	
	C. C.	**Limiting amino acid**	**C. C.**	**Limiting amino acid**	**C. C.**	**Limiting amino acid**
Chía S[1]	**0.73**	**MET + CYS**	**0.84**	**MET + CYS**	**1.08**	**MET + CYS**
Chía P[2]	**0.56**	**MET + CYS**	**0.64**	**MET + CYS**	**0.82**	**MET + CYS**

*Values obtained are based on the Table 3, (Salgado-Cruz, 2006). S: solvent extrraction[1], P: press extraction;[2]

DIETARY FIBER, SOLUBLE AND INSOLUBLE

Chia seed is considered an excellent source of fiber (composed of cellulose, hemicellulose and lignin) (Capitani et al., 2012), and also the meal and fibrous chia fractions showed a high content of total dietary fiber (TDF, 33.5 ± 2.7 g/100 gds), consisting mostly of insoluble (IDF, 25.4 ± 2.2 g/100 g dry solids) and soluble (SDF, 8.2 ± 0.8 g/100 g dry solids) fractions (Sandoval-Oliveros and Octavio Paredes-López, 2013).

Although these amounts might differ as observed in Table 2, the concentration of TDF and its corresponding components (SDF and IDF) were significantly higher in the chia fiber fractions samples than those reported for canola, soybean and flax meals (11.54, 3.50 and 5.27 g/100 g, respectively). In other research, the fiber-rich fraction was obtained by dry fractionation of the defatted flour showing that the fiber content increased from 26.5 g/100 g in the chia flour to 29.56 g/100 g in the chia fiber-rich fraction (FRF) (Table 5). These results are different from those presented in Table 2 due to the flour process used. This increase may not seem substantive, but the fiber content of the fiber rich fraction (FRF) is appropriate for a fibrous product, suggesting that the dry processing method employed here is feasible and improved nutritional characteristics. The relatively high IDF percentage in chia FRF suggests possible applications in light food products. It is considered also that the extracting process of dietary fiber residues could affect its functional properties (water-holding, oil-holding capacity, water adsorption capacity, water absorption capacity, organic molecule absorption capacity, emulsifying activity and emulsion stability).

It has also been reported that the fiber-rich fraction of chia shows high values of water-holding capacity (15.41 g/g), water absorption capacity (11.73 g/g) and organic molecule absorption capacity (1.09 g/g) while its oil-holding capacity (2.02 g/g) and water adsorption capacity (0.3 g/g) were low. This high hydration capacity may be due to the presence of chia seeds mucilage which can act as soluble dietary fiber. Mucilages have excellent water-holding properties (Muñoz et al., 2012) as the gel creates a physical barrier between carbohydrates and the digestive enzymes that break them down, thereby decreasing the conversion of carbohydrates into sugars while increasing the satiety sensation (Rubio, 2002; Muñoz et al., 2012). These physicochemical and functional properties turned chia byproducts into important ingredients for the manufacture of products such as desserts, drinks, breads, jellies, emulsions, cookies, among others (Capitani et al., 2012).

OIL, ANTIOXIDANTS AND FATTY ACID COMPOSITION

The seed has between 28 and 35 % oil content as seen in Table 2. The chemical composition of Chia seed (Salvia hispanica L.) shows that oil is an interesting source of polyunsaturated fatty acids (Table 6) (Ayerza and Coates, 2011; Ayerza, 1995). The fatty acids of chia oil are highly unsaturated, showing a iodine value of 197.68 cg/g. Black chia seed oil contains higher amounts of polyunsaturated fatty acids than other legume seed oils such as canola (110-126 cgI_2/g), safflower (136-148 cgI_2/g), olive (118– 139 cgI_2/g) and soybean (118–139 cgI_2/g). The peroxide value obtained for chia oil was 2.64 meq/Kg. This low value shows that natural antioxidants that have been identified such as γ-tocopherol, protects the oxidation of omega-3 fatty acids (Alonso-Calderón et al., 2013). In addition, chia seed oil is an excellent source of natural lipid antioxidants (flavonol glycosides, chlorogenic acid (7.1×10^{-4} mol/kg of seed), caffeic acid (6.6×10^{-3} m/kg) flavonol aglycones/kaempferol (1.1×10^{-3} m/kg), quercetin (2.0×10^{-4} m/kg), myricetin (3.1×10^{-3} m/kg); and caffeic acid (t.35 X 10^{-2} m/kg) and also it is rich in polyphenols, (González Jiménez et al., 2010; Taga y col., 1984; Reyes-Caudillo, Tecante, and Valdivia-López, 2008).

Table 5. Proximate composition of chia (Salvia hispanica) fiber-rich fraction (FRF) compared to defatted flour (g/100 g d.b.)

Component	[a]Chia FRF	[a]Defatted flour	[b]FRF_S	[b]FRF_P
Protein	28.14 ± 0.36	32.24 ± 0.17	35.32± 0.17	33.74 ± 0.05
Moisture	6.96 ± 0.02	6.87 ± 0.10	10.34± 0.03	10.00± 0.16
Crude fiber	29.56 ± 0.07	26.50 ± 0.28	32.84± 0.34	28.35± 0.80
Dietary fiber fractions of chia				
Total dietary fiber	56.46 ± 0.35	-----	52.26± 0.65	51.89± 0.20
Insoluble dietary fiber	53.45 ± 1.62	-----	47.51± 0.59	47.65± 0.18
Soluble dietary fiber	3.01 ± 0.28	-----	4.75± 0.06	4.33± 0.02
Nitrogen-free extract (NFE)	35.33 ± 0.38	33.72 ± 0.49	24.99± 0.56	21.02± 0.79
Fat	0.46 ± 0.00	0.45 ± 0.02	0.21± 0.05	10.85± 0.13
Ash	6.51± 0.04	7.09 ± 0.02	6.64± 0.03	6.04± 0.01

[a]Vázquez-Ovando et al., 2009; [b] Capitani et al., 2012.

FRF: Fiber- rich fraction ; FRF_S: solvent fiber- rich fraction; FRF_P: pressing fiber- rich fraction.

The fatty acid composition values of Chia seed oil is presented in Table 6, where the presence of α-linolenic, linoleic, oleic, palmitic and stearic fatty acids were detected, being predominant α-linolenic acid (64.2%- 54.2%) and showing that chia oil is a highly unsaturated material. Owing to its specific composition, chia oil appears unique among the common vegetable edible oils such as rape, corn, sunflower, soybean, olive, coconut oils, etc (Ayerza, 2011).

Vitamin and Mineral Content

Chia seed is a good source of vitamin B (B1, B2, B3) and vitamin A (Table 2) and a exceptionally source of calcium, magnesium and iron, being this content much higher than those expressed for coarse wheat, rice and sorghum meals (Ragaee, Abdel-Aal, and Norman, 2006). Regarding iron content, it was higher than sesame, but similar to barley (Egbekun and Ehieze, 1997; Capitani et al., 2012).

Potential Technological and Medicinal Benefits: Effect of Specific Seeds Components on Chronic Degenerative Diseases

The effect of specific chia components are summarized in Table 8. Studies carried out feeding rats with diets added with chia seed (*Salvia hispanica*L.) showed that the presence of α-linolenic acid and fiber had benefits upon dyslipidaemia and insulin resistance (IR) improving adiposity and normalizing hypertriacylglycerolaemia (Chicco et al., 2009). In other study, the use of chia seeds was very effective in increasing n-3 PUFA levels and may potentially be used to obtain meats with functional properties but only when supplemented with high levels of antioxidants (Meineri et al., 2010, Dalle Zotte and Szendrő 2011).

Another area in which chia seed is administered as an alternative to control certain diseases in humans is in animal feed. Incorporating chia seed in animal diets a decrement in its palmitic acid levels (up to 20.6% and 12.8% for white and dark meats, respectively) and in SFA (up to 17.5% and 12%) for white and dark meats respectively is observed (Ratanayake et al., 1989). Chia also seems to be a possible feedstock for healthier pork meat production for human due to the increase of n-3 α-linolenic acid content without negatively affecting animal production or meat taste (Coates and Ayerza, 2009). Productive performance and lipid composition of birds fed with 15 and 10% chia supplemented diets with whole chia flour led to the enrichment of the egg yolk and the chicken tissues with linolenic acid and a decrement in the percentage of palmitic acid (Salazar-Vega et al., 2009). This is of vital importance because the palmitic acid is considered as one of the most hypercholesterolemic compounds (Denke and Grundy, 1992; Salazar-Vega et al., 2009).

In a study carried out in normal Wistar rats, the addition of chia to their diet prevented the increase of the activity of lipogenic enzymes, observing that the high linoleic acid content of chia prevents the increment of hepatic lipogenesis by exerting hypolipidemic effects similar to those observed with ω-3 fatty acids of marine origin.

Table 6. Fatty acid composition (%) of chia seed

Location	Palmitic	Stearic	Oleic	Linoleic	Linolenic	P:S	n-6:n-3
Jaho	7.65[a]	3.50[a]	3.50[e]	19.25[e]	57.50[f]	6.89[e]	0.34[bc]
La Población	6.55[e]	2.95[b]	7.15[c]	20.30[c]	62.00[c]	8.66[a]	0.33[bcd]
Cerro Prieto	7.15[c]	3.00[b]	6.90[d]	18.35[f]	64.20[a]	8.14b[c]	0.29[e]
Yoshioka	7.60[a]	3.55[a]	10.30[b]	19.10[e]	58.55[e]	6.97[e]	0.33[cd]
La Invernada	7.40[b]	2.75[c]	6.75[d]	20.10[d]	62.20[c]	7.74[cd]	0.32[d]
LSD (0.05)	0.14	0.17	0.19	0.19	0.27	0.4	0.01
La Peña	7.25[c]	3.40[a]	3.40[e]	18.00[g]	63.20[b]	7.63[d]	0.29[e]
El Torrente	7.65[a]	3.55[a]	13.25[a]	20.50[b]	54.20[g]	6.67[e]	0.38[a]
C. del Valle	6.95[d]	2.75[c]	6.85[d]	21.05[a]	61.65[d]	8.53[ab]	0.34[b]
Santa Cruz*	7.40	2.69	7.10	18.70	63.60	8.15	0.29

Means in the same column with different letters are significantly different ($P < 0.05$). *Not replicated.
P:S; polyunsaturated:saturated ratio.
Adapted from: Ayerza and Coates, 2004.

Table 7. Fatty acids content (g/kg of total FA) in chia seed oil

	Peiretti and Gai (2009)	Ayerza (1995)	Ayerza and Coates (2004)	Coates and Ayerza (1996)	Heuer et al. (2002)
C_{16}	71 ± 0.64	62–71	66–77	64–79	76–87
C_{18}	33 ± 0.41	31–37	27–36	24–32	26–30
$C_{18:1n-9}$	60 ± 0.48	73–82	68–133	60–66	61–63
$C_{18:2n-6}$	188 ± 0.25	198–208	180–211	170–201	174–187
$C_{18:3n-3}$	641 ± 0.87	607–634	542–642	632–678	635–651
P/S[a]	7.9 ± 0.66	7.6–8.8	6.7–8.7	7.7–9.3	7.3–7.8
n-6/*n*-3	0.29 ± 0.00	0.31–0.33	0.29–0.38	0.25–0.32	0.27–0.29

[a]Polyunsaturated fatty acid/saturated fatty acid. (Peiretti and Meineri , 2009).

Replacing the dietary fat source, it failed to prevent the increment of the activity of key lipogenesis enzymes, maintaining the plasma levels and hepatic triglyceride concentration. These data are consistent with those reported by Poudyal in 2012, indicating that the administration of chia seeds to rats for 8 weeks in a rich carbohydrates and fats diet induces lipid redistribution, reduces hepatic steatosis and exerts a cardioprotective effect.

Regarding human studies, other researchers mentioned that supplementation of chia seeds to patients with type 2 diabetes mellitus for a long period of time (12 months) attenuated cardiovascular risk factors such as systolic blood pressure, while maintaining good glycemic and lipid control (Vuksan et al, 2007; Rossi et al, 2011).

On the other hand, it has been reported that chia consumption promotes the decrement of tumor weight and metastasis number and also inhibits growth and metastasis in murine mammary gland adenocarcinoma (Espada et al. 2007). Moreover in recent years an increased interest in public health benefits and the discovery of many traditional foods like seeds, vegetables, fish among other foods containing ingredients which are beneficial to the organism, has been in growth.

Therefore a study was made to determine the overall acceptability, sensory characteristics, functional properties, and nutrient content of cakes added with chia (*Salvia hispanica* L) gel as a replacement for oil or eggs, finding that up to 25% of oil or eggs can be replaced in cakes, yielding a more nutritious product with acceptable sensorial characteristics (Borneo et al., 2010). In other study was shown that the addition of 15 and 20% of chia flour to tortillas, promoted a significant increment in the protein, fat and fiber level without altering the sensorial properties while the *in vitro* digestibility of tortilla showed lowest values (Rendón-Villalobos et al., 2012). A similar study was carried out by adding amaranth and chia seed flours to prepared bread (Alvarado, 2011), observing an increment in the protein, fatty acids and fiber content.

Another interesting study is the production of biodiesel from chia seed together with sunflower seed and castor, resulting in a successful alternative for the production of biofuels (Mera, 2009; Wilson and Hildebrand, 2010).

Table 8. Active compounds identified in *Salvia hispanica* L. and their chemical structure

Active compounds in *Salvia hispanica* L.	Activities	Reference
Dietary fiber soluble and insoluble	The high fiber content may improve satiety, decrease energy intake, and promote weight loss glycemic control. Consume of fiber causes frequent peristaltic movements of the intestine, facilitates transit of the fecal bolus and intestinal distention, and thus aids in reducing the probability of intestinal tract disorders and constipation	Vuksan et al. 2007 *Rendón-Villalobos et al., 2012* Vázquez-Ovando et al., 2009
Omega 3 alpha-linolenic acid; ω-3 ALA (18 : 3) (*PUFA fatty acids*)	Essential fatty acid Benefits: lowering TG and cholesterol levels, which in turn results in low blood pressure and heart-related diseases.	Heuer et al., 2002, Jeong et al., 2010 Simopoulos et al., 2002

Table 8. (Continued)

Active compounds in *Salvia hispanica* L.	Activities	Reference
	Anti-inflammatory activity. cardioprotective and hepatoprotective activities by redistributing lipid away from visceral fat and liver; reduce the risk of suffering coronary heart diseases. antidiabetic action. protection against arthritis, autoimmune disease, and cancer	Poudyal et al., 2012 Mohd Ali et al. 2012 Alonso-Calderón et al ., 2013
Omega 6 linoleic acid; ω-6 LA (18 : 2) (*PUFA fatty acids*)	Essential fatty acid This FA has inflammatory, hypertensive, and thrombotic activities. Eicosanoid produced from LA has been associated with CVD and cancer It works inversely with ALA; thus a balanced ratio of ALA to LA is important in maintaining good health	Simopoulos et al., 2002 Betti et al., 2009 Dubois et al., 2007 Mohd Ali et al. 2012
Myricetin, Kaempferol, Quercetin,Caffeic acid (flavonols and phenolic acids)	Antioxidant, anti-inflammatory, anticancer, and Antithrombotic activities. The presence of antioxidants induce the reduction in the saturates deposit of in tissue	Jeong et al., 2010 Mohd Ali et al. 2012 Salazar-Vega et al., 2009
γ-tocopherol, campesterol, stigmasterol, β-sitosterol	The presence of β- sitosterol has been associated with prevention from prostatic and colon cancer and may help to alleviate the symptoms and urinary flow problems caused by an increase of the benign prostatic hyperplasia (BPH) and decrease cholesterol levels (HDL).	Alonso-Calderón et al ., 2013

Antinutritional Factors or Possible Negative Effects

Although *Salvia hispanica* was used for many years in the Aztecs diets in Mexico with no apparent side effects, there has been concern about that the high amount of n-3 PUFAs could alter the eicosanoids produced, leading to an adverse effect on clotting factors. High doses of n-3 PUFAs have been implicated in altered bleeding and clotting time (Harper and Jacobson, 2001; Vuksan et al., 2007). However, coagulation, liver enzymes, and kidney function were not significantly affected by the addition of *Salvia hispanica* in a supplementation study of conventional therapy with the novel grain (*Salvia hispanica L.*) (Vuksan et al., 2007). Factors in *Salvia hispanica L.* that may act against good glycemic control includes, high content of n-3 PUFAs, which in some studies has shown to increase A1C and fasting postprandial glycemia in type 2 diabetes (Mori et al., 2000). Mechanisms underlying the putative adverse effects of n−3 fatty acids on glycemic control includes an increment in hepatic glucose output, which may be related to an elevated flux of gluconeogenic precursors to the liver, increased plasma glucagon concentrations, changes in hepatic insulin or glucagon sensitivity, or decreased insulin secretion rates (Malasanos and Stacpoole, 1991; Kasim, 1993; Mori et al., 2000). However further studies are required to resolve these issues.

Conclusion

The agroindustrial and technological importance of chia remains in the use of all its main components, of which have been demonstrated to have economical value as sources of protein, nutraceutical compounds and even as raw materials for the production of hydrocolloids, being useful in several areas, such as food, livestock, pharmaceutical, and agricultural industries.

References

Ali, Q., Ashraf, M. and Anwar, F. (2010). Seed Composition and Seed Oil Antioxidant Activity of Maize Under Water Stress. *Journal of the American Oil Chemists' Society,* 87 (110), 1179-1187.

Alonso-Calderón, A., Chávez-Bravo, E., Rivera, A., Montalvo-Paquini, C., Arroyo-Tapia, R., Monterrosas-Santamaria, M., Jiménez-Salgado, T. and Tapia-Hernández, A. (2013). Characterization of Black Chia Seed (Salvia hispanicaL) and Oil and Quantification of β-sitosterol. *International Research Journal of Biological Sciences*, 2(1), 70-72.

Alvarado, R. D. I. (2011). Caracterización de la semilla del chan (Salvia hispanicaL.) y diseño de un producto funcional que la contiene como ingrediente. *Revista de la Universidad del valle de Guatemala,* 23, 43-49.

Ayerza, R. and Coates, W. (2004). Composition of chia (Salvia hispanica) grown in six tropical and subtropical ecosystems of South America. *Tropical Science*, 44, 131–135.

Ayerza, R. and Coates, W. (2009). Influence of environment on growing period and yield, protein, oil and α-linolenic content of three chia (Salvia hispanicaL.) selections. *Industrial Crops and Products,* 30(2), 321-324.

Ayerza, R. (2008). Chia as a new source of ω-3 fatty acids: nutritional comparison with other raw materials and its advantages when producing ω-3 enriched eggS. In: Wild-type food in health promotion and disease prevention: the Columbus Concept Editors, Meester, F. de;Watson, R. R. pp. 179-194.

Ayerza, R. (2009). The seed's protein and oil content, fatty acid composition, and growing cycle length of a single genotype of chia (Salvia hispanica L.) as affected by environmental factors. *Journal of Oleo Science*, 58, 347-354.

Ayerza, R. (2011). The seed's oil content and fatty acid composition of chia (Salvia hispanicaL.) var. Iztac 1, grown under six tropical ecosystems conditions. *Interciencia*, 36 (8).

Ayerza, R. (h) (1995). Oil Content and Fatty Acid Composition of Chia (Salvia hispanicaL.) from Five Northwestern Locations in Argentina. *Journal of the American Oil Chemists' Society*, 72 (9).

Ayerza, R., and Coates, W. (2005). Chia. Rediscovering a forgotten crop of the Aztecs. Tucson: The University of Arizona Press.

Baughman, W. F. and Jamieson, S.G. (1929). Chia Seed Oil. *Oil and Fat Industries*, 6 (9), 15-17.

Baytop, T., (1999). Therapy with Medicinal Plants in Turkey (Past and Present), vol. 2. Nobel Tıp Kitabevleri, Istanbul,142.

Beltrán-Orozco, M. C. and Romero, M. R. (2003). Chía, alimento milenario. Revista Industria Alimentaria, September/October, 20-29.

Bethesda, Md. (1991). Report of the join FAO/WHO Expert Consultation. Food and Agriculture Organization of the United Nations, Rome.

Betti, M., Perez, T. I., Zuidhof, M. J., and Renema, R. A. (2009). Omega-3-enriched broiler meat: 3. Fatty acid distribution between triacylglycerol and phospholipid classes. *Poultry Science*, 88 (8), 1740-1754.

Borneo, R., Aguirre, A. and León, A. E. (2010). Chia (Salvia hispanicaL) Gel Can Be Used as Egg or Oil Replacer in Cake Formulations. *Journal of the American Dietetic Association,* 110(6), 946-949.

Brummitt, R.K. (1992). Vascular plant families and genera. Royal Botanical Gardens, Kew, UK.

Cahill, J. P. and Ehdaie, B. (2005). Variation and heritability of seed mass in chia (Salvia hispanicaL.). *Genetic Resources and Crop Evolution*, 52, 201–207.

Cahill, J. P. (2003). Ethnobotany of Chia, Salvia hispanica L. (Lamiaceae). *Economic Botany*, 57, 604-618.

Cahill, J. P. (2004). Genetic diversity among varieties of Chia(Salvia hispanicaL.). *Genetic Resources and Crop Evolution,* 51,773-781.

Capitani, M. I., Spotorno, V., Nolasco, S. M., Tomás, M. C. (2012). Physicochemical and functional characterization of by-products from chia (Salvia hispanicaL.) seeds of Argentina. *LWT - Food Science and Technology*, 45, 94-102.

Chicco, A. G., D'Alessandro, M. E., Hein, G. J., Oliva, M. E. and Lombardo, Y. B. (2009). Dietary chia seed (Salvia hispanicaL.) rich in a-linolenic acid improves adiposity and normalises hypertriacylglycerolaemia and insulin resistance in dyslipaemic rats. *British Journal of Nutrition,* 101, 41-50.

Coates, W. and Ayerza, R. (2009). Chia (Salvia hispanica L.) seed as an n-3 fatty acid source for finishing pigs: effects on fatty acid composition and fat stability of the meat and internal fat, growth performance, and meat sensory characteristics. *Journal of Animal Science,* 87(11), 3798-3804.

Cornejo-Tenorio, G., and Ibarra-Manríquez, G. (2011). Diversidad y distribución del género Salvia (Lamiaceae) en Michoacán, México. *Revista Mexicana de Biodiversidad*, 82, 1279-1296.

Dalle Zotte, A. and Szendrő Z. (2011). The role of rabbit meat as functional food. *Meat Science*, 88,319-331.

Deng, W., Jeng, D-S, Toorop, P. E., Squire, G. R. and Iannetta, P. P. M. (2012). A mathematical model of mucilage expansion in myxospermous seeds of Capsella bursa-pastoris (shepherd's purse). *Annals of Botany*, 109, 419-427.

Denke, M. A. and Grundy, S. M. (1992). Comparison of effects of lauric acid and palmitic acid on plasma lipids and lipoproteins. *American Journal of Clinical Nutrition,* 56(5), 895-898.

Di Sapio, O., Bueno, M., Busilacchi, H., Quiroga, M. and Severin, C. (2012). Caracterización Morfoanatómica de Hoja, Tallo, Fruto y Semilla de Salvia hispánica L. (Lamiaceae). *Boletín Latinoamericano y del Caribe de Plantas Medicinales y Aromáticas,* 11 (3), 249-268.

Dubois, V., Breton, S., Linder, M., Fanni, J., and Parmentier, M. (2007). Fatty acid profiles of 80 vegetable oils with regard to their nutritional potential. *European Journal of Lipid Science and Technology,* 109(7), 710-732.

(EC) No 258/97 (2009). Commission, E.U. Authorizing the placing on the market of chia seed (Salvia hispanica) as novel food ingredient under Regulation of European Parliament of the Council. *Official Journal of the European Union*, C, 7645.

(EC) No 258/97 (2011) Application for the authorization of Chia Seed from Salvia hispanicaL. for consumption as a food and as an ingredient in additional food groups. Submitted pursuant to Regulation (EC) No 258/97 of the European Parliament and of the Council of 27th January 1997 concerning novel foods and novel food ingredients.

(EC) No 258/97 (2013) Commission Implementing Decision: authorising an extension of use of Chia (Salvia hispanica) seed as a novel food ingredient under Regulation (EC) No 258/97 of the European Parliament and of the Council. *Official Journal of the European Union,* 24(1).

EFSA-Q-2005-059 (2005) Opinion of the Scientific Panel on Dietetic Products, Nutrition and Allergies on a request from the Commission related to the safety of chia (Salvia hispanicaL.) seed and ground whole chia seed as a novel food ingredient intended for use in bread. *The EFSA Journal,* 278, 1-12.

Egbekun, M. K., and Ehieze, M. U. (1997). Proximate composition and functional properties of full fat and defatted beniseed (Sesamun indicum L.) flour. *Plant Foods for Human Nutrition,* 51, 35-41.

Espada, C. E., Berra, M. A., Martinez, M. J., Eynard, A. R., and Pasqualini, M. E. (2007). Effect of chia oil (Salvia hispanica) rich in ω-3 fatty acids on the eicosanoid release, apoptosis and T-lymphocyte tumor infiltration in a murine mammary gland adenocarcinoma. *Prostaglandins, Leukotrienes and Essential Fatty acids,* 77, 21-28.

FAO/WHO. (1989) Protein quality evaluation. Report of the Joint FAO/WHO. Expert consultation. Bethesda, Md., USA.

González Jiménez, F. E. (2010). Caracterización de compuestos fenólicos presentes en la semilla y aceite de chía (Salvia hispánica L.), mediante electroforesis capilar. Master of science thesis, ENCB-IPN, México.

Gören, A. C., Kilic, T., Dirmenci, T., Bilsel, G. (2006). Chemotaxonomic evaluation of Turkish species of Salvia: Fatty acid compositions of seed oils. *Biochemical Systematics and Ecology,* 34, 160-164.

Harper, C. R. and Jacobson, T. A. (2001). The fats of life:the role of omega-3 fatty acids in the prevention of coronary heart disease. *Archives of Internal Medicine,*161, 2185-2192.

Hernández Gómez, J. A. and Miranda Colin, S. (2008). Caracterización morfológica de Chía (Salvia hispanica L.). *Revista Fitotecnia Mexicana*, 3,105-113.

Heuer, B., Yaniv, Z. and Ravina, I., (2002). Effect of late salinization of chia (Salvia hispanica), stock (Matthiola tricuspidata) and evening primrose (Oenothera biennis) on their oil content and quality. *Industrial Crops and Products*, 15(2), 163-167.

Hsu, H., Vavak, D., Satterlee, L., Miller, G. A. (1977).Multienzime technique for estimating protein digestibility. *Journal Food Science*, 42, 1269-1279.

INIFAP (1995). Instituto Nacional de Investigaciones Forestales y Agropecuarias. México:informe nacional para la conferencia técnica internacional de la FAO sobre los recursos }filogenéticos. Pagina Web:http://66.102.7.104/search?q=cache:vUggKB9JEgJ: www.fao.org/ag/AGP/AGPS/Pgrfa/pdf/mexico.pdf+ch%C3%ADa,+tepech%C3%ADa+o +chinetlacoloandhl=es

INNSS Instituto Nacional de Nutrición Salvador Subirán. (2002). Composición de Alimentos Mexicanos, composición química, aminoácidos y cálculo de aportes dietarios.

Ixtaina, V. Y., Vega, A., Nolasco, S. M., Tomás, M. C., Gimeno, M., Bárzana, E. and Tecante, A. (2010). Supercritical carbon dioxide extraction of oil from Mexican chia seed (Salvia hispanicaL.). *Journal of Supercritical Fluids*, 55(1), 192-199.

Jamboonsri, W. (2010). Improvement of New Oil Crops for Kentucky. University of Kentucky Doctoral Dissertations. Paper 120. http://uknowledge.uky.edu/ gradschool _diss/120.

Jamboonsri, W., Phillips, T. D., Geneve, R. L., Cahill, J. P., Hildebrand, D. F. (2012). Extending the range of an ancient crop, Salvia hispanica L.-a new ω-3 source. *Genetic Resources and Crop Evolution,* 59,171-178.

Jeong, S. K., Park, H. J., Park, B. D. and Kim, I. H. (2010). Effectiveness of topical chia seed oil on pruritus of end-stage renal disease (ESRD) patients and healthy volunteers. *Annals of Dermatology*, 22(2), 143-148.

Kasim, S. E. (1993). Dietary marine fish oils and insulin action in type 2 diabetes. *Annals of the New York Academy of Sciences*, 683, 250–257.

Kelly, A. and Stanley, C. A. (2001). Disorders of Glutamate Metabolism Mental Retardation and Developmental Disabilities. *Research Reviews*, 7, 287-295.

León, J. (2000). Botánica de los cultivos tropicales. Third Ed. San José CR, IICA.

Lu, Y., and Foo, L. Y. (2002). Polyphenolics of salvia. *Phytochemistry*, 59, 117-140.

Malasanos, T. H. and Stacpoole, P. W. (1991). Biological effects of omega-3 fatty acids in diabetes mellitus. *Diabetes Care*, 14, 1160-79.

Matt, B. (2008). Chia: The Ord Valley´s new super crop. ABC Rural. http://www.abc.net.au/ rural/content/2008/s2367335.htm.

Meineri, G., Cornale, P., Tassone, S., and Peiretti, P. G. (2010). Effect of Chia (Salvia hispanica L.) seed supplementation on rabbit meat quality, oxidative stability and sensory traits. *Italian Journal of Animal Science*, 9, 45−49.

Mera. (2009). Cultivos Energéticos Alternativos. Ed. By., Recalde Posso E R. and Durán Altisent J M, CIITTOL Centro Iberoamericano de Investigación y Transferencia de Tecnología en Oleaginosas. ISBN: ISBN-978-9978-375-01-03.

Miranda, F. (1978). La vegetación de Chiapas. Primera parte. Ediciones del Gobierno del estado. Sección autográfica. Departamento de Prensa y Turismo. Tuxtla Gutiérrez, México, pp. 311-312.

Mohd Ali, N., Keong Yeap, S., Yong Ho, W., Kee Beh, B., Wei Tan, S. and Guan Tan S. (2012). The Promising Future of Chia, Salvia hispanica L. *Journal of Biomedicine and Biotechnology*, ID 171956, 9. doi:10.1155/2012/171956.

Monroy-Torres, R., Mancilla-Escobar, M. L., Gallaga-Solórzano, J. C., Medina-Godoy, S., Santiago-García, E. J. (2008). Protein digestibility of chia seed Salvia hispanicaL. *Revista Salud Pública y Nutrición (RESPYN)*, 9(1).

Mori, T. A. and Beilin, L. J. (2001). Long-chain omega 3 fatty acids, blood lipids and cardiovascular risk reduction. *Current Opinion in Lipidology*, 12, 11-17.

Muñoz, L. A., Cobos, A., Diaz, O., Aguilera, J. M. (2012). Chia seeds: Microstructure, mucilage extraction and hydration. *Journal of Food Engineering*, 108, 216–224.

NMI. (2010). "National Measurement Institute." Retrieved 18 February 2010, 2010, from http://www.measurement.gov.au/Pages/default.aspx.

Peiretti, P. G. and Meineri, G. (2008). Effects on growth performance, carcass characteristics, and the fat and meat fatty acid profile of rabbits fed diets with chia (Salvia hispanicaL.) seed supplement. *Meat Science*, 80(4), 1116-1121.

Poudyal, H., Panchal, S. K., Waanders, J., Ward, L. and Brown, L. (2012). Lipid redistribution by α-linolenic acid-rich chia seed inhibits stearoyl-CoA desaturase-1 and induces cardiac and hepatic protection in diet-induced obese rats. *Journal of Nutritional Biochemistry,* 23(2), 153-162.

Ragaee, S., Abdel-Aal, M., and Norman, M. (2006). Antioxidant activity and nutrient composition of selected cereals for food use. *Food Chemistry*, 98, 32-38.

Ratanayake, W. M. N., Ackman, R. G., and Hulan, H. W. (1989). Effect of redfish meal enriched diets on the taste and n-3 PUFA of 42-day-old broiler chickens. *Journal Science Food Agriculture*, 49,59-74.

Rendón-Villalobos, R., Ortíz-Sánchez, A., Solorza-Feria, J. and Trujillo-Hernández, C. A. (2012). Formulation, Physicochemical, Nutritional and Sensorial Evaluation of Corn Tortillas Supplemented with Chía Seed (Salvia hispanicaL.). *Czech Journal of Food Sciences,* 30(2), 118-125.

Reyes-Caudillo, E., Tecante, A. and Valdivia-López, M. A. (2008) Dietary fiber content and antioxidant activity of phenolic compounds present in Mexican chia (Salvia hispanicaL.) seeds. *Food Chemistry*, 107, 656–663.

Rodríguez Vallejo, J (1992). Historia de la agricultura y de la fitopatología, conreferencia especial a México. Colegio de Post-graduados en Ciencias Agrícolas,Ciudad de México, México.

Rossi, A., Oliva, M. E., Ferreira, M. R., Chicco, A. and Lombardo, Y. B. (2011). La semilla de Salba (Salvia hispánica L.) dietaria previene la esteatosis hepática y dislipidemia inducida en ratas normales por una dieta rica en sacarosa. *Revista FABICIB*, 15 ,47-59.

Rovati, A., Escobar, E. and Prado, C. (2012). Particularidades de la semilla de chía (Salvia hispánica L.). *EEAOC - Avance Agroindustrial,* 33 (3), 39-43.

Rubio, M. A., 2002. Implicaciones de la fibra en distintas patologías. *Nutrición Hospitalaria,* 2, 17–29.

SAGARPA-SIAP (2011) http://www.siap.gob.mx/aagricola_siap/icultivo/index.jsp

Salazar-Vega, M. I., Rosado-Rubio, J. G, Chel-Guerrero, L. A., Betancur-Ancona, D. A. and Castellanos-Ruelas, A. F. (2009). Composición en ácido graso alfa linolénico (?3) en huevo y carne de aves empleando chía (Salvia Hispánica L.) en el alimento. *INTERCIENCIA*, 34 (3).

Sandoval-Oliveros, M. R. and Paredes-López, O. (2013). Isolation and Characterization of Proteins from Chia Seeds (Salvia hispanica L.). *Journal Agricultural and Food Chemistry*, 61,193-201.

Simopoulos, A. P. (2002). Omega-3 fatty acids in inflammation and autoimmune diseases. *Journal of the American College of Nutrition*, 21 (6), 495-505.

Taga, M. S., Miller, E. E., and Pratt, D. E. (1984). Chia seeds as a source of natural lipid antioxidants. *Journal of the American Oil Chemist's Society*, 61, 928-931.

Ting, I. P., Brown, J. H., Naqvi, H. H., Estilai, A., Kummamoto, J. and Matsumura, M. (1990). Chia: a potential oil crop for arid zones. In: Naqvi, H. H., Estilai, A. and Ting, I. P. (eds) New Industrial Crops and Products, pp. 197–200. Tucson, AZ: University of Arizona,

Ulubelen, A., (2000). III. Chemical constituents, terpenoids in the genus Salvia. In: Kintzios, S.E. (Ed.), Medicinal and Aromatic Plants. C.H.I.P.S. Sage. The Genus Salvia, Harwood Academic Publishers, 55-68.

Vázquez-Ovando, A., Rosado-Rubio, G., Chel-Guerrero, L., Betancur-Ancona, D. (2009). Physicochemical properties of a fibrous fraction from chia (Salvia hispanica L.).*LWT - Food Science and Technology,* 42,168-173.

Vuksan, V., Whitham, D., Sievenpiper, J. L., Jenkins, A. L., Rogovik, A. L., Bazinet, R. P.Vidgen, E.; Hanna, A. (2007). Supplementation of conventional therapy with the novel grain Salba (Salvia hipanica L.) improves major and emerging cardiovascular risk factors in type 2 diabetes. *Diabetes Care*, 30, 2804-2810.

Walker, J. B., Sytsma, K. J., Treutlein, J. and Wink, M. (2004). Salvia (lamiaceae) is not monophyletic: Implications for the systematics, radiation, and ecological specializations of salvia and tribe mentheae. *American Journal of Botany*, 91(7), 1115-1125.

Watson, G. (1938). Nahuatl word in American English. *American Speech*, 13, 108-121.

Webster, T., Rippin, L., Morison, J., Herr, A., Abel, N., Taylor, B., Clark, E. and Stone, P. (2009). 10 Irrigated agriculture: development opportunities and implications for northern Australia. Northern Australia Land and Water Science Review full report. Chapter 10.

Wilson, R. F. and Hildebrand, D. F. (2010). Engineering Status, Challenges and Advantages of Oil Crops. Chapter 8. University of Kentucky, Veterans Drive, Lexington KY 40546-0312, USA. DOI 10.1007/978-3-642-13440-1_8, # Springer-Verlag Berlin. http://www.ibce.org.bo/exportemos/exportemos35.pdf

PART 2. SEEDS AS BYPRODUCTS

In: Seeds as Functional Foods and Nutraceuticals
Editors: R. Mora-Escobedo, J. De J. Berrios et al.
ISBN: 978-1-62808-489-4

Chapter 11

PHYSICOCHEMICAL COMPOSITION, BIOLOGICAL ACTIVITIES AND TECHNOLOGICAL APPLICATIONS OF AVOCADO SEEDS

Obed Osorio-Esquivel[1], Lidia Dorantes Álvarez[2] and Alicia Ortiz-Moreno[2,*]

[1] Instituto de Salud Pública, Universidad de Chalcatongo,
Avenida Universidad S/N,
Chalcatongo de Hidalgo, Oaxaca, México
[2]Departamento de Ingeniería Bioquímica,
Escuela Nacional de Ciencias Biológicas,
Instituto Politécnico Nacional, Carpio y Plan de Ayala S/N,
Colonia Santo Tomas, Delegación Miguel Hidalgo, Distrito Federal, México

ABSTRACT

Avocados (*Persea Americana*) are an important commercial tropical fruit, mainly classified into four strains: West, Indian, Guatemalan, West Indian-Guatemalan hybrid, and Mexican. Avocado pulp composition is well-known, therefore it is marketed worldwide as a functional food. However, the fresh kernel seed constitutes 10-13% weight of the fruit that is discarded after processing or human consumption raising a considerable waste problem. However, avocado seeds are a rich source of fiber, phytosterols, triterpenes, fatty acids, furanoic acid, abscisic acid, catechins, procyanidins, and hydroxycinnamic acid. As a result of their composition, avocado seeds exhibit higher antioxidant capacities than pulp. Several pharmacological studies reported that avocado seed extracts are useful for the treatment of diabetes, inflammation and gastrointestinal irregularity. In a new report,avocado seed was concluded to have a hypolipidemic effect in mouse models. Besides, the potential to use it as a source of natural colorant, and effective sorbent has been considered. Thus, this chapter summarizes the most recent investigations about avocado seed, pointing out their phytochemical composition, biological activities and technological applications.

* Tel: 52 5557296000 Ext. 62372. Fax: 62359. E- mail: ortizalicia@hotmail.com.

1. Introduction

Avocados (*Persea americana* Lauraceae) are an important tropical and subtropical fruit grown worldwide. Botanically, avocados are classified into four strains: West Indian, Guatemalan, West Indian-Guatemalan hybrid and Mexican [1]. Although avocado fruits are native to southern Mexico, nowadays, they are cultivated in places as far from America as Australia, South Africa, Spain, Turkish or India. Avocado is rich in unsaturated fatty acids, fiber, vitamins B and E, and other nutrients. The Hass variety is the most commonly cultivar grown commercially. It has been reported that avocados are known with other common names include "alligator pear", "avocado", "ahuacate", "aguacate" and "palta" among others [2]. A unique feature of avocados is that the fruits mature on the tree and ripen after harvest. The ripening process takes 5–7 days at room temperature. The fruits are ripe when they yield to gentle pressure [3].

1.1. Cultivars

In general, small to medium-sized fruits are the best for commercial production and especially for metropolitan markets. Large fruits are suitable for local commercialization. A smooth, thin or fairly thin, pliable, and green skin is preferred by the consumers. The flesh should be virtually fibreless, of pleasant flavor and low oil content. The seed must be small and tight so as not to bruise the flesh during handling and shipping. The seed coats ought to adhere to the seed and not to the cavity. The fruit should ship well and stand cold storage.

Avocados are classified mainly into four strains.Tables 1, 2, 3, 4 and 5 summarize the most known strains and cultivars of *Persea Americana* which are important due their commercial role in avocado industry. Altogether, about 70 species in the genus *Persea* occur in warmer regions of North, Central and South America, and 80 species east and southeast Asia [2]. Therefore, new cultivars are developed frequently that may have special resistance to wheatear condition, plagues or land conditions such as Guatemalan - Mexican hybrids that includes the following cultivars: "Hass", "Fuerte", "Rincon", "Ryan", "Sharwil" and "Susan".

1.1.1. West Indian Strain

Table 1. West Indian strain and main varieties belong to the race

Cultivars	Shape	Size	Skin characteristic	Pulp	Seed
Butler	Pear	Medium-large	Smooth	-	Medium size
Fuchs	Pear to oblong	Medium	Smooth	Pale greenish-yellow flesh Oil level from 4 to 6 %	Loose seed
Pollock	Oblong to pear	Large	-	Oil content 3 to 5%	Large size
Simmonds	Oblongoval to pear	Large	Smooth, light Green	Good flavor Oil content 3 to 6%	Medium size
Trapp	Round to pearshape	Medium to large	Smooth	Golden-yellow flesh, oil content 3-6%	Large seed, loose in cavity
Cultivars	Shape	Size	Skin characteristic	Pulp	Seed
Waldin	Oblong to oval	Medium to large	Smooth skin	Pale to greenish-yellow Oil content 5 to 10%	Medium to large size, and tight

Source: Morton, [4].

Table 2. Guatemalan strain and its main varieties

Cultivars	Shape	Size	Skin	Pulp	Seed
Dickinson	Oval to obovate	Small to médium	Dark-purple with large maroon dots, rough, thick, and granular	Good quality	Small to medium size
Edranol	Pear	Medium	Olive-green, slightly rough, thin leathery	High quality and nutty flavor, oil content 15-18%	Small size
Hazzard	Pear	Medium	Rough and fairly thin	Good quality flesh Oil content 15 to 34%	Small size
Linda	Elliptical	Large	Rough, dull-purple	Yellow flesh Oil content 10 to 14%	Small size
Macarthur	Pear	Large	Thin	Sweet, nutty and watery flavor Oil content 13-16.7%	Medium to large size
Reed	Rounded	Medium to large size	Slightly rough	Cream-colored with faintly nutty flavor, and excellent quality	Small size
Schmidt	Pear	Medium tolarge	Rough	Pale-yellow flesh Oil content 12 to 16%	Medium size
Sharpless	Slender-pear	Large to very large	Slightly rough, greenish-purple to dark-purple, thick and granular	High quality and good flavor	Small size
Solano	Obovate to oval	Medium to large	Smooth, bright-green, medium-thick, and granular	Fair quality	Small and tight size
Taylor	Obovate to pear	Small to medium	Rough, and fairly thin with small yellow dots	Excellent quality and flavor Oil content 12 to 17%	Medium size
Tonnage	Pear	Medium tolarge	Dark-green, rough, and thick	Flesh green near skin, rich in flavor Oilcontent 8 to 15%	Medium size
Wagner	Rounded toobovate	Small to médium	Slightly rough	Light yellow, oil content 16 to 20%	Large size

Source: Morton, 1987 [4].

1.1.2. Guatemalan Strain

This strain includes more than 20 cultivars; some of them are displayed in Table 2. The Guatemalan varieties or strain are native to highlands of Central America, Ecuador and Mexico.The large fruits have rough or smooth, gritty skin. They ripen from nine to fourteen months after flowering and have seeds which are small and held tightly by the flesh. The trees are cold resistant and many varieties are used commercially. Guatemalan strain includes Dickinson, Edranol, Hazzard, Linda, Macarthur, Reed, among other.

1.1.3. West Indian-Guatemalan Hybrids

West Indian-Guatemalans have major importance in the Florida avocado industry, representing more than half of the more than 20 major and minor commercial cultivars grown in this state today. Prominent cultivars past and present are shown in Table 3.

1.1.4. Mexican Strain

Mexican strain; thin skin and tender, flesh of high oil content, up to 30%. The foliage has a pronounced anise-like odor; the tree is more cold resistant than those of the other races or hybrids, thriving near Puebla, Mexico, at 500 ft (1,800 m) above sea-level. Main cultivars are summarized in Table 4.

1.1.5. Guatemalan-Mexican Hybrids

Guatemalan - Mexican hybrids include one of the most important cultivar which is Hass avocado due their commercial value. Avocados belongs to this cultivar are produced in several countries such as USA, Chile, and Mexico. Hass avocado are commercial important due higher amount of pulp than other cultivars. Table 5 shows the main cultivars of Guatemalan-Mexican hybrids. Hass and Fuerte are the most important cultivars of this strain. Avocado Hass and Fuerte are shown in Figures 1 and 2.

1.2. Production

The 2010 world total production of avocado was 3.84 million metric tons (MMT), which represents about 92% increased as compared to 1990. Mexico is the leading producer with 1.10 million metric tons (approximately 29% total world production), followed by Chile, the Dominican Republic, Indonesia, and Colombia.

Table 3. West Indian-Guatemalan hybrid and its main cultivars

Cultivars	Shape	Size	Skin	Pulp	Seed
Bonita	Obovate	Medium	Slightly rough	Oil content 8 to 10%	Medium size
Booth 1	Round-obovate	Medium-large	Smooth, medium thick, and brittle	Pale, oil content 8 to 12%	Large size
Booth 7	Round obovate	Medium	Slightly rough, thick and brittle	Oil content 7 to 14%	Medium size and tight
Collinson	Broad-obovoid to elliptical	Large	Smooth	Excellent flavor Oil content 10 to 16%	Medium size, and tight
Cultivars	Shape	Size	Skin	Pulp	Seed
Grande	Pear	Large	Rough, green to purplish	-	Medium size
Hall	Pear	Large	Smooth, fairly thick	Deep-yellow, oil content 12 to 16%	Medium large size and tight

Cultivars	Shape	Size	Skin	Pulp	Seed
Hickson	Obovate	Medium tosmall	Sligltly rough, thick, brittle	Fair to good quality Oil content 8 to 10%	Small size, and tight
Simpson	Obovate-elliptical	Large	Slightly rough and thick but no brittle	Pale, oil content 10 to 14%	Medium-large, size and tight
Winslowson	Round-oblate	Large	Smooth	Pale, oil content 9 to 15%	Medium size

Source: Morton, [4].

Table 4. Mexican strain and main cultivars

Cultivars	Shape	Size	Skin	Flesh	Seed
Duke	Elongated	Small (150-200 g)	-	Good quality Oil content 14.5%	-
Gottfried	Pear	Medium	Smooth, and purple	Excellent quality Oil content 9 to 13%	Medium size
Mexicola	Pear	Very small	Black	Excellent flavor	Large size
Northrop	-	Small 100-150 g	Nearly black	Good quality Oil content 26%	Medium size
Puebla	-	Medium	Smooth, and purple	Good flavor Oil content nearly 20%	Medium to large size
Zutano	Pear	Medium-small	Light green, very thin, and leathery	Oil content 15 to 22%	Medium size

Source: Morton, [4].

Table 5. Guatemalan-Mexican strain and main cultivars

Cultivars	Shape	Size	Skin	Pulp	Seed
Hass	Pear to ovoid	Medium	Tough, leathery, dark-purple, fairly thin	Good flavor Oil content 18 to 22%	Small size
Fuerte	Pear	Small to medium	Slightly, rough, with many small yellow dots	Green flesh near to skin Oil content 12 to 17%	Small size and tight
Rincon	Pear	Small to medium	Smooth, leathery and fairly thin	Buttery, oilcontent 15 to 26.5%	Medium size
Ryan	Pear	Medium 226-230g	Medium-rough	Fair quality	Largesize
Susan	Pear	Medium 227-283g	Light-green, smooth, thin	Pale cream-color	Large size

Source: Morton,[4].

Credit: photo by Dr. Alicia Ortiz-Moreno.

Figure 1. Hass cultivar belongs to Guatemalan-Mexican hybrids strain.

Credit: photo by Dr. Alicia Ortiz-Moreno.

Figure 2. Fuerte cultivar belongs to Guatemalan-Mexican hybrids strain.

Table 6. Market Forecast: 2008 - 2012 World's Avocado Production, Kton

Countries	Years					Market Share
	2008	2009	2010	2011	2012	
Mexico	1,257	1,289	1,321	1,353	1,385	29.5%
Indonesia	323	340	357	374	391	8.3%
United States	246	252	257	263	269	5.7%
Colombia	256	271	287	302	318	6.8%
Brazil	218	227	236	244	253	5.4%
Chile	227	241	255	269	283	6.0%
Dominican Rep.	193	202	212	221	230	4.9%
Peru	140	146	152	158	163	3.5%
China	118	123	129	135	140	3.0%
Ethiopia	92	93	93	94	95	2.0%
World Total	4,113	4,259	4,405	4,551	4,697	100.0%

Source: Market Research Analyst[5].

Avocado world production in 2011 was estimated at 4.55 MMT, produced mainly in Latin America and the Caribbean zones. The production was estimated to reach 4.7 MMT in 2012.

Detailed list of main avocado's producer are displayed in Table 6. The estimate of the global avocado market in 2012 was about $7.5 billion in retail prices [5].

Due to the high economic importance of avocado fruit in Mexico, the food industry is showing a remarkable interest in processing and enhancing the value of this crop. Industrial processing of avocados generates a large amount of peels and seeds as wastes, causing a serious environmental problem. Recently, exploiting the phytochemicals content of these wastes may lead to new products and add value to the avocado industry.

2. Chemical Composition of *Persea americana* Seed

Interest in by-products from avocado has been a trend topic in last decade. Recent reports have focused on chemical composition of seeds from different cultivars. Avocado (*Persea americana* Mill var. Hass) seed mainly contains carbohydrates, protein and fat and, 34.8±3.4 g dietary fiber/100 g dry weight [6].

Moreover, Rodríguez-Carpena et al., [13] informed the chemical composition of two varieties of avocados: "Fuerte" and "Hass". They results are similar in both varieties; moisture is the major component, followed by fat and protein contents. Carbohydrates content is not reported.

Results of both studies are displayed in Table 7. The lipid analysis by gas chromatography/mass spectrometry indicates the presence of palmitic acid (21.3%), palmitoleic acid (1.6%), stearic acid (2.2%), oleic acid (24.1%) and linoleic acid (27.6%) in avocado seed [7].

Table 7. Chemical composition of "Hass" and "Fuerte" avocado seeds

Cultivar	Moisture	Fat	Protein	Carbohydrates	Ash
Hass[d]	55.76±4.34 [a]	1.39±0.54 [a]	2.19±0.38 [a]	NR	0.70±0.14 [a]
Hass[e]	4.0±0.08 [b]	4.38±0.8 [b]	4.74±0.01 [b]	79.10±0.8[b,c]	2.2±0.14 [b]
Fuerte[d]	52.69±1.49 [a]	1.52±0.83 [a]	2.22±0.46 [a]	NR	0.83±0.21 [a]

[a]Results expressed as g/100 g wet weight. [b]Results expressed as g/100 g dry weight.
[c]Value calculated as the difference to a total of 100%. [d]Rodríguez- Carpena et al., [13],
[e]Pahua-Ramos et al., [6], NR No reported.

3. Occurrence of Functional Compounds in Avocado Seeds

3.1. Total Phenolic Content

Phenolic compounds are responsible for the high free radical inhibition activity of fruits which indicates that there is a correlation between phenolic and flavonoids content with percentage of radical inhibition [8]. Table 8 shows differences in concentration in similar cultivars may be due different solvents employed for extractions, extraction time, and procedure conditions (temperature, pressure, speed, and vacuum, among other variables). In the preliminary phytochemical analyses performed by Giffoni-Leite et al., [7] revealed the presence of flavonoids, anthocyanins, condensed tannins, alkaloids and triterpenes in methanol extracts from avocado seed. The authors also reported that sterols and triterpenes were detected in the hexane extract. Phenolic compounds are concentrated mainly in avocado pericarp and seed and, in minor proportion in the flesh. Tannins have been identified in aqueous extract of avocado seed [9]. Recently,Hurtado-Fernandez et al.,[10] reported the phenolic content in avocado fruits from "Hass", "Lamb-Hass", and "Rugoro" varieties.

Table 8. Total phenolic content and antioxidant capacities in avocado seeds of different cultivars

Cultivar	TPC (mg GAE/g)[a]	Solvent extraction
Slimcado[d]	19.2 ± 3.3 [a]	acetone:water:acetic acid (70:29.7:0.3)
Simmonds[d]	40.2 ± 1.9 [a]	
Lorreta[d]	31.5 ± 2.2 [a]	
Choquette[d]	33.4 ± 1.5 [a]	
Booth 7[d]	33.4 ± 5.0 [a]	
Booth 8[d]	35.7 ± 0.8 [a]	
Tonnage[d]	33.1 ±0.9 [a]	
Hass	51.6 ± 1.6 [a, d]	acetone:water:acetic acid (70:29.7:0.3)
	60.82 ± 14.4[b, e]	Acetone
	16.99 ± 4.08[b, e]	Ethylacetate
	35.11 ± 9.88[b, e]	Methanol
	219.4 ± 4.5[b,c, f]	methanol:ethyl acetate and water
Fuerte	69.12± 16.99[b, e]	Acetone
	20.29 ± 7.15[b, e]	Ethylacetate
	41.64 ± 10.48[b, e]	Methanol

[a]Fresh weight, [b] Dry weight, [c] Dry extract.
[d]Wang et al., 2010, [e]Rodríguez-Carpena et al., [13],[f]Dabas et al., [12].

Hurtado-Fernandez et al.,[10] evaluated the capability of different solvents as extraction mediums such as acetone, methanol, ethanol, and ethyl acetate. They informed that the pulp of three avocado varieties contains common compounds, such as quinic acid, succinic acid, pantothenic acid, *p*-coumaroyl-D-glucose, abscisic acid, pentadecylfuran, avocado furan, and oleic acid [10]. However, phenolic level and profile compounds are different in seeds; Wang and others (2010) used acetone:water:acetic acid (70:29.7:0.3, v/v/v) to extract phenolic compounds from avocado Hass.

They reported values of 51.6 mg/g GAE fresh weight for the seed of avocado Hass, whereas the corresponding values for pulp and peel were (4.9 mg/g GAE and 12.6 mg/g GAE fresh weight). Phenolic content in seed is about 4-fold and 10-fold higher values than in peel or pulp respectively. The authors studied also other avocado cultivars, and seeds contain higher phenolic content than pulps or peels. Total phenolic content in different seeds cultivars ranged from 19.2 to51.6 GAE mg/g. It was reported that "Simmonds" avocado seed has the highest phenolic content and antioxidant activity [11]. Dabas et al.,[12]reported also that the phenolic content in two avocado seeds extracts was found to be 219.4±4.5 mg/g GAE (colored extract) and the uncolored extract was found to be 283.2±5.8 mg/g GAE. The authors evaluated also the efficiency of different solvent mixes to extract phenolic compounds and determined that methanol:ethyl acetate mix is better extraction medium than water. In addition the phenolic content in colored extract was 22.5% lower than in uncolored one. Thus, there is a correlation among color development and phenolic content in crushed avocado seed, this behavior depends of polyphenoloxidase enzyme activity. No changes in phenolic content are reported with peroxidase enzyme activity [12].

Rodrìguez-Carpena et al., [13] reported that seeds contain less total phenolic levels than peel in two avocado seeds ("Hass" and "Fuerte"), opposite to most reports. Recently, Pahua-Ramos et al., [6] informed similar total phenolic levels in dry avocado seed and antioxidant capacity of 173.3 μmol TE/g seed dry weight in "Hass" cultivar. Results of different studies are summarized in Table 8.

3.2. Phenolic Profile of Avocado Seed

Avocado seeds have gained much attention due their phenolic composition. Recently, it has been reported that several compounds which are important in antioxidant and biological activities are in avocado seeds. Catechin and epicatechin has been isolated as major active compound in methanolic extract of avocado seeds [2]. Recently, Rodriguez-Carpena et al., [13] reported a large variety of phenolic compounds in seeds from two avocado varieties ("Hass" and "Fuerte"). These phenolics have been classified into five groups: catechins, hydroxybenzoic acids (OH-B), hydroxycinnamic acids (OH-C), flavonols, and total procyanidins. The avocado seeds are rich in catechins ("Hass", 237.8 mg/100 g dry weight and "Fuerte", 96.7 mg/100 g dry weight), procyanidins (the sum of dimers, oligomers, and polymers) ("Hass", 4592 mg/100 g dry weight and "Fuerte", 876.9 mg/100 g dry weight), and OH-C ("Hass", 282.7 mg/100 g dry weight and "Fuerte", 72.4 mg/100 g dry weight). Seeds from "Hass" and "Fuerte" varieties contain small amounts of flavonols (1.7 mg/100 g dry weight and 2.1 mg/100 g dry weight, respectively). The chlorogenic and the neochlorogenic acids are the dominant OH-C in avocado seeds. The derivates from hydroxybenzoic acids (p-coumaric, caffeic, ferulic, and sinapic) are found only in avocado seed "Fuerte" (1.0 mg/100 g dry weight) [13]. Rodriguez-Carpena et al., [13]concluded that avocado seeds from the 'Hass" cultivar has significantly higher amounts of catechins, procyanidins, and OH-C than avocado seed from the "Fuerte" cultivar. Pahua-Ramos et al., [6] identified seven phenolic compounds in methanolic extract from avocado seed ("Hass" variety): protocatechuic acid (128.18 μg/g dry weight) kaempferide (107.42 μg/g dry weight) and vanillic acid (28.67μg/g dry weight). In addition, small amounts of clorogenic acid, syringic acid, rutin, and kaempferol are contained in avocado seed [6].

3.3. Procyanidins Content

Procyanidins are the oligomers and polymers of flavanols and are one of the major polyphenol compounds in foods. The sum of total procyanidins has been reported by Rodríguez-Carpena et al.,[13], they results (see Figure 1) are in agreement with reports of Wang et al.,[11]. B-type procyanidin oligomers (monomers to octamers) and polymers have been identified by Wang et al., [11] in seeds from eight avocado cultivars, using mass spectrometry and liquid chromatography. About 40% of procyanidins in the seeds are polymers with a degree of polymerization above 10. They findings suggest that a small percentage of procyanidins in avocado seeds contained A-type linkage. Procyanidins content in avocado seeds is similar to those found in natural cocoa powder and dark chocolates. Procyanidin monomer levels are similar in eight cultivars investigated ("Slimcado", "Simmonds", "Loretta", "Choquette", "Booth 7", "Booth 8", "Tonnage", and "Hass").

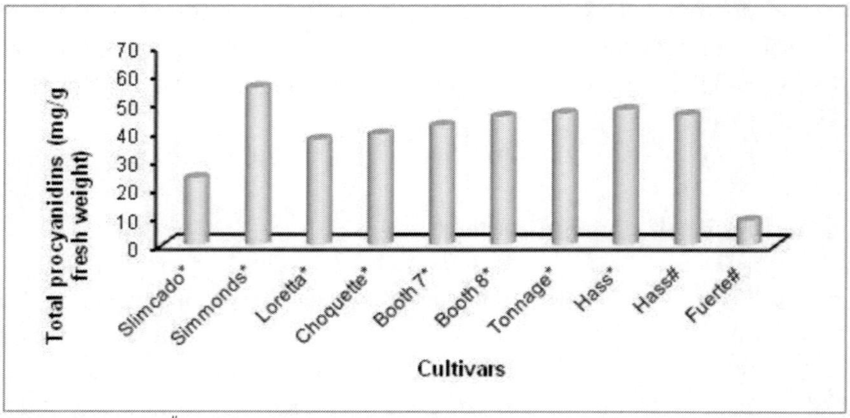

Source: *Wang et al., [11]; #Rodríguez-Carpena, et al., [13].

Figure 3. Total procyanidins content in different avocado seeds.

Procyanidin dimers are mainly found in "Booth 8" and "Hass" cultivars (5.8 ± 0.02 mg/g fresh weight and 4.5 ± 0.0 mg/g fresh weight, respectively). "Tonnage" (12.5 ± 0.3 mg/g fresh weight), "Simmonds" (11.2 ± 0.4 mg/g fresh weight) and "Hass" (10.6 ± 0.3 mg/g fresh weight) cultivars seeds contain the highest procyanidin 4-6mers levels. Procyanidins polymers are found in most of the cultivars being mainly in "Simmonds" (26.6 ± 0.6 mg/g fresh weight), "Hass" (18.6 ± 0.2 mg/g fresh weight) and "Booth 8" (18.3 ± 0.3 mg/g fresh weight) cultivars [11]. According to Wang et al.,[11], the "Simmonds" cultivar (55.6 ± 1.8 mg/g fresh weight) is very rich in total procyanidins followed by "Hass" (47.7 ± 0.6 mg/g fresh weight), "Tonnage" (46.4 ± 1.5 mg/g fresh weight) and "Booth 8" cultivars (45.3 ± 1.2 mg/g fresh weight) results of other five cultivars are displayed in Figure 3.

3.4. Carotenoids Content

Carotenoids are phytocompounds found in high levels in tropical exotic fruit tissue. These compounds are important since they play an important role in human health by acting as biological antioxidants, protecting cells and tissues from damaging effects of free radicals and singlet oxygen and are used as natural colorants in the food industry [8, 14-15]. Wang et al., [11] indicated that carotenoid content in seeds is much lower than other phytocompounds such as total polyphenols in different avocado seeds. According to Wang et al.,[11]. Hass cultivar contains the highest level of carotenoids (6.3 μg/g fresh weight), followed by Choquette, and Booth 7 (2.1 μg/g fresh weight). Carotenoids content in Slimcado, Simmons, Loretta, Booth 8, and Tonnage cultivars were similar (<2 μg/g fresh weight). Seed from Slimcado cultivar contains the lowest level of carotenoids (<1 μg/g fresh weight) [11].

3.5. Chlorophyll Content

Chlorophylls levels in avocado seeds are not as important as polyphenols, however Wang et al.,[11] determined the content of chorophyll a, b and total chlorophylls in avocado seed from eight cultivars. The authors found that chlorophyll content in Hass avocado is much higher (41.2 μg/g fresh weight) than in seeds of non-Hass avocados which contain less amounts of chlorophyll (<30 μg/g fresh weight).

Seed from "Hass" cultivar contains equal levels of chlorophylls a and b (21.0 μg/g and 20.2 μg/g fresh weight, respectively). Results of other cultivars are displayed in figure 4. The seed from "Slimcado" cultivar contains the lowest level of chlorophylls a and b (0.1 μg/g and 0.1 μg/g fresh weight, respectively). The total chlorophylls content in "Choquette" and "Tonnage" seeds are the same (3.0 μg/g fresh weight), while Booth 7 and Booth 8 cultivars are also similar (2.1 μg/g fresh weight, and 2.6 μg/g fresh weight, respectively) [11].

4. ANTIOXIDANT ACTIVITY

The intense antioxidant activity of phenolics and chlorophylls has been closely linked to the health-promoting effects of avocado fruit and oil [13].

Antioxidant activities from avocado seeds extracts are related to their high procyanidin content, otherwise antioxidant activity due pigments content indicates that pigments are not the major antioxidant compounds in avocado seed [11]. In agreement with this report it has been informed that antioxidant activity of peel and seed extracts give considerably more intense antioxidant power than pulp [13]. The solvent extraction and avocado seed varieties have stronger influence in antioxidant activity value. Rodríguez-Carpena et al., [13] indicated that acetone extract is more suitable for CUPRAC assay, followed by methanol and ethyl acetate counterparts.

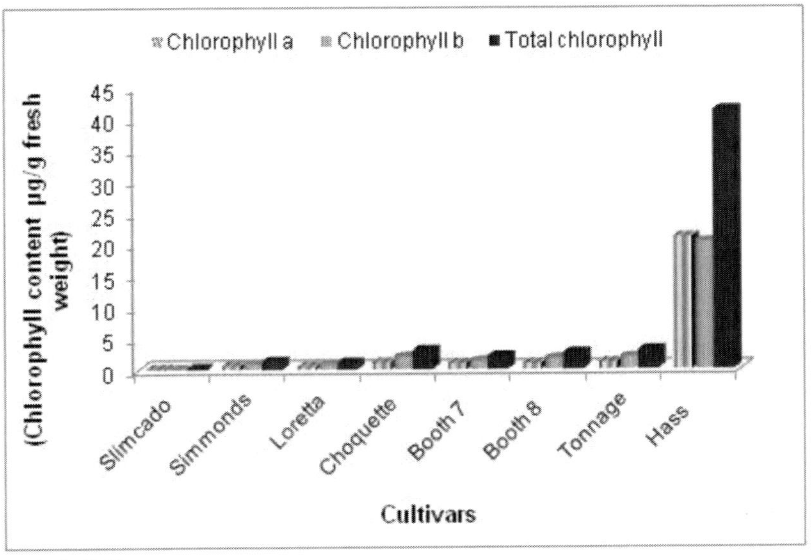

Source: Wang et al.,[11].

Figure 4. Clorophylls a, b, and total chlorophyll content in avocado seeds from eight cultivars.

In addition, acetone and methanol extracts of "Fuerte" seed exhibit a higher antioxidant potential than extracts from "Hass" variety.

Results found with ABTS radical assay are similar to those found in the CUPRAC assay.

Acetone extract of seeds has the greatest peroxil radical scavenging ability in ABTS method. Otherwise, intermediate values are reported for methanol extracts from "Fuerte" and "Hass" seeds cultivars whereas the ethyl acetate extracts displays the lowest values. The acetone extract from "Fuerte" seed shows the highest antioxidant activity against the DPPH radical.

"Fuerte" seed exhibits higher antioxidant activity than "Hass" variety with CUPRAC and ABTS assays instead of DPPH assay. Even, there are differences in antioxidant values; the total phenolic content and particular phenolics such as catechins, procyanidins and OH-C (hydroxycinnamic acids) levels are no significantly different between seeds from the two cultivars [13]. Results of several reports in antioxidant activity of avocado seeds from different cultivars are displayed in Table 9.

5. Biological Activity

Several reports have associated avocado seeds extracts or seed powder with biological activity such as hypolipidemic activity, hypocholesterolemic activity, and antimicrobial activity and, hypogliceamic activity, among others.

5.1. Antimicrobial Activity of Avocado Seeds

Aztec and Maya cultures have used infusions of avocado seeds to treat mycotic and parasitic infections since ancient times. Also, it has been reported that seeds extracts are effective against diabetes, inflammation and gastrointestinal disorders [12]. It has been reported, that 1,2,4-trihydroxyheptadec-16-ene, 4-acetoxy-1,2-dihydroxyheptadec-16-ene, 1,2,4-trihydroxyheptadec-16-yne, 4-acetoxy-1,2-dihydroxyheptadec-16-yne, 1,2,4-trihydroxynonadecane among other compounds, purified from *P. americana* seeds exhibited moderate activity against epimastigotes and trypomastigotes[2, 16]. Other compounds such as quercitrin showed virustatic effects by inhibiting HIV syncytium formation and viral p24 antigen formation.

The methanolic extract of avocado seeds possess antioxidant activity in an AMVN-induced methyl linoleate peroxidation assay [2]. Therefore, the extracts obtained from avocado seeds (*Persea americana* Mill.) and their constituents represent possible candidates for use as alternative dengue control agents, and also act as antifungal agents against pathogenic yeasts.

Giffonni-Leite et al., [7] tested two extracts (methanol and hexane) from avocado seeds; the hexane extract was able to inhibit *Artemia salina* at LC_{50} of 2.37 mg/mL.Notably, methanol and hexane extracts have larvicidal effects, being the most effective hexane extract.

This effect was attribute to phytochemical compounds such as: 1,2,4-trihydroxy-nonadecane, 1,2,4-trihydroxy-heptadec-16-ene, and 1,2,4-trihydroxy-heptadec-16-yne from unripe avocado fruits, that showed toxicity on *Artemia salina and Aedesa egypti* larvae. β-sitosterol and 1,2,4-trihydroxy-nonadecane were detected in the hexane extract from *Persea americana* seeds and may be responsible for its larvicidal activity.

Table 9. Antioxidant activity in different avocado seeds cultivars

Cultivar	ABTS (mmol TE/g)	ORAC (mmol TE/g)	DPPH (mmol TE/g)	CUPRAC (mmol TE/g)
Slimcado[f]	NR	0.229[a]	0.128[a]	NR
Simmonds[f]	NR	0.459[a]	0.240[a]	NR
Lorreta[f]	NR	0.299[a]	0.159[a]	NR
Choquette[f]	NR	0.348[a]	0.157[a]	NR
Booth 7[f]	NR	0.319[a]	0.188[a]	NR
Booth 8[f]	NR	0.368[a]	0.207[a]	NR
Tonnage[f]	NR	0.464[a]	0.162[a]	NR
Hass[g]	NR	0.428[a,f]	0.164[a,f]	NR
	158.290[a,c]	NR	0.130[a,c]	275.36[a,c]
	21.57[a,d]	NR	17.78[a,d]	58.00[a,d]
	78.93[a,e]	NR	66.24[a,e]	141.67[a,e]
Fuerte[g]	69.12[a,c]	NR	167.50[a,c,g]	353.43[a,c]
	20.29[a,d]	NR	27.80[a,d,g]	96.09[a,d]
	41.64[a,e]	NR	94.27[a,e,g]	184.42[a,e]

[a]Fresh weight, [b] Dry weight. [c] Acetone (70%) extraction, [d] Ethyl acetate extraction, [e] Methanol (70%) extraction. [f] Wang et al.,[11], [g] Rodríguez-Carpena et al., [13].

The methanol extract contains phenolic compounds such as flavonoids and anthocyanins, for which antimicrobial activity has been reported, and alkaloids with several actions such as antiviral, antimalarial and antiamoebic activity [7].

Giffoni-Leite et al., [7] investigated also the antifungal potential of avocado seed extracts against three yeast *Candida* spp, *Cryptococcus neoformans* and *Malasseziapachydermatis* which are of great importance for human and veterinary medicine. The avocado seed extracts (inhibitory concentration for hexane extract ranged from 0.312 to 1.25 mg/mL and for methanol extract from 0.125 to 0.625 mg/mL) inhibited grown of *Candida* species. The hexane extract inhibitory concentration values ranged from 0.031 to 0.625 mg/mL and, methanol extract from 0.321 to 0.625 mg/mL for *Malasseziapachydermatis* strains. *Cryptococcus neoformans* strains can be inhibited at low hexane extract concentration (0.312 – 0.625 mg/mL) and methanol extract (0.08 mg/mL - 0.156 mg/mL). [7]. Antimicrobial activity of ethanolic extracts from three avocado cultivars ("Shepard", "Hass" and "Fuerte") has been confirmed by Chia et al., [17]. They tested seed extracts against four Gram-positive bacteria, six Gram-negative bacteria, and one yeast. The ethanol extracts show antimicrobial activity from 104.2 to 416.7 μg/mL concentrations toward both Gram-positive and Gram-negative bacteria (except *Escherichia coli*).

The minimum concentration required to inhibit *Zygosaccharomyces bailii* was 500 μg/mL of ethanol extract. No inhibition by ethanol extract was reported against *Penicillium* spp. and *Aspergillus flavus*. The seed of three avocado cultivars show activity against yeast, Gram negative and gram-positive bacteria [17].

Similar results have been found recently by Rodriguez-Carpena et al., [13]. The authors tested different extracts obtained with avocados seeds from two varieties "Fuerte" and "Hass" against several pathogens and spoilage microorganisms commonly found in meat products. Their findings suggest that avocado extracts are promising sources of potential antimicrobial compounds and may be efficient natural additives for extending the shelf-life of fresh food products. This study concludes that Gram-positive bacteria are generally more sensitive than Gram negative bacteria.

This effect is attributed to the fact that Gram-negative bacteria have an extra protective outer membrane, thus they are usually more resistant to antibacterial agents than their Gram positive counterparts. The highest inhibitory effect is observed against *Bacillus cereus* and *Listeria monocytogenes* reported in Gram-positive bacteria strains, whereas *Escherichia coli* is the most sensitive among Gram-negative bacteria.

Pseudomonas spp. was found to be the most resistant bacteria and, *Aspergillus niger* was the most resistant to all extracts. Notably, extracts from the "Fuerte" variety display antimicrobial activity in more strains than extracts from the "Hass" cultivar. Both extracts from "Fuerte" and "Hass" do not show antimicrobial activity against *Pseudomonas* spp. Avocado seed extract from "Hass" cultivar do not show activity against *E.coli*. Authors suggest that the differences in the antimicrobial activity of avocado extracts could be due to the nature of the antimicrobial substances present in the extracts and their mechanisms of action on the tested microorganisms [13].

5.2. Treatment of Hypertension

Avocado seed extracts have been used by herbalists in Nigeria in folk medicine to management the hypertension. Anaka et al., (2009) informed that the aqueous seed extract of *P. americana* reduces blood pressure and heart rate in Sprague Dawley rats. The authors suggest that the mechanism involved in the blood pressure lowering effect of the extract appears to be cardio dependent as heart rate. This effect is due polyphenol composition in aqueosus seeds extracts of *P. americana* Imafidon and Okunrobo, [18] administered methanolic extracts on albino rats.

The methanolic extracts of *Persea americana* seeds (Lauraceae) reduce blood pressure of hypertensive rat treated at doses of 200 mg/kg 500 mg/kg and 700 mg/kg body weight. Supplementation of albino rats with aqueous extract reduces cholesterol, glucose, urea and sodium levels in hypertensive rats.

However, protein and albumin levels significantly increase, while potassium level does no change. This study demonstrates that methanolic extract from *Persea americana* seed has anthypertensive, hypoglycaemic and hypocholesterolemic activities at long-term administration period (five weeks) [18].

5.3. Hypoglycemic Activity

Diabetes mellitus is a primary disorder of carbohydrate metabolism, which generally involves absolute or relative insulin deficiency and/or insulin resistance and ultimately leads to hyperglycemia. Avocado seed (*Persea americana*) has been used in traditional medicine in treatment of diabetes and its possible role on pancreatic tissue. Edem et al.,[20] investigated the hypoglycemic activity and its possible role on pancreatic tissue of aqueous extracts from avocado seed in Wistar rats. Test groups were made diabetic with intra-peritoneal injection of alloxan and treated with 300 mg and 600 mg/kg body weight of avocado seed extract. They results show that in diabetic rats supplemented with extracts, blood glucose levels decrease by 73.26-78.24.

In normal rats supplemented with seeds extract, blood glucose levels decrease by 34.68-38.9%. These results suggest both hypoglycemic and antihyperglycemic effects of the avocado pear seed extracts. [19]. This study, suggests that avocado pear seed may have some chemical components that exert regenerative effects on beta cells leading the stimulation of these cells to produce more insulin (pancreatotrophic action) or may have some insulin-like substances. The effect on the islet cells of diabetic rats increases when dose is increased. Excitingly, findings in this research support the traditional use of avocado pear seed for controlling hyperglycemia in diabetics [20].

In addition, Koffi et al., [21]evaluated the supplementation of aqueous extract from seeds of *Persea americana* indiabetic-induced rabbits. The extract was oral-administered in order to be similar to the traditional method. The oral administration of the aqueous extract reveals that it has glycemic properties. The glycemic effect dependents of dose administered. The glycaemia of rabbits treated with *P americana* extract at 10 g/l, does no vary. However, the glycaemia of rabbits treated with the extract at 20 g/l and 30 g/L decreases the glycaemia at about 2 g/L and 1 g/L respectively. In any case, a weak resumption of the hyperglycemia is reported, after the stop of the treatment [21].

5.4. Hypolypidemic Activity

Cholesterol is a lipid, waxy steroid found in the cell membrane. It can dissolve and travel in the water-based blood stream at exceedingly small concentrations. Since cholesterol is insoluble in blood, it is transported in the circulatory system within lipoproteins. Abnormal high dietary cholesterol which leads to hypercholesterolemia is strongly associated with cardiovascular diseases because it promotes atherosclerosis [22-23]. The methanolic (methanol at 70%) extract of *P. americana* was evaluated on hypercholestreolemic-induced rats by Asaolu et al., 2010.

Plasma levels of total cholesterol (TC), triglycerides (TG), very low density lipoprotein cholesterol (VLDLC) and low density lipoprotein cholesterol (LDLC) reduce significantly in rats treated with different concentrations (at dose of 50, 100, 200 and 300 mg/kg) of methanolic extract. Also, a positive effect on high density lipoprotein cholesterol (HDLC) levels was observed in hypercholestrolemic rats treated with methanolic extract. These results point out that methanolic extract of *Persea americana* seed may be used as an effective supplement in hyperlipidemic patients [23]. Similar effect is reported in Pahua-Ramos et al., (2012) study. They evaluated the supplementation of avocado seed flour at doses at 125, 250 and 500 mg/kg body weight on hypercholesterolemic mice.

In the report treatment of mice with avocado seed flour reduces the elevated levels of TC by 33% and LDL-C by 39.4%. In addition, treatment with 250 mg/kg body weight reduces TC and LDL-C by 34 and 39%, respectively, and treatment with 500 mg/kg body weight reduces the TC and LDL-C levels by 36 and 41%, respectively.

The hypocholesterolemic effect has been associated with crude fiber in avocado seed that possess a cholesterol-lowering effect. [6]. These studies together indicate that methanol extract or seed flour of avocado can be used as an effective supplement in mice and rats for treating hyperlipidemia.

6. TECHNOLOGICAL APPLICATIONS

6.1. Snack Manufacturing

Avocado seeds are considered waste material, and also an environmental problem for the avocado industry. One of the most important commercial; cultivars is "Hass" avocado which contains about 12% seed respect to the total fruit weight. Recently, new ways to management the waste has been published. One of the most remarkable works has pointed out its use as principal material to produce snacks. Avocado seed powder can be blended with milled corn at 40-60%, and then this mix is extruded. The mixture yield was 8.4% and the appearance and adhesion are good. The snack obtained with 100% of avocado powder is very similar to commercial one. Thus, it is viable to produce snacks with avocado seed powder alone or combine with other flours [24].

6.2. Sorbent Material

Adsorption has proven to be one of the most effective physicochemical methods for textile wastewater treatment. The development, characterization and application of adsorbents prepared from avocado kernel seeds was done by Elizalde-González et al.,[25]. They obtained a series of environmentally friendly adsorbents by simple activation of avocado kernel seeds waste to dye removal.

The main components identified in avocado kernel seeds absorbents are silicon and potassium. The sorbent material has elevated oxygen levels, and after carbonization the surface retains acidic groups, mainly phenolic moieties, also increases specific activated surface. The sorbents of the avocado kernel seeds absorb higher amounts of a dye basic class compare to activated carbon prepared from corncob and plum kernels [25].

6.3. Natural Color

Nowadays, natural colors have gained much interest in industry, due their health benefits and consumer preference. Avocado (*Persea americana*) seeds produce an orange color in water due polyphenol oxidase enzyme activity. The enzyme acts on polyphenols from the seed producing the orange color. This color is more intense and stable at ranged pH from 2.0 to 11.0.

The color extract is stable at this pH value and in the presence of oxygen, but it is no stable at low pH and anaerobic conditions. Also, the color is stable in solution at -18 °C for 2 months, but at higher temperatures color is unstable. This study suggests that the extract may be a potential source of natural colorant and could be used in products stored at low temperature [12].

Conclusion

Avocado seed is an important plant material that has been investigated around the world. Its chemical composition and phytochemicals constituents have an important role in health. Also, there are several investigations about technological applications which may help to intensify the effort to use it in more practical products. This tendency is very trendy with environmental friendly technologies that may help to management this waste material. However, further studies are needed to identify the active principles in *P. americana* seeds, mechanisms, compounds stability, and clinical trials to be assure the safety consumption. In the future, the use of avocado seed as natural treatment of hypercholesterolemia, hyperglycemia, and hyperlipidemia appears possible. Consequently these applications may increase the commercial value of avocado fruit and subproducts.

References

[1] Evans, E. and NaLampang, S. World, US and Florida avocado situation and outlook. *Food and Resource Economics Department*, 2009. 639: p. 7-22.

[2] Ding, H., Chin, Y.-W., Kinghorn, A. D., and D'Ambrosio, S. M., Chemopreventive characteristics of avocado fruit. *Seminars in Cancer Biology*, 2007. 17: p. 386–394.

[3] Ozdemir, F. and Topuz, A., Changes in dry matter, oil content and fatty acids composition of avocado during harvesting time and post-harvesting ripening period. *Food Chemistry*, 2004. 86(1): p. 79-83.

[4] Morton, F. J., *Avocado*, in Fruits of Warm Climates. 1987: Miami, FL. p. 91-102.

[5] Market Research Analyst. Avocado Market Forecast 2008 - 2012: Production, Exports, Imports. 2012 [February 10, 2013]; Available from: http://www.Marketresearch analyst.Com/2008/01/14/Avocado-Market-Forecast-2008-2012-Production-Exports-Imports/.

[6] Pahua-Ramos, M. E., Ortiz-Moreno, A., Chamorro-Cevallos, G., Hernández-Navarro, M. D., Garduño-Siciliano, L., Necoechea-Mondragón, H., and Hernández-Ortega, M., Hypolipidemic effect of avocado (*Persea americana* Mill) seed in a hypercholesterolemic mouse model. *Plant Foods for Human Nutrition*, 2012. 67: p. 10–16.

[7] Giffoni-Leite, J. J., Salles-Brito, E. H., Aguiar-Cordeiro, R., Nogueira-Brilhante, R. S., Costa-Sidrim, J. J., Medeiros-Bertini, L., de Morais, S. M., and Gadelha-Rocha, M. F., Chemical composition, toxicity and larvicidal and antifungal activities of *Persea americana* (avocado) seed extracts. *Revista da Sociedade Brasileira de Medicina Tropical*, 2009. 42(2): p. 110-113.

[8] Ayala-Zavala, J. F., Vega-Vega, V., Rosas-Domínguez , C., Palafox-Carlos, H., Villa-Rodriguez, J. A., Wasim-Siddiqui, M., Dávila-Aviña, J. E., and González-Aguilar, G.

A., Agro-industrial potential of exotic fruit byproducts as a source of food additives. *Food Research International*, 2011. 44: p. 1866–1874.

[9] Anaka O. N., Ozolua R. I., and Okpo S. O., Effect of the aqueous seed extract of *Persea americana* mill (*Lauraceae*) on the blood pressure of sprague-dawley rats. *African Journal of Pharmacy and Pharmacology*, 2009. 3(10): p. 485-490.

[10] Hurtado-Fernández, E., Carrasco-Pancorbo, A., and Alberto Fernández-Gutiérrez, A., Profiling LC-DAD-ESI-TOF MS method for the determination of phenolic metabolites from avocado (*Persea americana*). *Journal of Agricultural and Food Chemistry*, 2011. 59: p. 2255-2267.

[11] Wang, W., Bostic, T. R., and Gu, L., Antioxidant capacities, procyanidins and pigments in avocados of different strains and cultivars. *Food Chemistry*, 2010. 122: p. 1193–1198.

[12] Dabas, D., Elias, R. J., Lambert, J. D., and Ziegler, G. R., A Colored Avocado Seed Extract as a Potential Natural Colorant. *Journal of Food Science*, 2011. 76(9): p. C1335-C1345.

[13] Rodríguez-Carpena, J.-G., Morcuende, D., Andrade, M.-J., Kylli, P., and Mario Estévez, M., Avocado (*Persea americana* Mill.) phenolics, *in vitro a*ntioxidant and antimicrobial activities, and inhibition of lipid and protein oxidation in porcine patties. *Journal of Agricultural and Food Chemistry,* 2011. 59: p. 5625–5635.

[14] Rufino, M., Alves, R., de Brito, E., Pérez-Jiménez, J., Saura-Calixto, F., and Mancini-Filho, J., Bioactive compounds and antioxidant capacities of 18 non-traditional tropical fruits from Brazil. *Food Chemistry*, 2010. 121: p. 996−1002.

[15] Oreopoulou, V. and Tzia, C., Utilization of plant by-Products for the recovery of proteins, dietary fibers, antioxidants, and colorants. Utilization of by-products and treatment of waste in the food industry, V. Oreopoulou and W. Russ, Editors. 2007, Springer US. p. 209-232.

[16] Abe, F., Nagafuji, S., Okawa, M., Kinjo, J., Akahane, H., Ogura, T., Martínez-Alfaro, M. A., and Reyes-Chilpa, R., Trypanocidal constituents in plants 5. Evaluation of some Mexican plants for their trypanocidal activity and active constituents in the seeds of *Persea Americana*. *Biological and Pharmaceutical Bulletin*, 2005. 28: p. 1314-1317.

[17] Chia, R., Wah, T., and Dykes, G. A., Antimicrobial activity of crude epicarp and seed extracts frommature avocado fruit (*Persea americana*) of three cultivars. *Pharmaceutical Biology (Formerly International Journal of Pharmacognosy)*, 2010. 48(7): p. 753−756.

[18] Imafidon, E. K. and Okunrobo, O. L., Biochemical evaluation of the tradomedicinal uses of the seeds of *Persea americana* Mill., (Family: Lauraceae). *World Journal of Medical Sciencies*, 2009. 4(2): p. 143-146.

[19] Yadev J. P., Saini S., Kalia A. N., and Dangi A. S., Hypoglycemic activity of ethanolic extract of *Salvadora oleoides* in normal and alloxan – induced diabetes rats. *Indian Journal Pharmacology*, 2008. 40(1): p. 23-27.

[20] Edem D., Ekanem I., and Ebong P., Effect of aqueous extracts of alligator pear seed (*Persea americana* Mill) on blood glucose and histopathology of pancreas in alloxan-induced diabetic rats. *Pakistan Journal of Pharmaceutical Sciences*, 2009. 22(3): p. 272-279.

[21] Koffi, N., Ernest, A. K., and Dodiomon, S., Effect of Aqueous Extract of *Persea Americana* Seeds on the Glycemia of Diabetic Rabbits. *European Journal of Scientific Research*, 2009. 26 (3): p. 376-385.

[22] Durrington, P., Dysplipidaemia. *The Lancet*, 2003. 362(9385): p. 717-731.

[23] Asaolu, M. F., Asaolu, S. S., Oyeyemi, O. A., and Aluko, B. T., Hypolipemic effects of methanolic extract of *Persea americana* seeds in hypercholestrolemic rats. *Journal of Medicine and Medical Sciences*, 2010. 1(4): p. 126-128.

[24] Olaeta, J. A., Schwartz, M., Undurraga, P., and Contreras, S., Use of Hass avocado (*Persea americana* Mill.) seed as a processed product. Proceedings VI World Avocado Congress, 2007.

[25] Elizalde-González, M. P., Mattusch, J., Peláez-Cid, A. A., and Wennrich, R., Characterization of adsorbent materials prepared from avocado kernel seeds: Natural, activated and carbonized forms. *Journal of Analytical and Applied Pyrolysis*, 2007. 78 p. 185–193.

In: Seeds as Functional Foods and Nutraceuticals
Editors: R. Mora-Escobedo, J. De J. Berrios et al.
ISBN: 978-1-62808-489-4

Chapter 12

Cactus Seeds: A Source of Biocompounds with Functional Activity

***Alicia Ortiz-Moreno*[1], *Gisela Palma-Orozco*[2,3] *and Obed Osorio-Esquivel*[4,*]**

[1]Departamento de Ingeniería Bioquímica, Escuela Nacional de Ciencias Biológicas, Instituto Politécnico Nacional, México D.F., México

[2]Departamento de Ciencias Naturales, Universidad Autónoma Metropolitana-Cuajimalpa, México D.F., México

[3]Departamento de Bioprocesos, Unidad Profesional Interdisciplinaria de Biotecnología, Instituto Politécnico Nacional, México D.F., México

[4]Instituto de Salud Pública, Universidad de Chalcatongo, Chalcatongo de Hidalgo, Oaxaca, México

Abstract

Plants of the Cactaceae family are distributed from Canada to Argentina and are mainly consumed as fresh fruits. Recently, these fruits have been processed into jam, jellies, juice, liquor, and other items. However, only the flesh is used, while the seeds are discarded. These by-products have prompted much interest due their special compositional characteristics, such as their fatty acid profile and their phenolic, flavonoid, carotene, and tocopherol contents, as well as their insoluble dietary fiber content. Cactus seeds are an important source of essential fatty acids, such as linoleic and linolenic acids, which are the biosynthetic precursors of eicosanoids. It has been demonstrated that consumption of seeds from *Opuntia ficus indica* and *Opuntia joconostle* provides beneficial biological effects, such as a hypolipidemic effect in hypercholesterolemic rat and mouse models. Seeds from prickly pears exhibit an excellent antioxidant capacity due to their total phenolic content. For these reasons, cactus seeds could potentially be commercialized as natural sources of biocompounds,

---

* Tel.: 52 95320399; Ext: 899; E-mail: obed2004@hotmail.com.

oils, and fiber for supplementation of food for animals or humans. Thus, this chapter contains valuable information about the origin and taxonomy, physicochemical composition, nutritional value, economic situation, future trends, and effects on some chronic ailments revealed by several studies of *Opuntia* seeds.

1. INTRODUCTION

The Cactaceae family includes over 1,000 species; some of the most important species are *Opuntia ficus indica* (Figure 1), *Opuntia leucotricha*, *Opuntia streptacantha,* and *Opuntia streptacantha,* among others. These plants grow wild in arid and semiarid regions, where the production of more succulent food plants is severely limited. It is known that their cladodes are consumed by local and international populations. Cacti possess crassulacean acid metabolism, making them highly tolerant to drought due to their very efficient water use. Therefore, cacti are suitable for sustainable agriculture in semiarid and arid lands worldwide. The characteristics of the main species are described in Table 1.

Cladodes have been investigated as a possible treatment for gastritis, hyperglycemia, arteriosclerosis, diabetes, and prostatic hypertrophy. The fruits are used for the manufacture of food products, such as juices, alcoholic beverages, and jams, as sources of natural liquid sweetener and for the production of fungal protein via solid substrate fermentation. Cactus juices and cladodes have been extensively studied. However, there are few reports about cactus seeds, which are usually discarded [1].

The 2010 total worldwide production of prickly pears (*Opuntia ficus indica*) was 500,000 metric tons. Mexico produced approximately 79.4%, followed by Italy (12.2%), South Africa (3.7%), Chile (2%), and other countries 2.8% [3].

Mexico is the leading producer of prickly pears with a cultivated area of 74,000 ha, followed by South Africa, Italy, Chile, Argentina, and Peru [4]. Other countries with a presence in the international market are listed in Table 2.

Recently, the international scientific community has shown interest in *Opuntia joconostle* and *Opuntia matudae* fruits. These fruits are known as xoconostles, which are sour prickly pears due to their sour flavor.Production is currently over 10,000 metric tons per year. The interest is due to these fruits being a new source of biocompounds. The fruits possess high antioxidant activity, previously reported in different parts of the *O. joconostle* sour prickly pear, such as the pericarp (62.96±0.5%), mesocarp (42.27±0.5%), and endocarp (51.70±0.55%), which is related to the phenolic and flavonoid composition [5]. *Opuntia joconostle* grows mainly in the central area of Mexico [6]. A plant produces 60 rounded fruits with a diameter of 4–5 cm that can be stored in a dry environment for several weeks. Despite the fruit's characteristics, its use and production are limited to certain geographical areas. *Opuntia joconostle* and *O. matudae* seeds are discarded as waste because they are hard and it is not possible to chew them. However, this by-product could be used to obtain other subproducts. These seeds represent approximately 5.83% of the total fruit weight.

Information about *Opuntia* seeds is still scarce; therefore, the objective of this chapter is to summarize the most recent studies concerning the chemical composition, metabolite identification and biological activity of these seeds.

Table 1. Main species of the Cactaceae family

Species	Characteristics
Opuntia leucotricha	The most widely distributed species, which grows at altitudes of between 1,500 and 2,500 m where the annual rainfall ranges from 220 to 450 mm.
Opuntia streptacantha	It is present in large areas of Mexico (Zacatecas and San Luis Potosí).
Opuntia robusta	It grows in association with *Opuntia leucotricha* and *Opuntia streptacantha.* It is widely distributed throughout the states of Zacatecas, San Luis Potosí, Guanajuato, Aguascalientes and Jalisco in Mexico.
Opuntia cantabrigiensis	It is a shrub with well-defined stems and abundant spines, used mainly to feed goats. It has a wide distribution at altitudes from 1,500 to 2,200 m, with relatively low densities on calcareous soils.
Opuntia rastrera	It is a shrub with a creeping habit, less than 1 m tall, widely used to feed dairy cattle. It grows well in a wide range of soils, from deep to shallow and rocky and calcareous.
Opuntia imbricata	It exhibits great variability. It grows well in relatively poor soils and is a typical invader plant in poorly managed grasslands. It is used to feed goats and sheep after burning off the spines *in situ*.
Opuntia microdasys	It is a short shrub without true spines but loaded with short bristles, which are blown by the wind and can easily blind animals. It grows in calcareous soils and is used to feed animals under extreme drought conditions.
Opuntia violacea	It is recognized by its purple-colored cladodes. A short shrub (< 1 m) that grows in deep clay soils in areas with only 200 mm of annual rain, hot summers (up to 45°C) and cold winters (-8°C). Its forage quality is poor.
Opuntia rufida	It is a shrub that reaches approximately 2 m and grows in calcareous clay soils on slopes or in deep valleys. Its cladodes are larger than those of *Opuntia microdasys,* and its quality as forage is also poor.

Mondragon-Jacobo and Pérez-González, 2001 [2].

Table 2. Worldwide areas of *Opuntia ficus indica* cultivation

Country	Cultivated area (ha)
Mexico	74,500
USA	250
Colombia	500
Peru	1,000
Chile	1,000
Argentina	1,000
Italy	2,500
Israel	300
North Africa: Morocco, Algeria, Libya, Tunisia, Egypt	20,000
South Africa	4,500

Flores, 2006 [4].

Credit: Photo by Dr. Clemente Gallegos.

Figure 1. *Opuntia ficus indica* var. Rojo Pelon cladodes and prickly pears.

The results of a recent investigation of the fatty acid profile, antioxidant activity and lethal dose-50 of *Opuntia joconostle* seed oil has been included at the end of the chapter.

2. Characteristics of the Main Species of *Opuntia*

2.1. *Opuntia Ficus Indica*

Opuntia ficus indica possesses crassulacean acid metabolism, which makes them exceptionally tolerant to drought, allowing for very highly efficient water use. These properties make it suitable for sustainable agriculture in semiarid and arid lands [7]. The plant is consumed as a fresh vegetable (cladodes) in some regions where it is produced, such as Mexico, Spain, Sicily, and the coasts of Southern Italy, Morocco, Algeria, Egypt, Saudi Arabia, Israel, Palestine, Chile, Brazil, Turkey, and northern Africa, as well as in Eritrea and Ethiopia. *Opuntia ficus indica* produces sweet fruits that are called cactus pears and are also known as prickly pears (Fig. 1). These barrel-shaped fruits are 5–10 cm long and 4–9 cm in diameter. The fruits can have white/yellowish or reddish/purple flesh, depending on the variety, and they contains many tiny hard seeds of 5 mm diameter that are brown in color, embedded in the flesh, and are generally ingested [8].

In the past decade, the fruits of *Opuntia* species have become popular for the manufacture of food products. Juice pulp and cladodes have been extensively studied due to their commercial value. However, information about the composition, uses, and benefits of the seeds remains scarce [1].

The seeds are formed of two different tissues, the endosperm and the pericarp, in a ratio of 1:9, respectively. Reports have indicated that prickly pear seeds are mainly composed of polysaccharides, such as cellulose and hemicellulose [9], and reserve proteins such as albumin (molecular weight of 6.5 kDa) [1]. Cactus seeds are an important source of minerals and sulfur-containing amino acids (methionine and cysteine). Table 3 presents the general composition of *Opuntia ficus indica* seeds.

Matthäus and Ōzcan (2011) indicated that the oil content of the seeds of *Opuntia ficus indica* from different locations and provinces of Turkey ranged from 5.0% to 14.4%. Nassar (2008) reported that prickly pear seed flour obtained from *Opuntia ficus indica* contains 10.43% crude oil. Therefore, considering the oil content, cactus seeds could be interesting to commercialize the oil [12].

The fatty acid composition of the oils from *Opuntia ficus-indica* seeds is shown in Table 3. In many plants, the seed oil composition differs from that of the photosynthetic tissue lipids. There is a wide variation in the reported fatty acid composition of prickly pear seeds. The cactus seed oil contained common fatty acids, including saturated fatty acids (palmitic and stearic), and unsaturated fatty acids (oleic, vaccenic, and linoleic). The major fatty acids found in cactus seed oils were palmitic, oleic, and linoleic acids. Stearic and vaccenic acids were found at lower levels. The palmitic acid contents of the seed oils ranged between 10.6% and 12.8%. Whereas the oleic acid contents varied between 13.0% and 23.5%, the linoleic acid contents ranged from 49.3% to 62.1%. Additionally, the stearic acid content was found to be 3.3% to 5.4%, and that of vaccenic acid ranged from 5.0% to 6.3%. The main fatty acid found in *O. ficus indica* seeds was linoleic acid, followed by oleic, palmitic, vaccenic, and stearic acid. The fatty acid profile is dependent on the location where the cacti were grown.

Table 3. Chemical composition and fatty acid profile of *Opuntia ficus indica* seeds

Composition	g/kg
Crude oil	61.9
Protein	9.4
Crude fiber	507.4
Ash	12.3
Carbohydrate	409
Tocopherols	0.039 – 0.5
Fatty acids	
Myristic (14:0)	1.9 – 1.9
Palmitic (16:0)	132.1 – 156
Palmitoleic (16:1)	14.4 – 18.5
Stearic (18:0)	33.1 – 47.9
Oleic (18:1)	210.5 – 256
Linoleic (18:2)	522.5 – 577.6
Linolenic (18:3)	2.9 – 9.7
Arachidic (20:0)	4.2 – 6.6
Behenic (22:0)	2.1 – 3.0

Coskuner and Tekin, 2003; Matthäus and Özcan, 2011 [10, 11].

The chemical composition of the seeds is affected by climatic factors, soil type, genetic factors, and season of the year, among other conditions. Prickly pear seed oil is similar to sunflower and grape seed oils; therefore, prickly pear seeds may be an interesting natural source of edible oil due their high concentration of essential fatty acids [11].

Tocopherols are important components of the unsaponifiable fraction of vegetable foods. In addition to fatty acids, seed oils are excellent sources of vitamin E. Tocopherols are biologically highly active natural antioxidants. Tocopherols are important due to their biochemical functions, such as protecting polyunsaturated fatty acids from peroxidation. In most cactus seed oils, the total content of tocopherols ranged between 3.9 mg/100 g and 50.0 mg/100 g. Generally, the β-tocopherol content of *Opuntia* seed oils varied between 3.9% and 50.0% [11, 13].

2.2. *Opuntia Monacantha* Haw

Opuntia monacantha Haw. belongs to the *Cactaceae* family and grows in semiarid regions. The plant is from South America and is mainly cultivated in Argentina, Paraguay, Uruguay, and Brazil. This cactus has been introduced and naturalized in tropical and subtropical regions throughout the world.

Cladodes from *Opuntia monacantha* Haw. are more fibrous than *Opuntia ficus indica*. The unripe fruit is green/yellowish, but when ripe, it takes on a reddish purple color. It is ovoid shaped, 5–7.5 cm long, and 4–5 cm in diameter. The fruit is processed into jams, jellies, and alcoholic beverages. The seeds are smooth, pale brown in color, irregularly elliptical shaped (4 × 3 mm), and embedded in the reddish flesh. The chemical composition of *Opuntia monacantha* Haw cladodes has been reported as follows: water (91.1%), ash (15.0%) protein (5.4%), fiber (18.5%), and lipid (1.4%), but information about the fruit and seeds is still unavailable [14]. Flavonoids, kaempferol, and isorhamnetin were also found in cladodes of *O. monacantha Haw* [15].

2.3. *Opuntia Stricta* (Haw.) Haw

Opuntia stricta (Haw.) Haw. is a succulent creeping plant belonging to the *Cactaceae* family (Figure 2). This species is native to the Caribbean region, the tropical and subtropical regions of North America, and South America.

The cladodes of *Opuntia stricta* (Haw.) Haw. are not consumed. The flesh and skin of the fruits are red to purplish-red, and the fruits are barrel shaped, 4–6-cm long, and 2.5–4 cm in diameter. The seeds are round, 4–5 mm in diameter, pale brown, with a slightly irregular surface, and are embedded in the flesh. The fruit juice is consumed or made into jams, jellies, and alcoholic beverages. This fruit is a potential source of betacyanin and pigments, which can be used as a natural red-purple food colorant. The composition of *Opuntia stricta* (Haw.) Haw. fruit is shown in Table 4.

Ennouri et al. (2005) reported that the oil content of *O. stricta* seeds is 11.05%. The seeds contain 3.9% protein, 1.64% ash, and 95% dry matter. The refractive and iodine indices were 1.469 and 91.6, respectively. The fatty acid composition of *O. stricta* oil is similar to that of other Opuntia seeds.

Credit: Photo by Dr. Clemente Gallegos.

Figure 2. *Opuntia stricta* Haw.

Table 4. Chemical composition of *Opuntia stricta* Haw. Fruit.

Composition	
Water[a]	82.9
Lipid[a]	0.1
Protein[a]	0.8
Calcium[b]	230
Copper[b]	0.1
Iron[b]	0.4
Magnesium[b]	43
Phosphorus[b]	15
Sodium[b]	700
Zinc[b]	0.7
Niacin[b]	0.1
Vitamin C[b]	1
Betacyanin[b]	80

[a] g/100 g of fruit, [b] mg/100 g of fruit [16, 17].

Linoleic acid is the major component (74.8%), followed by oleic (12.8%) and palmitic acids (7.21%).

Both lauric and myristic acids were detected in *O. stricta* oil at low levels. *O. stricta* oil is highly unsaturated – approximately 88.0%. *Opuntia stricta* consists of approximately 18–20% seeds with respect to the peeled fruits.

2.4. Pitaya

The pitaya is the fruit of several cactus varieties that belong to the *Cactaceae* family, which are native to Mexico, Central America, and South America. Presently, it is cultivated in many Asian countries, such as Vietnam, Taiwan, the Philippines, and Malaysia. In Asia, the pitaya is often called ‘‘dragon fruit” because of its bright red skin and the green overlapping fins covering the fruit [18].

There are different varieties of pitaya, which can be distinguished from one another by either the color of the flesh skin (exocarp) and/or the color of the soft flesh center (mesocarp), which contains the seeds, i.e., white flesh and pink skin (*Hylocereus undatus*), white flesh and yellow skin (*Selenicereus megalanthus*), red flesh and pink skin (*Hylocereus monacanthus*), or red or wine flesh and red skin (*Hylocereus polyrhizus*) [18].

Separation and recovery of the seeds is difficult due to the flesh composition (cellulose, hemicellulose, and simple saccharide polymers), which snugly envelope every seed. Pitaya seeds may also contain oil like that of most seed oils. Pitaya seed oils exhibit a significantly high content of linoleic and linolenic essential fatty acids and their isomers [18]. Information on only two pitaya varieties, *H. polyrhizus* and *H. undatus,* is available. The data are summarized in Table 5.

2.5. *Opuntia joconostle*

The genus *Opuntia* includes approximately 1,500 species of cacti, and many of them that grow in arid or semiarid climates produce sweet (cactus pear) or acidic fruits known as xoconostles, which are considered valuable vegetable foods in Latin America. Xoconostle (*Opuntia joconostle*) (Figure 3) fruit is a piriform berry that exhibits an apical depression or receptacle.

It is composed of the pulp skin (pericarp), the flesh (mesocarp), and a core (endocarp) wherein the seeds are tightly packed together in a mucilaginous structure [5].

This fruit can remain in the plant for several months without deteriorating, and it can even be kept for several weeks in a dry and cool environment without losing its flavor, color, or moisture.

The flesh is the edible part of the fruit and is used as a condiment in the Mexican diet; like the prickly pear, it is also used to produce candies, jam, and beverages. It contains absorbable sugars, dietary fiber, ascorbic acid, polyphenols, carotenoids, and betacyanin pigments that have been related to its health benefits such as its hypoglycemic and hypolipidemic action, as well as its antioxidant properties [5, 21].

The xoconostle seeds are considered potential food by-products because they exhibit a high content of unsaturated fatty acids, particularly polyunsaturated fatty acids, which are associated with a reduced risk of developing cardiovascular, inflammatory, and autoimmune diseases [22].

Table 5. Chemical composition of *H. polyrhizus* seeds (red flesh) and *H. undatus* seeds (white flesh)

Composition	*H. polyrhizus*	*H. undatus*
Moisture[a]	89.4	85.4
Proteins[a]	0.5	0.4
Crude fiber[a]	0.3	0.5
Ash[a]	0.5	0.4
Carbohydrates[a]	9.2	13.2
Fatty acids		
Myristic (14:0)[b]	0.19	0.19
Palmitic (16:0)[b]	16.53	12.78
Palmitoleic (16:1)[b]	0.94	0.64
Stearic (18:0)[b]	5.57	4.67
Oleic (18:1)[b]	26.8	24.43
Linoleic (18:2)[b]	48	55.63
Linolenic (18:3)[b]	1.3	1.18
Arachidic (20:0)[b]	0.49	0.35
Antioxidants		
α-tocopherol[c]	31.9	24
γ-tocopherol[c]	11.6	12.7
Protocatechuic acid[c]	0.93	0.93
p-coumaric acidc	0.78	0.79
p-hydroxybenzoic acid[c]	0.66	0.72
Vanillic acid[c]	0.64	0.70
Gallic acid[c]	0.25	0.20
Caffeic acid[c]	0.08	0.71
Syringic acid[c]	0.08	0.21
β-sitosterol[c]	676	548
Campesterol[c]	252	198
Stigmasterol[c]	106	72

[a] g/100 g of fruit,
[b] g/100 g oil,
[c] mg/100 g of fruit [19, 20].

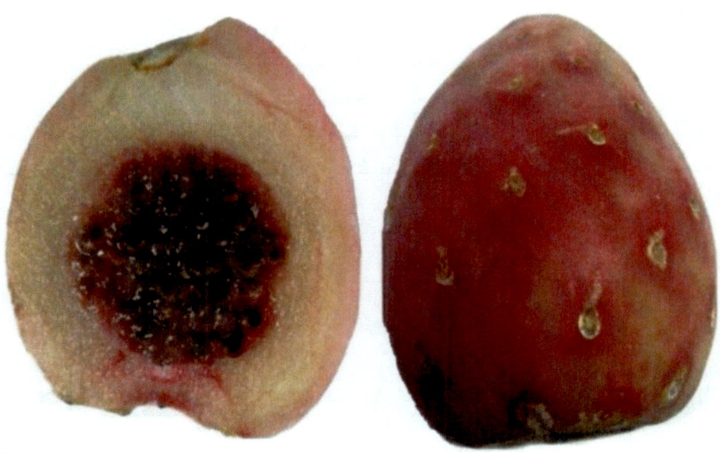

Credit: Photo by Dr. Obed Osorio-Esquivel.

Figure 3. *Opuntia joconostle* var. Weber.

At least two varieties of xoconostle fruits *Opuntia joconostle* F.A.C. Weber ex Diguet, cv. Cuaresmeño and *Opuntia matudae* Scheinvar, cv. Rosa (Table 6), have been studied.

Vitamin C has not been reported in xoconostle seeds [24]. Vitamin E and tocopherols (α, β, γ, and δ- tocopherol) are antioxidant nutrients that play important roles in health by inactivating the free radicals produced through normal cellular activity and due to various stressors.

Vitamin C is also believed to protect against degenerative processes, such as cancer and cardiovascular diseases [25]. Considering their nutritional composition and antioxidant properties, xoconostle fruits and their seeds could be attractive products for the food industry because of their high nutritional value, being a rich source of polyunsaturated fatty acids, tocopherols, and dietary fiber.

Table 6. Chemical composition of the most common xoconostle seeds

Composition	*Opuntia joconostle*, cv. Cuaresmeño	*Opuntia matudae*, cv. Rosa
Moisture[a]	73.95	60.44
Protein[a]	3.12	3.45
Fat[a]	2.45	3.52
Soluble sugars[a]	0.95	1.47
Total dietary fiber[a]	19.22	30.17
Insoluble fiber[a]	18.85	29.04
Ash[a]	0.54	0.85
Fatty acids		
Myristic (14:0)[b]	0.04	0.04
Palmitic (16:0)[b]	12.35	9.42
Palmitoleic (16:1)[b]	0.41	0.31
Stearic (18:0)[b]	3.28	2.04
Oleic (18:1)[b]	9.72	7.82
Linoleic (18:2)[b]	72.42	79.15
Linolenic (18:3)[b]	0.36	0.41
Arachidonic (20:0)[b]	0.26	0.19
Behenic (22:0)[b]	0.19	0.04
Antioxidants		
α-tocopherol[c]	0.09	0.19
β-tocopherol[c]	0.01	0.02
γ-tocopherol[c]	3.07	6.43
δ-tocopherol[c]	0.06	0.08
Total tocopherols[c]	3.23	6.71
Total phenolics[d]	50.43	59.48
Total flavonoids[e]	24.18	58.40

[a] g/100 g of fruit; [b] relative percentage fatty acids; [c] mg/100 g of fruit; [d] mg of gallic acid equivalents/g of extract.

[e] mg of catechin equivalents/g of extract; [23].

3. ANTIOXIDANT PROPERTIES

Phenolic compounds are capable of protecting the body from damage caused by free radicals. Consequently, there is an increasing interest in natural sources of antioxidants, such as fruits and seeds [26]. Recently, Osorio-Esquivel et al. reported a value for the total phenol content of *Opuntia joconostle* seeds of 47.85±1.28 mg gallic acid equivalents/g on a wet basis (51.71 mg gallic acid equivalents/g of dry weight), which is similar to that reported by Morales et al. (2012).

The antioxidant properties of *Opuntia joconostle* and *Opuntia matudae* have been evaluated based on the DPPH radical scavenging capacity, the reducing power (Fe^{3+} into Fe^{2+}), and the inhibition of lipid peroxidation using the β-carotene–linoleate model system that assesses the neutralization of the linoleate-free radical and other free radicals formed in the system, which attack the highly unsaturated β-carotene molecules [27]. The high antioxidant capacity of *O. joconostle* seeds is due to their high phenolic and flavonoid contents.

The seeds of *O. joconostle* fruits presented high DPPH-scavenging activity (EC_{50}: 1.53 mg/mL) and reducing power (EC_{50}: 0.27 mg/mL): those of *O. matudae* fruits also showed high DPPH-scavenging activity (EC_{50}: 1.88 mg/mL) and reducing power (EC_{50}: 1.58 mg/mL) [23] Notably, Osorio-esquivel et al. reported a higher antioxidant value for the *O. joconostle* methanolic extract, 4.38±0.11 mg/ml [26].

The seeds contain high levels of phenolic compounds, which give them their antioxidant power. The authors reported that the methanolic extract of *O. joconostle* seeds inhibited the formation of DPPH radicals by 49.76±0.49%, similar to the value reported by Morales et al. (2012).

γ-tocopherol is the main tocopherol present in both varieties (*Opuntia joconostle*, cv. Cuaresmeño and *Opuntia matudae*, cv. Rosa. Tocopherols act as antioxidants because of their capacity to scavenge the lipid peroxyl radicals of unsaturated lipid molecules, preventing the propagation of lipid peroxidation [25].

4. BIOLOGICAL ACTIVITY

4.1. Proteinase Inhibitory Activity

Plant proteinase inhibitors have different biological functions. These functions include preservation of storage proteins and defense against insects and other pathogens. The interest in plant proteinase inhibitors is due to their potential utilization as tools in different research areas. These proteins are thought to have possible applications in the development of antiviral agents, as well as in the treatment of disorders involving proteinase defects. Proteinases have the ability to degrade extracellular matrices. In the case of cancer progression, this step is required for the processes of invasion and metastasis. Torres-Castillo et al. (2009) focused on the identification, purification, and partial characterization of a low molecular weight trypsin inhibitor of *O. streptacantha* seeds using a combination of chromatographic procedures. The inhibitory effect of partially purified *Opuntia streptacantha* trypsin inhibitor type 2 was evaluated using *Prostephanus truncatus*, *Periplaneta americana*, *Acheta* sp., and *Gryllus* sp., the main intestinal proteinase of which belongs to the serine type. The proteinase activities of *Callosobruchus maculatus*, *Tribolium castaneum*, *Acanthoscelides obtectus*, and *Gnathocerus maxillosus* are of the cysteine type. The authors concluded that the *Opuntia streptacantha* trypsin inhibitor type 2 inhibited only the trypsin-like proteinases that are present in *P. truncatus*, *P. americana*, *Acheta sp*, and *Gryllus sp*.

One of the possible applications of proteinase inhibitors is their use as molecular markers. Torres-Castillo et al. determined that environmental conditions have no effect on the relative amount and distribution of these proteins, at least in the species analyzed, raising the possibility of using them as biochemical markers in studies of evolutionary diversity. Moreover, their effect in controlling some insect pests was studied, and the results showed no effect on the gut proteinases of certain insects known to feed on *Opuntia* plants. This result raised the possibility of some type of adaptive resistance of the evaluated insects.

The trypsin inhibitor of *O. streptacantha* was found to be a 38-residue peptide with high pH- and temperature-tolerance. Although the biological role of the trypsin inhibitor type 2 of *O. streptacantha* remains unknown, it had no effect on the gut proteinases of the insects collected from infested *Opuntia* plants [28].

4.2. Effect of *Opuntia* Seeds on the Lipid Profile

Several studies have shown that high levels of LDL-cholesterol increase the risk of heart attacks, whereas a high concentration of HDL-C or a low LDL/HDL ratio may protect against coronary heart disease.

Studies of the effect of dietary supplementation with cactus seeds on liver parameters are scarce. However, there are a few reports about the importance of supplementation with seeds, seed oils, or seed methanolic extracts that have shown their promising beneficial effects on lipid profile parameters. Recently, Osorio-Esquivel et al. (2012) reported the beneficial effect of supplementation with a methanolic extract of *Opuntia joconostle* seeds in a mouse model. Their findings suggested that biocompounds in seeds might play an important role in reducing total cholesterol, triglyceride, and LDL-cholesterol levels. The authors also indicated that groups treated with different doses of the methanolic extract showed a significant decrease in the atherogenic index compared with the hypercholesterolemic group. Treating mice with the methanolic extract at 1, 2, and 5 g/kg/day of body weight reduced their total cholesterol and LDL-cholesterol levels compared with those of the hypercholesterolemic group. The total cholesterol reduction ranged from 0.44% to 43.95% in the treated groups. The HDL-C level increased in the groups treated with the methanolic extract.

This study established that there is a dose-dependent effect on the total cholesterol and LDL-C levels and the atherogenic index. Their results showed that the HDL-C levels in the hypercholesterolemic-induced groups treated with the methanolic extract were similar to those of the normocholesterolemic group. Supplementation containing alkaloids, flavonoids, and glucosanates contributes to the hypolipidemic effect. It is suggested that methanolic extract from *Opuntia joconostle* seeds may reduce lipid levels via an antioxidant mechanism. *Opuntia joconostle* seeds contain quercetin, rutin, and cathechin, which may inhibit lipid oxidation by acting as antioxidants. Table 7 shows the most representative results of this investigation.

Table 7. Serum lipid parameter values in the control and hypercholesterolemic groups and in groups treated with 1 or 5 g/kg body weight of a methanolic extract of *Opuntia joconostle* seeds after 6 days of treatment

Parameters	Control	Hypercholesterolemic group	Methanolic extract: 1 g/kg/day body weight	Methanolic extract: 5 g/kg/day body weight
Total cholesterol	25.00	77.80	60.83	43.60
LDL-C	10.69	62.19	47.87	32.36
HDL-C	13.65	6.97	12.32	10.98
Triglycerides	1.47	0.88	1.30	1.16
Atherogenic index	0.87	10.67	5.19	2.94

Osorio-Esquivel et al., 2012.

Prickly pear seeds and oil exhibited hypoglycemic and hypercholesterolemic effects. These benefits may be due to their fatty acid composition and antioxidant contents [29].

Ennouri et al. (2007) reported the effects of supplementation with cactus pear oil or seeds (*Opuntia ficus indica*) on the serum and liver parameters of adult male rats. Supplementation with cactus seeds did not have a significant effect on the weights of the pancreas, kidney, or liver, whereas the groups fed cactus pear oil exhibited a significant increase in the weights of the liver and pancreas. The authors observed that both of the supplemented diets significantly decreased the atherogenic index compared with that of the control diet group. Reduction of the serum cholesterol levels was observed only in the groups supplemented with cactus pear oil. Therefore, the results of this study indicated that supplementation with the oil or seeds of the cactus pear may decrease the atherogenic risk factors of rats.

Moreover, the serum triglyceride level increased by 93% in the group treated with cactus pear oil, whereas in the group treated with cactus pear seeds, the serum triglyceride level decreased by 53% compared with the control group. The atherogenic index was significantly decreased in rats fed diets supplemented with cactus pear oil or seeds. This study demonstrated that cactus pear seeds exert very important hypolipidemic effects compared with cactus pear oil in rats fed a diet supplemented with those ingredients. Thus, the use of these seeds in hypocholesterolemic and hypolipidemic therapies may be a natural alternative option to treat these ailments, and it should be studied in rats with other pathologies. Some of the results of this study are shown in Table 8.

Table 8. Serum lipid parameter values in control, cactus pear oil-treated, and cactus pear seed-treated groups after 9 weeks of treatment

Parameter	Control	Cactus pear oil	Cactus pear seeds
Total cholesterol	0.66	0.43	0.76
HDL cholesterol	0.36	0.35	0.48
HDL cholesterol/total cholesterol (%)	0.55	0.81	0.64
Triglycerides	0.717	1.38	0.333
Atherogenic index	0.822	0.505	0.564

Ennouri et al., 2007.

5. Measured Trends in an *Opuntia Joconostle* Seed Study

5.1. Methods

As mentioned previously, interest in *Opuntia joconostle* has prompted investigations into different technological and biological applications of it. Thus, in this study, we determined some of the physical characteristics of both its seeds and the seed oil, such as the seed oil's acute oral toxicity and fatty acid profile. *Opuntia joconostle* fruits were purchased at a local market in Mexico City.

The fruits were identified and authenticated as number 15721 in the Herbarium of the Centro Medico Nacional of the Instituto Mexicano del Seguro Social. The seeds were manually separated from the fruits and washed with distilled water to eliminate the mucilaginous material. Then, the seeds (Fig. 4) were dried at room temperature for 1 week. Later, they were ground in a mill and maintained in a dry, cool environment.

Credit: Photo by Dr. Obed Osorio-Esquivel.

Figure 4. *Opuntia joconostle* var. Weber seeds.

5.1.1. Oil Extraction

Two-hundred grams of powdered seeds were accurately weighed and were extracted for 9 h using a total volume of 2,500 mL of hexane in a Soxhlet extractor. Then, we evaporated the solvent phase using a rotary evaporator at 35°C.

Finally, the oil was dried using nitrogen gas. The oil was kept in amber-colored containers at -20°C prior to analysis [29].

The refractive index of the oil was measured at 20 ºC using a refractometer (Atogo model NAR-2T, WA, U.S.A.). The saponification and iodine index assays were conducted according to the Mexican Standards [7, 30, 31].

5.1.2. Fatty Acid Analysis

Fatty acid methyl esters (FAMEs) were produced according to the method prescribed by the International Olive Oil Council with modifications [32]. FAMEs were prepared in the presence of 2 N potassium hydroxide in methanol. Briefly, 0.2 mL of methanolic KOH (2 N) was added to 0.1 g of *O. joconostle* seed oil in 2 mL of hexane. The mixture was shaken vigorously by vortexing for 30 s and left to stratify until the upper layer became clear. The hexane layer containing the methyl esters was decanted and maintained for no more than 12 h at 5 °C. This solution was diluted 1:100 (v/v) with fresh hexane immediately before injection into a gas chromatograph. A capillary gas chromatograph (HP 6890A; Agilent Technologies, Inc., Wilmington, DE) equipped with a flame-ionization detector (FID) was used for the quantitative analysis of the fatty acids. The fatty acid methyl ester sample (0.2 μL) was injected using an auto injector (HP G2613A) [33]. The injection port was packed with glass wool and maintained at 220°C. The FID was set to 250 °C. The oven temperature was held at 110°C for 1 min after the injection and increased at 10 °C/ min until a maximum temperature of 230°C was obtained.

The carrier gas pressure was held at 82.7 kPa consistently. A 25-m-long column (HP-FFAP) with a 0.32-mm internal diameter and 0.5-mm film thickness was used for the free fatty acid (FFA) analysis. The column was heated to 238°C for 11 min to purge the volatile compounds and to recondition it for a subsequent injection. The gas chromatograph was operated in a split injection mode with a split ratio of 1:1. The carrier gas used was 99.99% pure helium. The carrier gas velocity was 44.8 cm/s. The split flow was 2.73 mL/min. The chromatograms were integrated using HP Chemstation data analysis software. Quantification of the FFAs was performed by comparing the peak areas with those of internal standards and by employing linear regression, curves prepared using authentic fatty acids (Sigma-Aldrich Co., St. Louis, MO).

5.1.3. Acute Oral Toxicity

Male 8-week-old CD-1 mice (*n*=24) with a mean weight of 28±2 g were obtained from the Centro de Investigación y Estudios Avanzados of the Instituto Politécnico Nacional (IPN) in Mexico. The mice were housed in plastic cages in a well-ventilated room maintained at 23–25ºC with a 12/12 light/dark cycle (light: 7 a.m.–7 p.m.). They had access to food and drinking water *ad libitum*. The mice were fed standard diet (Laboratory chow 5001 Lab Rodent Diet, PMI Nutrition International, Inc. Bienwood, MO, USA).

All of the procedures concerning animal treatment and experimentation were conducted in accordance with the Guiding Principles for the Care and Use of Animals, approved by the Laboratory Animal Care Committee of IPN, and were in compliance with the Official Mexican Standard NOM-062-ZOO-1999 [34] regarding technical specifications for the production, care, and use of laboratory animals.

Acute oral toxicity was evaluated in two steps according to Lorke's method [35]. In the first step, 4 groups of 3 mice each were used in the experiment. The control group received the vehicle (distilled water), and the test groups received arithmetic doses of 10, 100, and 1000 mg/kg body weight of the seed oil. These doses were chosen following the recommendations for the method. The experimental animals were fasted for 12 h prior to oil administration. They were monitored continuously for 3 h thereafter for any signs of toxicity, such as diarrhea, depression or aggressiveness, hypermotility, and mortality. After this period, the animals were provided food and water *ad libitum*. The dead animals in each group were counted within 48 h following the administration of the oil, and the LD_{50} values were determined using the Lorke's methodology. The surviving animals were monitored daily for 14 days for changes in their body weight and food or water consumption. At the end of the experiment, the kidneys, livers, and testicles were dissected and weighed. The results of two experiments are summarized in Table 9.

All of the physicochemical data are reported as the mean values of 3 replications. The statistical analyses were conducted using Sigma Stat version 3.5 software (Jandel Corp., San Raphael, CA, USA). The acute oral toxicity data were analyzed using a one-way ANOVA and Dunn's test using Sigma Stat version 3.5 software. The data are reported as the mean ± standard error $P< 0.001$ was considered statistically significant.

5.2. Results

5.2.1. Physicochemical Characteristics

The *Opuntia joconostle* seed oil was a light brown color and had a characteristic odor. This oil remained liquid at room temperature. The physicochemical characteristics of the crude oil are shown in Table 9. These values were similar to the results reported by other authors for different *Opuntia* spp. seed oils.

The weight of 100 *O. joconostle* seeds was similar to the weight reported for *O. ficus indica* seeds (1.38 g) and was lower than that of *O. stricta* seeds (1.69 g) [29].The refractive and iodine indices of *Opuntia joconostle* seed oil are comparable to those of the oils obtained from O. *ficus indica* and *O stricta* seeds (1.475 and 101.5, and 1.469 and 91.6, respectively). Similar to most vegetable oils, *O. joconostle* seed oil has a relatively high iodine value, reflecting a high degree of unsaturation. Its iodine index value was identical to that of melon seed oil (116) but lower than that of grape seed oil (132). The density of *O. joconostle* seed oil at 20°C was similar to seed oils from *O. ficus indica* (0.903), *O. stricta* (0.919), and soybean (0.921). The saponification index of *O. joconostle* seed oil is higher than that reported for *O. ficus indica*, *O. stricta*, and *Ximena americana* seed oils (169, 174, and 182.3, respectively) but lower than that of *Haematostaphis. barteri* seed, melon seed and grapeseed oils (213, 248, and 194, respectively) [33].

Table 9. Characteristics of *Opuntia joconostle* var. Weber seed oil.

Characteristic[a]	Value
Weight of 100 seeds (g)	1.33 ± 0.001
Oil content (%)	10.50 ± 0.06
Dry matter (%)	89.96 ± 0.15
Iodine value	116.4 ± 0.51
Refractive index[b]	1.471 ± 0.22
Saponification index	191.1 ± 2.8
Density[b]	0.837 ± 0.03

Mean values of three determinations, [b] at 20°C.

5.2.2. Fatty Acid Composition

Figure 5 shows the GC chromatogram of the lipid components.

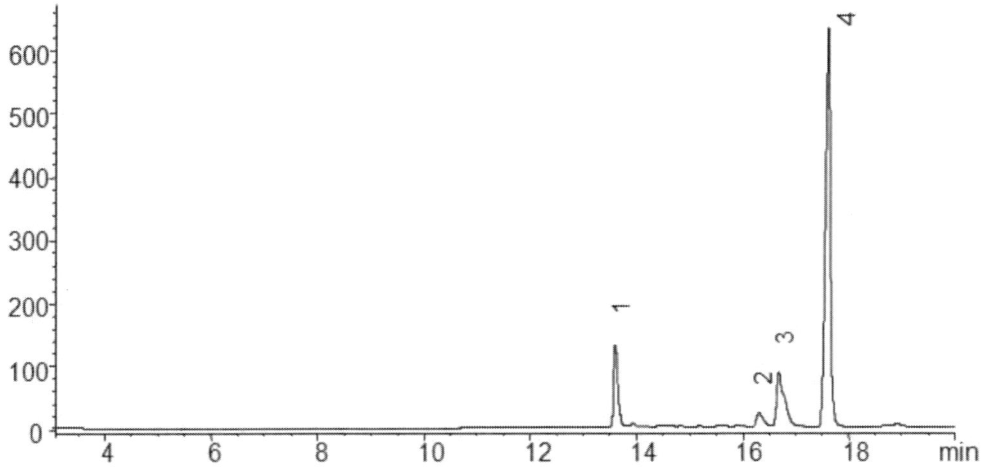

Figure 5. Chromatogram showing the fatty acid composition of *Opuntia joconostle* seed oil. Peaks: 1 = palmitic acid (C16:0), 2 = stearic acid (C18:0), 3 = oleic acid (C18:1), and 4 = linoleic acid (18:2).

The fatty acid composition of *O. joconostle* seed oil is not significantly different from those of the other *Opuntia* seed oils previously reported by other authors.

In all of these extracted oils, linoleic acid is the main fatty acid. It has been reported that the unsaturated oil of *Pinus pinea* L seeds confers some important dietary properties [36]. In fact, it is recognized that unsaturated fatty acids may affect the physical properties of membranes, such as their fluidity and permeability. Linoleic acid was the major fatty acid in *O. joconostle* seed oil (72.52 ± 0.03%), followed by oleic (13.36 ± 0.02%), palmitic (11.21 ± 0.11%), and stearic acids (2.79 ± 0.12%). This fatty acid profile is similar to that of sunflower oil (oleic: 19%, linoleic: 68%, and linolenic: 1%) and grapeseed oil (oleic: 15%, linoleic: 73%, and linolenic: 1). Recently, the fatty acid profiles of *O. joconostle* and *O. matudae* seed oils were reported and were found to be very different from our results. Other authors found at least 19 fatty acids [23], whereas we found 4 main fatty acids and traces of others that were not identified. However, similar levels of linoleic acid were reported (72.49 and 79.15% in the Cuaresmeño and Rosa cultivars, respectively). The recommended intake of linoleic acid for male adults is approximately 14–17 g per day, and that of α-linolenic acid is 1.6 g per day, whereas those for females are lower, 11–12 g per day of linoleic acid and 1.1 g per day of α-linolenic acid. The differences in composition could be explained by different climatic and soil conditions, times of harvest, and postharvest conditions. Linoleic fatty acid is indispensable for the growth of healthy human skin. The human organism can transform a series of long fatty acid chains, which are the precursors of eicosanoids, a family of compounds containing 20 carbons, including prostaglandins and leukotrienes. Prostaglandins have an important role at the vessel level and in blood coagulation.

5.2.3. Acute Oral Toxicity

Acute oral toxicity was evaluated in two steps; in the first step, we evaluated 10, 100, and 1000 mg/kg b.w. doses. Then, according to Lorke's methodology, higher doses were evaluated. Table 10 summarizes the results of this experiment. In the second step, we observed no toxicological symptoms, including writhing, diarrhea, depression, aggression, or hypermotility in the groups administered *O. joconostle* seed oil at any of the doses.

Table 10. Daily food and water intake and the relative liver and kidney weights of CD-1 mice administered different doses of *O. joconostle* seed oil

Group	Dose (mg/kg)	Daily Food Intake (g/mouse/day)	Daily Water Intake (g/mouse/day)	LBW (%)	KBW (%)
Control: 1st step	-	6.1±9.34	34.5±5.01	6.10±0.31	1.45±0.01
Opuntia joconostle seed oil	10	6.5±9.5	30.23±10.3	5.92±0.52	1.47±0.01
	100	7.5±9.5	33.45±5.35	5.98±0.81	1.46±0.01
	1000	24.65±8.32	36.19±4.67	5.90±0.37	1.44±0.01
Control: 2nd step	-	18.63±1.98	35.5±5.01	6.10±0.31	1.44±0.007
Opuntia joconostle seed oil	1600	15.29±2.34[a]	20.66±4.53[a]	5.89±0.62	1.46±0.001
	2900	19.1±2.14	33.07±3.26	5.95±0.71	1.57±0.018[a]
	5000	14.51±2.45[a]	23.66±6.29[a]	5.89±0.27	1.44±0.011

Data are expressed as the mean ± standard deviation; n=3, LBW, liver-to-body weight ratio; KBW, kidney-to-body weight ratio; [a] $P \leq 0.001$, significant difference with respect to the control group.

All of the animals survived to the end of the 14-day study period. The weight gain in the control group was similar (≈ 20%) to those of the treated groups. As shown in Table 10, the daily food intake and water consumption of the groups treated with 1600 and 2900 mg/kg b.w. of *O. joconostle* seed oil were significantly different from those of the control group, whereas there was no significant difference in the group administered 5000 mg/kg b.w. ($P<0.001$). The relative weights of the livers, kidneys, and testes were not significantly different from those of the control group. Based on these results, the oral LD_{50} for *O. joconostle* seed oil was determined to be > 5000 mg/kg b.w. The oral LD_{50} of *O. joconostle* seed oil in mice indicated that it possessed low toxicity. This information is important because of the possibility of commercializing this oil to take advantage of its fatty acid profile and phytocompound contents.

Opuntia joconostle seeds possess an important antioxidant activity. Their fatty acid profile is also important due to the high polyunsaturated fraction. Moreover, their high dietary fiber content makes them good ingredients for designing new foodstuffs. Moreover, no toxicity was found in mice administered high doses of *O. joconostle* seed oil.

Our study demonstrated that *Opuntia joconostle* seeds are a good source of free fatty acids and a good material for creating new products.

CONCLUSION

Opuntia seeds are a natural source of compounds such as fatty acids, polyphenols, vitamins D and A, tocopherols, and dietary fiber. Their hypolipidemic activity has been proven by dietary supplementation with seed powder, seed oil, and/or a methanolic extract of the seeds of some *Opuntia* varieties. The geographical distribution of the *Cactaceae* family allows most commercial varieties to be cultivated to take advantage of their nutriment composition while increasing the value of these crops. In the future, we expect that new investigations and chemical characterizations will elucidate the metabolite profile and biological activities of these compounds. *Opuntia* seeds may also be a by-product for the enrichment of food for humans or animals.

REFERENCES

[1] Ennouri, M., Fetoui, H., Hammami, M., Bourret, E., Attia, H., and Zeghal, N., Effects of diet supplementation with cactus pear seeds and oil on serum and liver lipid parameters in rats. *Food Chemistry*, 2007. 101: p. 248-253.

[2] Mondragon-Jacobo, C. and Pérez-González, S., eds. *Cactus (Opuntia spp.) as forage*. 2001, Food and Agriculture Organization of the United Nations pp. 29-36: Rome, Italy. 29-36.

[3] Ñacari, A. (2010) Tuna, prodigio olvidado. Revista Agronoticias.

[4] Flores-Valdez, C. A., Superficies de tuna por país. Mercado de la Tuna en el Mundo. Chapingo, Estado de México 2006.

[5] Osorio-Esquivel, O., Ortiz-Moreno, A., Álvarez, V., Dorantes-Álvarez, L., and Giusti, M., Phenolics, betacyanins and antioxidant activity in *Opuntia joconostle* fruits. *Food Research International*, 2011. 44: p. 2160–2168.

[6] Pimienta-Barrios E., Méndez-Moran, L., Ramirez-Hernandez, C., García de Alba-García, E., and Domínguez-Arias, M., Effect of Xoconostle (Opuntia joconostle Web.) fruit consumption on glucose and seric lipids. *Agrociencia*, 2008. 42: p. 645-653.

[7] Mizrahi, Y., Raveh, E., Yossov, E., Nerd, A., and Ben-Asher, J., New fruit crops with high water use efficiency. In: Janick, J., Whipkey, A. (Eds.), Creating Markets for Economic Development of New Crops and New Uses. Creating Markets for Economic Development of New Crops and New Uses, ed. J. In: Janick, Whipkey, A. (Eds.), Creating Markets for Economic Development of New Crops and New Uses. Alexandria, VA.: ASHS Press 2007.

[8] Lim, T. K., Opuntia ficus-indica, in Edible Medicinal and Non-Medicinal Plants. 2012: Springer Netherlands. p. pp. 660-682.

[9] Habibi, Y., Mahrouz, M., and Vignon, M. R., Isolation and structure of D-xylans from pericarp seeds of Opuntia ficus-indica prickly pear fruits. *Carbohydrate Research*, 2002. 337(17): p. 1593-1598.

[10] Coşkuner, Y. and Tekin, A., Monitoring of seed composition of prickly pear (*Opuntia ficus-indica* L) fruits during maturation period. *Journal of the Science of Food and Agriculture*, 2003. 83: p. 846-849.

[11] Matthäus, B. and Özcan, M. M., Habitat effects on yield, fatty acid composition and tocopherol contents of prickly pear (*Opuntia ficus-indica* L.) seed oils. *Scientia Horticulturae*, 2011. 131: p. 95-98.

[12] Nassar, A. G., Chemical composition and functional properties of prickly pear (Opuntia ficus indica) seeds flour and protein concentrate. *World Journal of Dairy and Food Sciences*, 2008. 3(1): p. 11-16.

[13] Kemal-Eldin, A. and Andersson, R. A., A multivariate study of the correlation between tocopherol content and fatty acid compostion in vegetable oils. *Journal of American Oil Chemistry Society*, 1997. 74: p. 375-380.

[14] Lim, T. K., Opuntia monacantha, in Edible Medicinal and Non-Medicinal Plants. 2012, Springer Netherlands. p. 683-686.

[15] Valente, L. M. M., da Paixão, D., do Nascimento, A. C., dos Santos, P. F. P., Scheinvar, L. A., Moura, M. R. L., Tinoco, L. W., Gomes, L. N. F., and da Silva, J. F. M., Antiradical activity, nutritional potential and flavonoids of the cladodes of Opuntia monacantha (Cactaceae). *Food Chemistry*, 2010. 123(4): p. 1127-1131.

[16] Castellar R., Obon J. M., Alacid M., and Fernandez-Lopez J. A., Color properties and stability of betacyanins from Opuntia fruits. *Journal Agriculture Food Chemistry*, 2003. 51(9): p. 2772-2776.

[17] Brand-Miller J., James K. W., and Maggiore P., Tables of composition of Australian aboriginal foods. Canberra: Aboriginal Studies Press 1993.

[18] Ariffin, A., Bakar, J., Tan, C., Rahman, R., Karim, R., and Loi, C., Essential fatty acids of pitaya (dragon fruit) seed oil. *Food Chemistry*, 2009. 114: p. 561-564.

[19] Lim, H., Tan, C., Karim, R., Ariffin, A., and Bakar, J., Chemical composition and DSC thermal properties of two species of *Hylocereus cacti* seed oil: *Hylocereus undatus* and *Hylocereus polyrhizus*. *Food Chemistry*, 2010. 119: p. 1326-1331.

[20] Tapiero, H., Townsend, D., and Tew, K., Phytosterols in the prevention of human pathologies. *Biomedicine & Pharmacotherapy*, 2003. 57: p. 321-325.

[21] Paiz, R. C., Juárez-Flores, B. I., Aguirre Rivera, J. R., Cárdenas Ortega, N. M., Reyes Agüero, J. A., García Chávez, E., and Álvarez Fuentes, G., Glucose-lowering effect of xoconostle (*Opuntia joconostle* A. Web., Cactaceae) in diabetic rats. *Journal of Medicinal Plants Research,* 2010. 4: p. 2326-2333.

[22] Simopoulus, A. P., The importance of ratio of omega-6/omega-3 essential fatty acids. *Biomedical Pharmacotherapy*, 2002. 56: p. 365-379.

[23] Morales, P., Ramírez-Moreno, E., Sanchez-Mata, M. C., Carvalho, A. M., and C. F. R. Ferreira, I., Nutritional and antioxidant properties of pulp and seeds of two xoconostle cultivars (*Opuntia joconostle* F.A.C. Weber ex Diguet and *Opuntia matudae* Scheinvar) of high consumption in Mexico. *Food Research International*, 2012. 46: p. 279-285.

[24] Corral-Aguayo, R., Yahia, E. M., Carrillo-López, A., and González-Aguilar, G., Correlation between some nutritional components and the total antioxidant capacity measured with six different assays in eight horticultural crops. *Journal of Agricultural and Food Chemistry*, 2008. 56: p. 10498–10504.

[25] Traber, M. G., Vitamin E regulatory mechanisms. *Annual Review of Nutrrition*, 2007. 27: p. 347-62.

[26] Osorio-Esquivel O., Ortiz-Moreno A., Garduño-Siciliano L., Álvarez B. V., and Hérnandez-Navarro M. D., Antihyperlipidemic effect of methanolic extract from *Opuntia joconostle* seeds in mice fed a hypercholesterolemic diet. *Plant Food for Human Nurtrition,* 2012. doi:10.1007/s11130-012-0320-2.

[27] Barros, L., Carvalho, A. M., Sá Morais, J., and Ferreira, I. C. F. R., Strawberry-tree, black thorn and rose fruits: Detailed characterization in nutrients and phytochemicals with antioxidant properties. *Food Chemistry*, 2010. 120: p. 247-254.

[28] Torres-Castillo, J. A., Mondragón Jacobo, C., and Blanco-Labra, A., Characterization of a highly stable trypsin-like proteinase inhibitor from the seeds of Opuntia streptacantha (O. streptacantha Lemaire). *Phytochemistry*, 2009. 70: p. 1374–1381.

[29] Ennouri, M., Bourret, E., Mondolot, L., and Attia, H., Fatty acid composition and rheological behaviour of prickly pear seed oils. *Food Chemistry*, 2005. 93: p. 431-437.

[30] NMX-F-408-S-1981, *Norma Mexicana NMX-F-408-S-1981.* Determinación de índice de Yodo por el método de Hanus: México.
[31] NMX-F-174-S-1981, Norma Mexicana *NMX-F-174-S-1981.* Determinación del índice de saponificación en aceites y grasas vegetales o animales: México.
[32] IOOC, I.O.O.C., *COI/T20/Doc.* 2001.
[33] Eromosele, O. C. and Eromosele, C. I., Fatty acid compositions of seed oils of Haematostaphis barteri and Ximena americana. *Bioresource Technology*, 2002. 82: p. 303-304.
[34] NOM-062-ZOO-1999, Norma Oficial Mexicana Especificaciones *NOM-062-ZOO-1999.* Técnicas para la produccion, cuidado y uso de los animales de laboratorio.: México.
[35] Lorke, D., A new approach to practical acute toxicity testing. *Archives of Toxicology*, 1983. 54(4): p. 275-287.
[36] Nasri, N., Khaldi, A., Fady, B., Triki, S. Fatty acids from seeds of *Pinus pinea* L.: Composition and population profiling. *Phytochemistry*, 2005. 66: p. 1729-1735.

In: Seeds as Functional Foods and Nutraceuticals
Editors: R. Mora-Escobedo, J. De J. Berrios et al.
ISBN: 978-1-62808-489-4

Chapter 13

NUTRACEUTICAL AND FUNCTIONAL PROPERTIES OF GRAPE SEEDS

María del Carmen Chaparro-Mercado[1], Evangelina García Armenta[2], Maribel Cornejo-Mazón[3], Ruth Pedroza-López[1] and Gustavo Fidel Gutiérrez-López[2*]

[1] Universidad Iberoamericana, Distrito Federal. Prolongación Paseo de la Reforma 880. Lomas de Santa Fe, México
[2] Escuela Nacional de Ciencias Biológicas, I.P.N. Departamento de Graduados en Alimentos. Carpio y Plan de Ayala, Col. Sto. Tomás, México
[3] Departamento de Biofísica, Escuela Nacional de Ciencias Biológicas, I.P.N. México

ABSTRACT

Besides giving rise to a new plant, seeds have excellent nutritional benefits, including high levels of dietary fiber, complex carbohydrates, bioactive peptides, and phytochemicals. Grapes (*Vitis vinifera* L.) belong to a group of the world's largest fruit crops. Most of these fruits are used for the production of table wine. Grape pomace (GP) is a waste product obtained during winemaking and it is composed of residual seeds and husks after winemaking. Treatment and disposal of winery waste represents a serious economic and environmental issue for wineries. Grape seeds and husks have been recognized for their high content of phenolic compounds. Humans have developed antioxidant systems to protect against free radicals. These systems include some antioxidants that are produced in the body and others that are obtained from the diet. The press residues of grape seed oil production are a rich source of polyphenolics with strong antioxidant activity. Future trends on grape GSE as antiviral agent are mentioned.

* Corresponding author: Tel. 525557296000 Ext. 62458, Email:gusfgl@gmail.com.

INTRODUCTION

Grapes (*Vitis vinifera* L.) belong to a group of the world's largest fruit crops. In 2007, the total global production of grapes exceeded 67 million metric tons (FAOSTAT 2011) and most of these fruits are used for the production of wine. Grape seeds and skins have been recognized for their high content of phenolic compounds such as gallic acid, monomeric phenolic compounds (catechin and epicatechin), and dimeric, trimeric, and tetrameric proanthocyanidins (Monagas et al., 2003).

Grape pomace (GP) is a waste product obtained during winemaking and it is composed of residual seeds and husks. During winemaking, approximately 20% of the total weight of the grape is waste that includes, stems, skins and seeds which constitute 40% to 60% of its dry weight, representing approximately 13 million tons of grape seed waste worldwide (Maier et al., 2009). Treatment and disposal of winery waste represents a serious economic and environmental issue for wineries. Frequently, the removing of the pomace is costly, and if it is not treated effectively, it can result on the generation of foul odors or ground water contamination (Moldes et al., 2003). Grape seed (*Vitis uinifera L.*) is a well-known oilseed crop containing typically 8-20% (w/w) of oil (Passos et al., 2009b) with recognized high quality due to its high level of unsaturated fatty acids, namely oleic and linoleic (Crews et al., 2006) and antioxidant-active compounds (Passos et al., 2010). Most of the polyphenols of grapes, extracted from wine, are originally present in the grape skins and seeds. As a result of processing, the duration of contact between the wine and the pomace has an effect on the total phenolic content (TPC) in the final wine (Sacchi et al., 2005; Ross et al., 2011).

Humans have antioxidant systems to protect against free radicals which include compounds that are produced in the body (endogenous) and others that are obtained from the diet (exogenous).

The first, include enzymatic defenses, such as *Se*-glutathione peroxidase, catalase, and superoxide dismutase, which metabolize superoxide, hydrogen peroxide, and lipid peroxides; thus, they prevent us from the formation of the toxic group HO•. Those that are exogenous, include non-enzymatic defenses, such as glutathione, histidine-peptides, iron-binding proteins: transferring and ferritin, dihydrolipoic acid, reduced CoQ10, melatonin, urate, and plasma protein thiols.

The last two provide the main contribution to the radical-trapping capacity of plasma. These defenses are complementary to each other, since they act against different species at different cellular compartments. However, despite these *defense antioxidants* which are able either to suppress free radical formation and chain initiation or to scavenge free radical and chain propagation, some ROS still escape to cause damage. However, the body antioxidant system is provided also by *repair antioxidants* (able to repair damage, and based on proteases, lipases, transferases, and DNA repair enzymes).

GP and seeds have been used to produce flour (GSF) which can then be utilized as a food ingredient. The inclusion of GSF in bakery products such as cookies, cakes, or some bread, provides a novel method to solve a waste disposal issue, besides the addition of extra health benefits (Kim et al., 2006).

SEEDS: ORIGIN AND TAXONOMY

The grape is one of the earliest domesticated fruit crops, has been widely cultivated for centuries and valued as a fruit and in the wine-production industry. The characteristic genomic patterns of grape, reported by the US Department of Agriculture cover around 1,000 varieties of the domesticated grape, *Vitis vinifera* subsp. *vinifera*, and its wild relative, *V. vinifera* subsp. *Sylvestris*.

High levels of genetic diversity have been maintained in *vinifera*, and it is consistent with a weak domestication problem followed by thousands of years of widespread vegetative propagation. The considerable genetic diversity within *vinifera*, however, is contained within a complex network of close pedigree relationships that has been generated by crosses among elite cultivars (Myles et al., *2011)*.

The first-degree relationships are rare between wine, grapes and among variety of grapes from geographically distant regions. Results suggest that although the substantial genetic diversity has been maintained, in the grape subsequent to domestication, there has been a limited exploration of this diversity which although caused a benefit by ensuring true breeding cultivars, it also is a drawback on the generation of unique cultivars through crosses. The grape fruit may suffer severe pathogen pressures and the long-term sustainability of the grape and wine industries will strongly rely on the exploitation of the grape's tremendous natural genetic diversity (Myles et al., 2011)

THE GLOBAL GRAPE SEEDS MARKETS

In the past, GP, particularly seeds, were considered a waste product. However a number of uses for this pomace and seeds have recently been explored, including the production of value-added products such as dietary supplements for disease prevention (Shrikhande, 2000), grappa production (Hang & Woodams 2008), laccase production (Moldes et al., 2003), and other applications.

Grape seed oil (GSO) has been traditionally recovered by hydraulic pressing and solvent extraction mainly with n-hexane (Passos et al., 2011). This is an important drawback of the process and may inclusive dictate its future viability. The yield of the extraction process might be significantly increased by mechanical treatment and thermal conditioning. For example, the milling of seeds, greatly improves extraction since enables cells break down and increases interfacial area for mass transfer (Silva et al., 2009). However, the usefulness of such mechanical treatments in industrial applications, might be questioned due to wetability problems, deficient solvent percolation and high pressure drop through the bed if fine powdered seed is used. Also, the application of thermal conditioning may degenerate the recovered oils (Passos et al., 2011).

Supercritical carbon dioxide is a suitable alternative for recovering seed oils; it transforms the products on high added value products due to their minimum degradation. The carbon dioxide at supercritical conditions is an environmentally friendly solvent, with high selectivity characteristics (Beckman, 2004). It achieves yields of extraction that are equivalent to those reported by conventional Soxhlet by using n-hexane (Passos et al., 2009a).

Agostini et al., (2012) studied the supercritical extraction from vinification residues in three varieties of *Vitis vinifera* (Moscato Giallo, Merlot, and Cabernet Sauvignon) and two varieties of *Vitis labrusca* (Bordô and Isabel). They found higher oil content in 2005 Bordô (15.40%) and in 2006 Merlot (14.66%). The higgest concentration of palmitic, stearic, and linoleic acids was observed in 2005Bordô and in 2006 Bordô, Merlot and Moscato Giallo. Bordô showed the highest concentration of oleic acid and α-tocopherol in both seasons too. For the equivalent of procyanidins, they did not found significant difference among the varieties from the 2005 harvest. In 2006, both varieties Isabel and Cabernet Sauvignon showed a value slightly lower than the other studied varieties. The concentration of total phenolics was higher in Bordô and Cabernet Sauvignon. These results were obtained for ground seeds (20/40 mesh) at 80°C, applying a CO_2 flow of 69 g/min and under 250 bar during 60 minutes of extraction. Supercritical fluid extraction (SFE) of grape seed oil in a semi-continuous mode was modeled and discussed in detail by Passos et al., (2011). An alternative to flavor oil release is the partial hydrolysis of the cell walls by means of appropriate enzymes cocktails (Dominguez et al., 1995; Fullbrook, 1983; Sineiro et al., 1998). This oil is rich in unsaturated fatty acids, in particular linoleic acid (Schieber et al., 2002), and has been recently utilized as a gourmet cooking oil.

Grape Seeds: Some Recent Findings

Ibrahim and Mohmed (2011) described the influence of administration of grape seeds as an antioxidant on the blue light effects induced in retinal sensitivity. The and β-waves of rabbit electrorentinogram (ERG) were measured after different periods (2, 7, 14 and 21 days) to blue light exposure. The results indicated that the blue light had an apparent effect on both α and β-waves. This effect is represented by a decrease in their amplitudes and an increment in their implicit time. This increment was carried out to have data on exposure period of blue light. However, the most noticiable feature in the ERG wave form variations appeared in case of the longest exposure time (21 days) where a negative ERG was recorded. This type of ERG reflects the non-vitality of the retina due to the photo-chemical damage of blue light.

After administration of grape seeds, a type of improvement in the ERG parameters was found towards control values but they were still lower than the control. However, no improvement appeared in the recorded negative ERG, and in this case the grape seed had no effects. Long exposure to blue light affected the high order of retinal neurons which was considered as a permanent damage. Therefore, the grape seeds might have a neuroprotective effect that help retinal neurons against photo-chemical damage which may induce retinal cell death. Hence, the consumption of grape seeds could help to protecting the retina.

Some reports indicate that grape seed extract (GSE) has several pharmacological benefits, including antioxidant, anti-inflammatory, anticarcinogenic, and antimicrobial properties. However, the effect of this inexpensive rich source of natural phenolic compounds on human enteric viruses has not been well documented (Su & D'Souza, 2011).

CHEMICAL COMPOSITION

In Italy and France grape seed oil (GSO) is obtained from grape seeds after wine processing. The GSO contains 75% linoleic acid, 15% oleic acid, 6% palmitic acid, 3% stearic acid, and 1% linolenic acid (Natella et al., 2002). The grape seed oil is rich in linoleic acid and α-Tocopherol (vitamin E); it has high nutritional value, and it can be used as a source of safe-to-eat vegetable oil. The grape seed oil has high amounts of linoleic acid (72–76%) compared with oils from sunflower seed (60–62%), corn (52%), olive (9–15%), soybean (54%), cotton (53%) , canola, and palm (9%), among others (Beveridge et al., 2005).

The most abundant fatty acid for most varieties of grape seeds is linoleic, which represents around the 65%, followed by the monounsaturated oleic acid with concentrations close to 25%. The total phytosterol concentrations were between 5179 and 5480 mg/kg, where the p-sytosterol represented more than the 66% in all grape seed oils. The ester l-buthyl-3-methylacetate was the most abundant in the volatile fraction compounds with concentrations of 5.4, 6.8 and 11.0 mg/kg for Syrah, Tintorera and various mixtures respectively. Other volatile compounds also present were the trans-2-hexenal (0.1 to 0.5 mg/kg), E-2-pentenal (3.1 to 4.2 mg/kg), hexenal (1.4 to 1.9 mg/kg) and heptanal (0.1 to 0.3 mg/kg). These compounds might be the responsible for the fruity flavor detected in all studied virgin oils. The α and γ isomers of the tocotrienols accounted for more than 80% of the tocochromanols present in the oils; whereas tocopherols represented only 10% The deep green color observed in all oil samples was associated to the presence of chlorophylls and other vegetable pigments (Bail et al., (2008). The GSO has a high level of antioxidant as vitamin E (60–120 mg/100 g), which makes the oil very stable. The chemical composition of grape seed extracts (GSEs) has been gradually elucidated (Yang, & Chien, 2000). Most antioxidants in GSE are proanthocyanidins, which are oligomers and polymers of polyhydroxy flavan-3-ol units, such as (+) catechin and (-) epicatechin (Hu et al., 2004).

PHYSICOCHEMICAL PROPERTIES

One of the final products of aging processes of edible oils is malondialdehyde (MDA). Niedworok & Bielaszka (2007) carried out a comparison of the effect of vitamins A and E on aging processes of edible vegetable oils. They determined MDA concentration of rapeseed oil, refined olive oil and grape seed oil with accelerated aging test Schaal Oven. In rapeseed oil, the highest concentration of MDA (1.59μM/dm3) was detected in the samples containing vitamin E and a mixture of vitamins A and E after 168 hours. The highest concentration of MDA in grape seed oil (about 1.54μM/dm^3), was observed in the samples with vitamin A after 72 hours.

EFFECT OF GRAPE SEEDS OIL COMPONENTS ON CHRONIC DEGENERATIVE DISEASES

Several epidemiological studies reported a protective effect of the consumption of fresh fruits and vegetables against cancer, heart disease, and brain stroke. Generally, high consumers of fruits and vegetables have a healthy life style, which might be an important factor for their resistance against chronic diseases. Therefore, fruits and vegetables play a preventive role due to they have a variety of constituents, including vitamins, minerals, fiber, and numerous phytochemicals, including flavonoids (Pietta, 2000).

Proanthocyanidin is an extract derived from grape seed (GSE), known for its preventive effects against cardiovascular diseases and uroinfections in women. However, the importance of studying these health beneficial effects remains on the elusive and incomplete understanding of uptake, metabolism and bioavailability of proanthocyanidins *in vivo* (Prasain, 2009).

Several studies have revealed the benefits of HDL effect of GSO and indicated the use of up to 45 mL of GSO in the daily diet as a substitute of cooking oil and within 2 weeks there was 13–14% increase in HDL blood levels. The antioxidant property is the mechanism of the hepatoprotective activity. The GSO exhibits a variety of interesting properties such as reducing platelet aggregation, prevents hypertension caused by sodium excess, and normalizes lesions caused from obesity and diabetes (Maheswari & Rao, 2005).

Maheswari & Rao (2005) studied the effect of oral administration of grape seed oil (GSO) against carbontetrachloride (CCl_4)-, they induced hepatotoxicity in rats and Histopathological examination of CCl_4-treated rat liver tissue revealed fatty degeneration, necrosis of the liver tissue, and fibrosis. Pretreatment with GSO exhibited protection and confirmed the results of biochemical studies. The results of our study indicated that simultaneous treatment with GSO protects the liver against CCl_4-induced hepatotoxicity. Simultaneous administration of GSO maintained the histological structure of liver though there was a slight congestion and regeneration of the liver tissue. The GSO offers several possibilities in the treatment of various liver disorders. This might be due to the high level of vitamin E, which was the mechanism of hepatoprotection. Further studies on any other models and extensive clinical trials are necessary to confirm these results.

The GSO exhibits a variety of interesting properties such as reducing platelet aggregation, it prevents hypertension caused by sodium excess, it normalizes lesions occurring from obesity and diabetes (Bagchi et al., 2002). Its antioxidant property is the mechanism of hepatoprotective activity (Bagchi et al., 2002).

SECONDARY METABOLITES

Recently, GSE has gained attention in the food industry due to its associated antimicrobial properties. Rhodes et al. (2006) showed that GSE at a concentration of 0.25 mg/ml decreased *Listeria monocytogenes* from 106 to 107 CFU/ml to undetectable levels within 10 min. Kao et al., (2010) found that GSE at 1 mg/ml could cause 99% inhibition in the growth of *Staphylococcus aureus*. GSE has also been shown to have antibacterial activity against many other epidemiologically significant food-borne bacterial pathogens such as *Escherichia coli* O157:H7, *Salmonella enterica* serovar Enteritidis, *Salmonella enterica* serovar Typhimurium, etc. Besides bacteria, GSE has also been shown to inhibit the replication and expression of human immunodeficiency virus type 1(HIV-1) (Nair et al., *2002*). The antiviral effect of GSE against human enteric viruses has not been explored (Su & D'Souza, 2011).

POTENTIAL HEALTH BENEFITS

Flavonoids are phenolic substances isolated from a wide range of vascular plants, with over 8000 individual compounds known. They act in plants as antioxidants, antimicrobials, photoreceptors, visual attractors, feeding repellants, and light screening (Ishii et al., 2000).

Maier et al., (2009) studied the polyphenolics content in seed oil of seven grape seed samples before (original seeds) and after (press residue) the oil recovery process (Tables 1 and 2). The results of this study confirm the press residues of grape seed oil production still to be a rich source of polyphenolics with strong antioxidant activity.

Table 1. Flavonoid composition of extracts from *Vitis vinifera* L. *cv.* 'Samtrot' grape seeds. Values are expressed as percentage of the total amount of flavonoids (Adapted from Maier et al., 2009)

Phytochemical	Content (%)
Procyanidins B1	13.92
Catechin	31.30
Procyanidins B2	17.85
Epicatechin	21.69
Epicatechin galato	13.38
Quercetin 3-o-glucoronide	1.0

Table 2. Flavonoid composition of extracts from *Vitis vinifera* L. *cv.* 'Samtrot' oil press residues. Values are expressed as percentage of the total amount of flavonoids (Adapted from Maier et al., 2009)

Phytochemical	Content (%)
Procyanidins B1	15.10
Catechin	45.88
Procyanidins B2	10.13
Epicatechin	24.38
Epicatechin galato	13.38
Quercetin 3-o-glucoronide	1.88

Many studies have suggested that flavonoids exhibit biological activities, including antiallergenic, antiviral, anti-inflammatory and vasodilating actions. However, most interest has been devoted to the antioxidant activity of flavonoids, which is due to their ability to reduce free radical formation and to scavenge free radicals (Pietta, 2000).

The capacity of flavonoids to act as antioxidants *in vitro* has been the subject of several studies in the past years. Important structure-activity relationships of the antioxidant activity have been established. The antioxidant *in vivo* efficacy of flavonoids is less documented than that *in vitro*, probably, because the limited knowledge on their uptake in humans. Most ingested flavonoids are extensively degraded to various phenolic acids, some of which still have a radical scavenging ability (Caccetta, et al., 2000)

Both, the absorbed flavonoids and their metabolites might display an *in vivo* antioxidant activity, which is evidenced experimentally by the increase of the plasma antioxidant status, the sparing effect on vitamin E of erythrocyte membranes and low-density lipoproteins, as well as the preservation of erythrocyte membrane polyunsaturated fatty acids. The ability of an antioxidant to either inhibit or promote oxidation can be very concentration dependent. This would be true for an antioxidant that could reduce pro-oxidant metals, where at low concentrations it could be prooxidative due to metal reduction and given that at high concentrations it could be antioxidative when enough radical scavenging capacity is present to overcome the prooxidant nature of the compound.

Grape seed extract and proanthocyanidins have been studied with respect to their potential health benefits. Recent reports have indicated a wide range of biological activities including antioxidant properties and cardioprotective and anti-inflammatory effects (Terra et al., 2007); anti-ulcer activity prevention of cataracts (Yamakoshi et al., 2002); improvement of insulin sensitivity (Pinent et al., 2008); and anti-carcinogenic activity.

Grape seed extract contains several health-promoting ingredients, such as protein, lipids, carbohydrates and polyphenols (which come mainly as flavonoids, known as bioflavonoids). The term flavonoid is used for a class of a plant chemical compounds known for its activity as highly potent antioxidants, their capability in protecting against oxidative and free radical damage. It has been demonstrated that the antioxidant power of polyphenols is 20 times more powerful than vitamin E and 50 times greater than vitamin C. It is known that flavonoids might help to prevent or delay the progression of eye diseases. Observational and clinical trials support this information. Kolb, et al., (2011) and Nakayama et al., (2011) proved neuroprotective effects of flavonoids on hypoxia-, glutamate- or oxidative stress–induced retinal ganglion cells death. The presence of a specific sugar side chain (rutinoside) might enhance neuroprotective activity. Rhone & Basu (2008) showed that lutein and zeaxanthin and their association reduce the risk of cataracts in healthy postmenopausal women and improving clinical features of Age-Related Macular (AMD) damage in a number of patients. Domalapalli & Neeraj (2007) found that oxidative free radicals, induced by ischemia or elevation of IOP, resulted in apoptosis and death of retinal ganglion cells and progressive vision loss in glaucoma. They found that neuroprotection could be achieved through the use of antioxidants, flavonoids and astrocyte cultures. Hanneken et al., (2006) identified a selected group of flavonoids that protected retinal pigment epithelium cells from oxidative stress–induced death with a high degree of potency and low toxicity. Dong et al., (2007) found that the flavonoids showed a strong increase of ocular blood flow and also produce marked positive effects on β-wave recovery, after ischemic action in rats.

Prasain (2009) carried out a special study in rats subjected to an oral intake of GSE (300 mg/kg, twice a day). Blood and urine of the rats were collected over a 24 h period. Monomeric catechins and their methylated metabolites, and proanthocyanidins up to trimers were detected in blood samples treated with GSE using LC-MS/MS and operating the multiple reactions on a monitored (MRM) mode. A new tetramethylated metabolite of dimeric proanthocyanidin (m/z 633) in GSE-treated urine was tentatively identified. Using LC-MS/MS, (+)-catechin and (-)-epicatechin were identified in the brain. These data suggested that GSE catechins cross the blood brain border and might be responsible for the neuroprotective effects of GSE.

Grape seed extract (GSE) has many pharmacological and health benefits that include antioxidant, cardioprotective, hepatoprotective, neuroprotective, anti-inflammatory, antidiabetic, anticarcinogenic, and antiaging effects (Nassiri-Asl & Hosseinzadeh (2009); Xia et al., (2010) Yadav et al., (2009)). GSE has gained attention in the food industry because it has been associated to antimicrobial effects.

GSE can also be used for the protection from viral contamination in coating ready-to-eat deli items, since GSE have the most potential activity after ingestion at 37°C. Thus, based on literature, GSE appeared to have a broad spectrum of activity, being effective against food-borne bacterial pathogens as well as food-borne viruses (Su & D'Souza, 2011).

Conclusion

The grape seed oil (GSO) offers several possibilities in the culinary and medical sectors. Is use is likely to continue growing and will, certainly offer new treatments for various physiological disorders. This might be due to the high levels of antioxidants present such as vitamin E and anthocyanins. The grape seeds oil might have a neuroprotective effect that help to the retinal neurons against photo-chemical damage which may induce retinal cell death. It has been shown that commercial GSE can reduce the titers of tested food-borne microorganisms and may be used as a novel and natural broad-spectrum alternative to increase food safety, extend shelf life, and improve food quality. GSE is available easily and inexpensively worldwide from diferent kinds of grapes. Its application for enhancing food safety would not require much labor, or sophisticated technology and have a global market for commercialization.

Some future works will focus on the application of GSE in the food industry and food processing environments to decrease viral contamination. The chemical compounds of GSE associated with antiviral activity have to be explored.

References

Agostini, F., Bertussi, R. A., Agostini, G., Atti dos Santos, A. C., Rossato, M. & Vanderlinde, R. (2012). Supercritical extraction from vinification residues: fatty acids, α-tocopherol, and phenolic compounds in the oil seeds from different varieties of grape. *Scientific World Journal*. Published online 2012 April 19. doi: 10.1100/2012/790486

Bail, S., Stuebigger, G., Krist, S., Unterweger, H. & Buchbauer, G. (2008). Characterization of various grape seed oils by volatile compounds, triacylglycerol composition, total phenols and antioxidant capacity. *Food Chemical*, *108*, 1122-1132

Bagchi, D., Ray, S. D., Bagchi, M., Preuss, H. G. & Stohs, S. J. (2002). Mechanistic pathways of antioxidant cytoprotection by a novel IH 636 grape seed proanthocyanidin extract. *Indian Journal Experimental Biology*, *6*, 717-26.

Bagchi, D., Bagchi, M., Stohs, S., Ray, S. D., Sen, C. K. & Preuss, H. G. (2002). Cellular protection with proanthocyanidins derived from grape seeds. *Annual NY Academic Science*, *957*, 260-70.

Beveridge, T. H. J., Girard, B., Kopp, T. & Drover, J. C. G. (2005). Yield and composition of grape seed oils extracted by supercritical carbon dioxide and petroleum ether: varietal effects. *Journal of Agricultural and Food Chemistry*, *53*(5), 1799–1804.

Beckman, E. J. (2004). Supercritical and near-critical CO_2 in green chemical synthesis and processing. *Journal of Supercritical Fluids*, *28*, 121-191.

Caccetta, R. A., Croft, K. D., Beilin, L. J. & Puddey, I. B. (2000). Ingestion of red wine significantly increases plasma phenolic acid concentrations but does not acutely affect ex vivo lipoprotein oxidizability. *American Journal Clinical and Nutrition*, *71(1)*, 67-74.

Crews, C., Hough, P., Godward, J., Brereton, P., Lees, M., Guiet, S. & Winkelmann, W. (2006). Quantitation of the main constituents of some authentic grape-seed oils of different origin. *Journal of Agricultural and Food Chemistry*, *54*, 6261-6265.

Domalapalli, M. & Neeraj, A. (2007). Oxidative Stress in Glaucoma: A Burden of Evidence. *Journal of Glaucoma*, *16*(3), 334-343.

Dong, A., Shen, J., Krause, M. & Hackett, F., Campochiaro, P. (2007). Increased expression of glial cell line-derived neurotrophic factor protects against oxidative damage-induced retinal degeneration. *Journal Neurochemistry*, *103*(3), 1041–1052.

FAOSTAT: Agricultural Production Domain online. (2011). Rome: *Food and Agricultural Organization of the United Nations*. Available from: http://faostat.fao.org/. Accessed: september, 3, 2012.

Hanneken, A., Lin, F., Johnson, J. & Maher, P. (2006). Flavonoids protect human retinal pigment epithelial cells from oxidative-Stress–Induced Death. 4 Invest. *Ophthalmological Vision Science.*, *7*(7), 3164-3177.

Hang Y, D. & Woodams, E. E. (2008). Methanol content of grappa made from New York grape pomace. *Bioresour Technologic*, *99*(9), 3923–5.

Hu, M., McClements, J. & Decker, E. A. (2004). Antioxidant activity of a proanthocyanidin-rich extract from grape seed in whey protein isolate stabilized algae oil-in-water emulsions. *Journal of Agricultural and Food Chemistry*, *52*, 5272-5276.

Ibrahim, E. A. & Mohmed, E. N. (2011). Neuroprotective Effects of grape seeds against photo-chemical damage–induced retinal cell death. *Nature and Science*, *9*(11), 83-89.

Ishii, K., Furuta, T. & Kasuya, Y. J. (2000). Mass spectrometric identification and high-performance liquid chromatographic determination of a flavonoid glycoside naringin in human urine *Agriculture Food Chemistry*, *48*, 56-59.

Kao, T. T., Tu, H. C., Chang, W. N., Chen, B. H., Shi, Y. Y., Chang, T. C. & Fu, T. F. (2010). Grape seed extract inhibits the growth and pathogenicity of *Staphylococcus aureus* by interfering with dihydrofolate reductase activity and folate-mediated one-carbon metabolism. *International Journal of Food Microbiology*, *141*, 17–27.

Kim, S. Y., Jeong, S. M., Park, W. P., Nam, K. C., Ahn, D. U. & Lee, S. C. (2006). Effect of heating conditions of grape seeds on the antioxidant activity of grape seed extracts. *Food Chemistry*, *97*(3), 472–9.

Kolb, H., Nelson, R., Fernandez, E. & Jones, B. (2011).Web vision. The organization of the retina and visual system. Copyright © web vision powered by word press. University of Utah Disclaimer Rhone, M. and A. Basu. (2008). Phytochemicals and age-related eye diseases. *Nutritional Review.*, *66*(8), 465-72.

Maheswari, M. U. & Rao, P. G. (2005). Antihepatotoxic effect of grape seed oil in rat. *Indian Journal Pharmacology*, *37*(3), 179-182.

Maier, T., Schieber, A., Kammerer, D. & Carle, R. (2009). Residues of grape (*Vitis vinifera L.*) seed oil production as a valuable source of phenolic antioxidants. *Food Chemical*, *112*, 551–9.

Myles, S., Boyko, A. R., Owens, C. L., Brown, P. J., Grassi, F., Aradhya, M. K., Prins, B., Reynolds, A., Chia, J. M., Ware, D., Bustamante, C. D. & Buckler, E. S. (2011). Genetic structure and domestication history of the grape. *Proceedings of the National Academic Sciences of U.S.A*, 1, *108*(9), 3530-3535

Moldes, D., Gallego, P., Rodriguez-Couto, S. & Sanroman, A. (2003). Grape seeds: the best lignocellulosic waste to produce laccase by solid state cultures of *Trametes hirsuta. Biotechnology Lett*, *25*, 491–5.

Montero, L., Herrero, M., Prodanov, M., Ibáñez, E. & Cifuentes, A. (2012). Characterization of grape seed procyanidins by comprehensive two-dimensional hydrophilic interaction × reversed phase liquid chromatography coupled to diode array detection and tandem mass spectrometry. *Analytical and Bioanalytical Chemistry*, *8*, 345-356.

Nair, M. P., Kandaswami, C., Mahajan, S., Nair, H. N., Chawda, R., Shanahan, T. & Schwartz, S. A. (2002). Grape seed extract proanthocyanidins downregulate HIV-1 entry coreceptors, CCR2b, CCR3 and CCR5 gene expression by normal peripheral blood mononuclear cells. *Biology Research*, *35*, 421–431.

Nakayama, M., Makoto, A., Chen, Y., Araie, M., Tomita, K., Yokotani, K. & Iwashina, T. (2011). Neuroprotective effects of flavonoids on hypoxia-,glutamate-, andoxidative stress–induced retinal ganglion cell death. *Molecular Vision*, *17*, 1784-1793.

Nassiri-Asl, M., & Hosseinzadeh, H. (2009). Review of the pharmacological effects of *Vitis vinifera* (grape) and its bioactive compounds. Phytotherminology. Research, *23*, 1197–1204.

Natella, F., Belleli, F., Gentili, V., Ursini, F. & Scaccini, C. (2002). Grape seed proanthocyanidins prevent plasma post prandial oxidative stress in humans. *Journal of Agricultural and Food Chemistry*, *26*, 7720-7725.

Navas, P. B. (2009). Chemical composition of the virgin oil obtained by mechanical pressing form several grape seed varieties *(Vitis vinifera L.)* with emphasis on minor constituents. *Archivos Latinoamericanos de Nutrición*, *59*(2), 214-219.

Passos, C. R., Silva, R. M., Da Silva, F. A., Coimbra, M. A. & Silva, C. M. (2010). Supercritical fluid extraction of grape seed (*Vitis vinifera L.)* oil. Effect of the operating conditions upon oil composition and antioxidant capacity. *Chemical Engineering Journal*, *160*, 634-640.

Passos, C. R., Silva, R. M., Silva, F. A., Coimbra, M. A. & Silva, C. M. (2009a). Enhancement of the supercritical fluid extraction of grape seed oil by using enzymatically pre-treated seed. *Journal of Supercritical Fluids*, *48*, 225-229.

Passos, C. R., Yilmaz, S., Silva, C. M. & Coimbra, M. A. (2009b). Enhaneement of grape seed oil extraction using a cell wall degrading enzyme cocktail. *Food Chemistry*, *115*, 48-53.

Pietta, P. G. (2000). Reviews: Flavonoids as Antioxidants. *Journal of Natural Products*, *63*, 1035-1042.

Prasain, J. K., Peng, N., Dai, Y., Moore, R., Arabshahi, A., Wilson, L., Barnes, S, Michael Wyss, J., Kim, H. & Watts, R. L. (2009). Liquid chromatography tandem mass spectrometry identification of proanthocyanidins in rat plasma after oral administration of grape seed extract. *Phytomedicine*, *16*(2-3), 233-43

Reverchon, E. & Marrone, C. (2001). Modeling and simulation of the supercritical CO_2 extraction of vegetable oils. *The Journal of Supercritical Fluids*, *19*,161-175.

Rhodes, P. L., Mitchell, J. W. Wilson, M. W. & Melton, L. D. (2006). Antilisterial activity of grape juice and grape extracts derived from Vitis vinifera variety Ribier. *International Journal of Food Microbiology*, *107*, 281–286.

Ross, C. F., Hoye, C. Jr. & Fernandez-Plotka, V. C. (2011). Influence of heating on the polyphenolic content and antioxidant activity of grape seed flour. *Journal of Food Science*, *76*, 6:C884-C890.

Shrikhande A. J. (2000). Wine by-products with health benefits. *Food Research International*, *33*(6), 469–74.

Silva, C. M., Passos, C. P., Coimbra, M. A. & Da Silva, E. A. (2009). Numerical simulation of supercritical extraction processes. *Chemical Product and Process Modeling (Article 9).*

Su, X. & D'Souza, D. H. (2011). Grape seed extract for control of human enteric viruses. *Applied and Environmental Microbiology*, *77*, 12: 3982-3987.

Xia, E. Q., Deng, G. F., Guo, Y. J. & Li, H. B. (2010). Biological activities of polyphenols from grapes. *International Journal Molecular Science*, *11*, 622–646.

Yadav, M., Jain, S., Bhardwaj, A., Nagpal, R., Puniya, M., Tomar, R., Singh, V., Parkash, O., Prasad, G. B., Marotta, F. & Yadav, H. (2009). Biological and medicinal properties of grapes and their bioactive constituents: an update. *Journal Medical Food*, *12*, 473–484.

Yang, Y. & Chien, M. (2000). Characterization of grape procyanidins using high-performance liquid chromatograph (mass spectrometry and matrix-assisted laser desorption/ionization time-of-flight mass spectrometry. *Journal of Agricultural and Food Chemistry*, *48*, *3990-3996.*

Thorsten, M., Schieber T., Kammerer D. R. & Carle R. (2009). Residues of grape (*Vitis vinifera* L.) seed oil production as a valuable source of phenolic antioxidants. *Food Chemistry*, *112*, 551–559.

In: Seeds as Functional Foods and Nutraceuticals
Editors: R. Mora-Escobedo, J. De J. Berrios et al.
ISBN: 978-1-62808-489-4

Chapter 14

CHEMICAL AND PHYSICAL PROPERTIES, STRUCTURE, BIOLOGICAL ACTIVITY, USES AND APPLICATIONS OF CASTOR SEEDS (*RICINUS COMMUNIS* L.)

***Stefany Cárdenas Pérez*[1], *María de Jesús Perea-Flores*[2], *José Jorge Chanona-Pérez*[1*], *Georgina Calderón-Domínguez*[1], *Vicente Garibay-Febles*[4], *Jorge Alberto Mendoza-Pérez*[3] *and Hilda María Hernández-Hernández*[1]**

[1]Departamento de Ingeniería Bioquímica, Escuela Nacional de Ciencias Biológicas, Instituto Politécnico Nacional, Plan de Ayala y Carpio s/n, Col Santo Tomas, D.F., México

[2]Centro de Nanociencias y Micro y Nanotecnologías, Instituto Politécnico Nacional, Unidad Profesional "Adolfo López Mateos", Luis Enrique Erro s/n, Zacatenco, D.F., México

[3]Departamento de Ingeniería en Sistemas Ambientales, Escuela Nacional de Ciencias Biológicas, Av. Wilfrido Massieu, Esq. Cda. Miguel Stampa s/n, U.P. Adolfo López Mateos, D.F., México

[4]Laboratorio de Microscopía Electrónica de Ultra Alta Resolución, Instituto Mexicano del Petróleo, Eje Central Lázaro Cárdenas N.152, Col. San Bartolo Atepehuacan, D.F., México

ABSTRACT

Castor bean (*Ricinus communis* L.) crops are grown across the world in tropical, sub-tropical and warm temperate regions; their farming is very important in many importing and exporting countries. Nowadays castor seed has received a special interest due to it being an important source of protein and oil with high potential to be used as animal's

[*] Corresponding author: jorge_chanona@hotmail.com; jchanona@ipn.mx.

feedstock and for biodiesel production. Nevertheless, beyond its applications in energy industry and the production of food for animals, the castor oil is one of the most useful plant oils, as its phytochemical compounds have been well known since antiquity and more than 700 industrial uses have been currently reported along with a yearly rising global demand. Thus, castor oil is used in pharmaceuticals, food and other industries. In addition, it is thesource of a number of useful oleochemicals. For this reason this chapter is aimed to present a brief recompilation about *Ricinus communis* L., covering from its origin, taxonomy, global production and marketing of the seed and its oil to its chemical and phytochemical characteristics. Its physical and microstructural properties, its storage and processing, the effects of castor seed components on some diseases, as well as some relevant medical and pharmaceutical uses and the future trends for improvement on its production and in the development of new products are also described.

ORIGIN AND TAXONOMY

Ricinus communis L. has several names, such as castor seed, castor bean, castor oil plant, higuerilla, mamona, palma and christi among others (Morris, et al., 2011; Falasca, et al., 2012). The plant is an evergreen shrub, fast growing in tropical regions. It is also a non-edible industrial oil seed crop usually containing about 42-59 oil percent and having a plethora of applications in petrochemicals, pharmaceuticals, cosmetics, textiles, chemicals, soap, leather, paints, varnishes, ink, nylon and plastic (Ramanjaneyulu, et al., 2013; Mhaske, et al., 2013). Castor oil has been traditionally used as laxative and also has medicinal and veterinary uses in the fields of obstetrics, dermatology etc. Nowadays castor oil is identified as a potential feedstock for biofuel production because of its proven technical and ecological benefits, and its recent utilization as bio-diesel has magnified its importance (Ogunniyi, 2006; Scholz and DaSilva, 2008; Pecina-Quintero, et al., 2013; Zapata, et al., 2012; Falasca, et al., 2012).

R. communis seeds were found in Egyptian tombs dating back to 4000 BC and the wild seed was mainly used as medicinal plant. The first historic register of this seed was referred in the Ebers Papyrus, an ancient Egyptian medical treatise date from 1552 BC. Then Herodotus (500 BC) reported that Egyptians used castor oil for purging purpose. In prehistoric timessmall seeded, indeterminate forms with dehiscent capsules were collected from arid climatic areas in territories of ancient Mesopotamia and Persia. Arabs did semi-nomadic cultivation of productive type wild forms with large seed, indehiscent capsules and dry or heat resistance. Also Iraq, Iran, Syria, Turkey and Armenia are considered to be areas of origin of *R. communis* in the West Asian region. Europeans introduced *R. communis* as a plantation crop in islands of Caribbean Sea and American continent. Castor was introduced in Brazil by Portuguese (Anjani, 2012)]. It is probably originated in tropical Africa, was then taken to India and China, and through gradual spreading it is now grown worldwide from temperate to tropical zones (Ogunniyi, 2006; Morris, et al., 2011). Today is not clear the site of origin of castor plant but four centers of origin has been reported, Ethiopian-East African region, Northwest and Southwest Asia and Arabian Peninsula, Sub continents of India, and China. Ethiopian-East African region is considered to be the most probable site of origin because of presence of high diversity in Ethiopia. (Salimon, et al., 2012; Ramanjaneyulu, et al., 2013)

The earlier taxonomic studies completed between 1930 and 1986 reported around 91 species, subspecies and varieties of genus *Ricinus*. It is now believed that all these species and subspecies are synonymous to *R. communis*. The FAO's Second Report on the State of the World's Plant Genetic Resources for Food and Agriculture (FAO, 2009) states that 17,995 castor germplasm accessions are being held with various institutes in the world, mainly from India, China, USA, Brazil, Russia, Ethiopia and others 52 Institutes around of world. Nevertheless, there are no consolidated reports describing the exact status of castor germplasm collections conserved in the world (Anjani, 2012).

The taxonomy classification of *R. communis* is shown in Table 1. *R. communis* is an herbaceous plant that can grow as a shrub of 1-3 m and the tree of 8-10 m, panicle inflorescence, monoecious flowers, the fruits are bristly capsules; it lives in the wild and grows on roadsides, edges of streams, gardens, farmland, rocky areas with low fertility and nutrients (**Figure 1a**). The fruit is a dehiscent capsule of three leaflets ovoid form, 1-2 cm of diameter, the surface is covered with thorns red before the ripening (San Miguel, 2006; Perea-Flores, et al., 2011) contains three and four ellipsoid seeds (from 10 to 17 mm of long), smooth, shining, the reddish-brown color, the seed as shown in Figure 1c. The stem is thickened and branched, the leaves are nearly orbicular foil from 10 to 60 cm of diameter, the divisions ovate-oblong to lanceolate, acute or acuminate, irregularly toothed edge-glandular; petiole is longer than the blade: glands between the sheet and the petiole.

Figure 1. Field crop of *Ricinus communis* in Michoacán-Mexico (a). Valva cluster of *R. communis* (b) and seeds of *R. communis* (c).

Table 1. Taxonomy of *R. communis* plant

Taxonomy	
Kingdom	Plantae
Subkingdom	Tracheobionta
Superdivison	Spermatophyta
Division	Magnoliophyta
Class	Magnoliopsida
SubClass	Rosidae
Order	Euphorbiales
Family	Euphorbiaceae
Genus	Ricinus
Species	Ricinus communis

Plants database, United States Department of Agriculture, USDA.

Table 2. Main producers of castor oil seeds

Rank	Country	Production (International price $1000)	Production (MT)	% Production $1000	% Production (MT)
1	India	907925	2339000	84.74	84.56
2	China	69426	180000	6.48	6.51
3	Brazil	45659	120166	4.26	4.34
4	Mozambique	21840	57429	2.04	2.08
5	Paraguay	4987	13000	0.47	0.47
6	Thailand	4409	11340	0.41	0.41
7	Others	17221	45211	1.61	2.20

FAO Stat, 2011.

GLOBAL MARKET OF *R. COMMUNIS* SEED

The share of *R. communis* seed is less than 0.15% of total world trade of oil seeds. Currently, the annual world yield of castor seeds is about 1.3 million tons, which corresponds to about 0.55 million tons of castor oil. Since the beginning of the 1970's, castor oil seed production increased continuously but, in some cases, subjected to yearly fluctuations of 20%, especially due to storm damage in the main producing regions. In 2008 about half of all the castor oil produced in the world was exported, with India dominating the market with a share of 80%. The FAO reported in 2011 that India produced over 80% of the castor oil in the entire world (FAO Stat, 2011). **Table 2** shows the main producers of castor oil seeds and its market value indicated in US dollars. The prices of castor oil vary considerably due to fluctuations in production and also due to speculation. The average price in the last decade was about 900 US$/metric tonne. The oil extracted from the *R. communis* seed already has a growing international market, assured by multiple uses, ranging from medicines and cosmetics to substituting petroleum in the manufacturing of biodiesel, plastics and lubricants. Approximately each hectare of castor oil bean plants growing in arid and semi-arid regions produces 350 - 650 kg of oil, which in turn produces 350-650 kg of biodiesel per hectare. (Biodiesel Technology, 2013). The imports and exports of castor oil in 2009 were estimated to be 268,500 tonnes and 289,500 tonnes, respectively. The major castor oil consuming countries are the European Union countries, USA and Japan. China and India are also emerging fast as major consumers. Demand for castor oil in the world is rising constantly at 3–5% per annum (Anjani, 2012). Castor oil derivatives potential has been estimated to be more than US$ 400 billions by 2012 (http://www.castoroil.in/).

FORMER SCOPE AND FUTURE TRENDS

The castor bean (*Ricinus communis*) has many industrial and medicinal values. However, it cannot be consumed by humans because of the presence of toxic ricin and allergens (Kim, et al., 2006; Dubois, et al., 2013). Castor oil has been used for coating fabrics and other protective coverings, in the manufacture of high-grade lubricants, transparent typewriter and printing inks, in textile dyeing, in leather preservation and in the production of 'Rilson', a polyamide nylon-type fiber (Salimon, et al., 2010). The uses of castor oil have changed over the years. The earlier uses of castor oil were for lamp oil, medicinal purposes and as a general industrial lubricant; for example sulfated castor oil or Turkey Red Oil were the first synthetic detergents after ordinary soap. Other forms of the oil became important for the leather, lubricants and other industries, even including medical applications where it has been employed as solvent of pure alkaloids and such solutions of atropine, cocaine, etc. used in ophthalmic surgery (Nielsen, 2011)]. Nowadays, there are many uses of castor oil and its derivatives such as: polyamide 11 (Nylon 11) engineering plastic, lubricating grease, coatings, inks, sealant, aircraft lubricants, surfactants, emulsifiers, encapsulators, plastic films, plasticizer for coatings, and components for shatterproof safety glass. It is an essential component in some artificial rubbers, in various descriptions of celluloid, and in the making of certain waterproof preparations and one of the largest uses is in the manufacture of transparent soaps. It also furnishes sebacic acid which is employed in the manufacture of candles and caprylic acid, which enters into the composition of varnishes. Castor oil has even made its way into cosmetics and related products. (Ogunniyi, 2006; Sathiskumar and Madras, 2011; Salimon, et al., 2012; Falasca, et al., 2012; Dubois, et al., 2013; Perdomo, et al., 2013). Table 3 summarized some applications for castor oil and its derivatives (Industrial Castor Oil, 2013; http://www.castoroil.in/).

Table 3. Some applications of castor oil and derivatives

Agriculture Organic Fertilizers **Food** Surfactants Viscosity Reducing Additives Flavorings Food Packaging	**Cosmetics & Perfumeries** Perfumery Products Lipsticks Hair Tonics Shampoos Polishes Emulsifiers Deodorants	**Textile Chemicals** Textile Finishing Materials Dyeing Aids Nylon, Synthetic Fibers & Resins Synthetic Detergents Surfactants, Pigment Wetting Agents
Plastics & Rubber Polyamide 11 (Nylon 11) Plastic Films Adhesives Synthetic Resins Plasticizers Coupling Agents Polyols	**Textile Chemicals** Textile Finishing Materials Dyeing Aids Nylon, Synthetic Fibers & Resins Synthetic Detergents Surfactants, Pigment Wetting Agents	**Paper** Flypapers Defoamer Water Proofing Additive **Electronics & Telecommunications** Polymers for Electronics & Telecommunications Polyurethanes Insulation Materials
Paints, Inks & Additives Inks Plasticizer for Coatings Varnishes Lacquers Paint Strippers Adhesive Removers Wetting & Dispersing Additives Finishing Materials	**Lubricants** Lubricating Grease Aircraft Lubricants Jet Engine Lubricants Racing Car Lubricants Hydraulic Fluids Heavy Duty Automotive Greases Fuel Additives Corrosion Inhibitors	**Pharmaceuticals** Antihelmintic Antidandruff Cathartic Emollient Emulsifiers Encapsulants Expectorant Laxatives & Purgative Surfactants

Adapted from http://www.castoroil.in.

The chemical structure of castor oil is of great interest because of the wide range of reactions it affords to the oleo chemical industry and the unique chemicals that can be derived from it. These derivatives are considerably superior to petrochemical products since they are from renewable sources, bio-degradable and eco-friendly. On the other hand, the increase in petroleum prices has prompted the study and development of renewable fuels worldwide. Currently, the mass production of biodiesel from crop species (canola, soybean, sunflower) on a commercial level is difficult; many of these crops are used for human consumption, and altering their use to include biodiesel production would create a supply chain shortage and a rise in food prices.

In this context, the search for renewable, biodegradable, nontoxic, sulfate and aromatic compound-free fuels has sparked interest in castor oil as a sustainable alternative that could partially replace petroleum-based diesel fuel (Pecina-Quintero, et al., 2013; Madankar, et al., 2013; Perdomo, et al., 2013)].

From an energy point of view, Castor oil is the best vegetable oil for producing biodiesel (Castor methyl ester) because it does not require heat and consequently neither the energy that is needed to transformed vegetable oils into biodiesel. However it has been reported that the biodiesel made from castor oil has a relatively high viscosity, making it unsuitable to be used as straight combustible, hence it is blended with biodiesel obtained from other vegetable oils.

In many countries castor is a crop that is still mostly cultivated by small farmers. It can be grown on marginal land, and can be converted in pure plant oil or biodiesel for local use. The economic viability depends in that case on various factors, such as high import (or transport) costs of fossil fuel, subsidies applied to the production of castor bio-fuel required to make it more competitive as compared to other cheaper fuels. For certain countries or regions, it might still be an attractive option. However with the present high price of castor oil (nearly double that of rapeseed oil), caused by its high value, as it is used for the production of a large range of products, it is unlikely that it will become a high volume feedstock for biodiesel or pure plant oil (Nielsen, 2011).

Despite of the interest of some countries to produce castor seeds, their adaption to the environmental conditions has been a problem. For instance, Mexico does not currently have any enhanced varieties of castor seeds that render high yields, (Pecina-Quintero, et al., 2013) although some varieties and hybrid have been evolved in the last four decades. However, to meet the tremendous global demand for castor oil, cultivars with further enhanced yield and oil percentage, disease and insect resistance and drought tolerance are needed (Anjani, 2012).

According to the Castor Oil website (http://www.castoroil.in), *R. communis* seed has not only received attention for biodiesel production, it is currently being investigated in many topics, such as the improvement of seed production or the development of new products such as oils of cosmetic grade, ricin toxicity, RCA (*Ricinus communis* agglutinin), epoxy resins, polyurethanes, antibacterial agents, production of sebacic acid and dimer acids from castor oil, energy production (biodiesel), improvement of castor oil productivity, complete studies in fat and oil, especially omega 3 and 6 fatty acids, castor derivatives with reduced toxicity, films and coatings from proteins of residual cake, adverse health effects of hydrogenated vegetable, genetic modification and improvement of oilseeds, among others. (Bradberry, 2007; Ferreira, et al., 2007; Chen, et al., 2004; Kim, et al., 2006; Sujatha, et al., 2008; Ali, et al., 2008; Madankar, et al., 2013; Perdomo, et al., 2013).

Table 4. Proximate composition of full-fat content, macro- and micro-mineral composition and amino acid profile in *R. communis* seed meal

Nutrients	Full-fat castor seed meal	Macro elements (mg/100g)	Full-fat castor seed meal	Aminoacid g/16gN	Castor seed cake
Dry matter (%)	92.97	Phosphorus	3.4	Glutamic acid	13.19
Crude protein (%)	20.78	Calcium	19.37	Arginine	8
Crude fat (%)	51.20	Magnesium	49.95	Aspartic acid	6.42
Crude fibre (%)	4.98	Potassium	21.89	Leucine	5.9
Total ash (%)	7.75	Sodium	74.32	Valine	5.46
Soluble Carbohydrate (%)	7.96	Copper	0.88	Lysine	4.11
		Iron	51	Alanine	3.96
		Manganese	1.62	Iso-leucine	3.09
		Cobalt	1.64	Threonine	3
				Proline	2.76
				Serine	2.13
				Methionine	2.06
				Histidine	1.38
				Glycine	0.56
				Cystine	0.53

Adapted from Annongu, et al., 2008.

Chemical Composition

R. communis seed contains a high level of lipids (50-70%) and proteins (25-30%); moreover it also contains minerals such as P, Ca, Na, K, Mg, Fe, Mn, Co and others. Seed components are shown in **Table 4**, where data of proximate composition of whole castor seed meal is presented. As it can be observed, castor seed contains high quantities of valuable nutrients, protein, crude fiber and high total ash content signifying the presence of high levels of macro- and micro-minerals. The presence of high nutrients suggests that castor seeds and cake may serve as alternative feedstuffs for livestock if is properly processed (Annongu, et al., 2008). Chemical amino acid analysis indicates that this seed is deficient in lysine, methionine, and tryptophan (Morris, et al., 2011). Research to obtain further methods of efficient processing and detoxification to enable full utilization of the castor seeds and cake as alternatives to conventional feedstuffs is being given consideration. (Annongu, et al., 2008).

The main fatty acids contained in castor bean oil are: ricinoleic acid (12-hydroxy-9-octadecaenoic acid $C_{18}H_{34}O$) that comprises close to 89% of the total fatty acid composition, linoleic 4.2%, oleic 3.2%, stearic 1%, palmitic 1%, dihidroxystearic 0.7%, linolenic 0.3%, and eicosanoic 0.3%. Thus, the unsaturated fatty acids portion in the seed represents about 97.5% of the total fatty acids composition (McKeon, et al., 2000; Scholz and DaSilva, 2008; Perdomo, et al., 2013; Perea-Flores, et al., 2011; Madankar, et al., 2013; Ramanjaneyulu, et al., 2013). Physically, castor oil is a high viscous, clear pale yellow liquid at room temperature (27°C) with no solid fat content at 0°C. This unusual high viscosity for a natural vegetable oil is ascribed to the hydroxyl groups in hydrogen bonding (Ogunniyi, 2006). On the other hand, the high percentage of ricinoleic acid (unsaturated fatty acid) in the oil keeps it in liquid form even at low temperature. Therefore the presence of hydroxyl groups and double bonds makes the Castor bean oil suitable for industrial application due to various chemical reactions that can be involved, but recently castor oil is identified as a potential feed-stock for biofuel production because of its proven technical and ecological benefits and guaranteed sale (Akpan, et al., 2006; Anjani, 2012).

Oil content for *R. communis* collections reported from Russia is reported to vary from 42 to 58%. For 36 *R. communis* seed collection from Brazil, a wide variability in the oil content, ranging from 39.6 to 59.5%, was observed. For 72 United States Department of Agriculture (USDA) castor accessions collected from 15 countries a range of 22–44% oil content was reported, while from 191 USDA castor germplasm collections the range reported was 44.8-56.5%. Oil content ranging from 36.6 to 53.8% was observed in six Nigerian collections. Indian germplasm collection at Directorate of Oilseeds Research (DOR) exhibited 28–55% oil content (Anjani, 2012).

Phytochemical studies carried out in *R. communis* seed indicated the presence of steroids, saponins, alkaloids, flavonoids, and glycosides. When analyzing dried leaves from castor the presence of alkaloids, ricinine (0.55%) and N-demethylricinine, as well as six flavones glycosides as kaempferol-3-O-β-D-xylopyranoside, kaempferol-3-O-β-D-glucopyranoside, quercetin-3-O-β-D-xylopyranoside, quercetin-3-O-β-D-glucopyranoside, kaempferol 3-O-β rutinoside and quercetin-3-O-β- monoterpenoids (1, 8-cineole, camphor and α-pinene) and a sesquiterpenoid (β-caryophyllene), gallic acid, quercetin, gentisic acid, rutin, epicatechin and ellagic acid are the major phenolic compounds isolated from were detected. The seeds with a contain 45% of oil also have glycosides of ricinoleic, isoricinoleic, stearic and dihydroxystearic acids, and lipases (Jena, et al., 2012).

Other important component found in all organs of castor is the alkaloid ricinine ($C_8H_8O_2N_2$). It can be detected early in the seedling stage. Ricinine concentration can be influenced by environmental factors, such as salinity. Ricinine concentration is high in young leaves but it decrease as long as the plant became older. This alkaloid is a human central nervous system stimulant that can induce seizures but also improves memory retention in low dose. It can be lethal if ingested in high doses (Salimon, et al., 2012).

Of all *R. communis* components the most known is the ricin, a potent cytotoxin that is a globular glycoprotein; it represents more of the 5% of the total protein contained in the *R. communis* seed. Ricin is only found in the endosperm of castor seed. Castor oil does not contain ricin because this protein is insoluble in oil, and any residual ricin is eliminated in the refining process. Ricin content varies among genotypes. The ricin concentration varied from 1.9 to 16 g kg^{-1} among 263 accessions from the USDA germplasm bank and from 3.5 to 32.2 among 20 accessions from Embrapa germplasm bank (Salimon, et al., 2012). This toxin has a molecular weight of 65 kDa. It consists of two polypeptides (A chain 32 kDa and B chain 34 kDa) linked by a single disulfide bond. The B chain is a lectin binded to galactose containing glycolipids and glycoproteins expressed on the surface of cells, facilitating the entry of ricin into the cytosol. The A chain inhibits protein synthesis by irreversibly inactivating eukaryotic ribosomes through removal of a single adenine residue from the 28S ribosomal RNA loop contained within the 60S subunit. This process prevents chain elongation of polypeptides and leads to cell death. Aqueous solutions of ricin are inactivated by heat (80 °C for 1 h); nevertheless, when ricin is purified is soluble in water and stable over a wide pH range. This ricin powder or crude preparations require higher temperatures or longer periods for inactivation. One molecule of ricin is sufficient to inactivate 1,500 ribosomes per minute leading to rapid inhibition of protein synthesis and cell death. The estimated lethal oral dose of ricin for human is 1,000–20,000 mg/kg body weight, which corresponds approximately to 8 beans (Greenwood, et al., 2005; Kim, et al., 2006; Pita, et al., 2004; Chen, et al., 2004; Bradberry, 2007; Mendes, et al., 2009; Morris, et al., 2011).

STRUCTURE AND PHYSICAL PROPERTIES OF PLANT AND SEED

Special identification features of the plant are shrub like herb, stems 1 to 4 meters high, branched, green to reddish or purple, leaves alternated, simple, long-stalked, 20 to 80 cm wide and palmately lobed with 5 to 11 long lobes which are toothed on the margins, green or reddish. The fruit is a cluster of seeds covered with fleshy spines and contains three to four celled capsules in each cell containing a mottled seed. The seeds contained in the pods are about 1.5 to 2.5 cm across, elliptical glossy, black or white or usually mottled with grey, black, brown and white (Duffus and Slaugther. 1985; Perea-Flores, et al., 2011). The seed is ovoid with a prominent hilum or caruncle (**Figure 2a**). Castor bean seeds have physiological dormancy and are a classical system for studying endosperm reserve breakdown, especially for lipid and protein mobilization.

Other important data of castor seeds are its physical properties; these could be useful for the design of equipment of separation, leaning, storage, transportation and processing of seeds. Relevant information about physical properties of castor seed collected in Michoacán Mexico is shown in **Table 5**.

Table 5. Physical properties *R.communis* seed

Physical properties	Seed
Bulk Density (r_b) (kg/m^3)	538
Density (r_t) (kg/m^3)	1458
Porosity (e) (%)	63
Average Unit Volume (Vi) (mm^3)	406.67
Length (mm)	13.5
Width (mm)	8.7
Thickness (mm)	6.1
Sphericity (ϕ)	0.67
Geometric mean diameter (D_g) (mm)	8.95
Superficial area (S) mm^2	251.54
Width of the head (mm)	281.97

Adapted from Perea-Flores, et al., 2011.

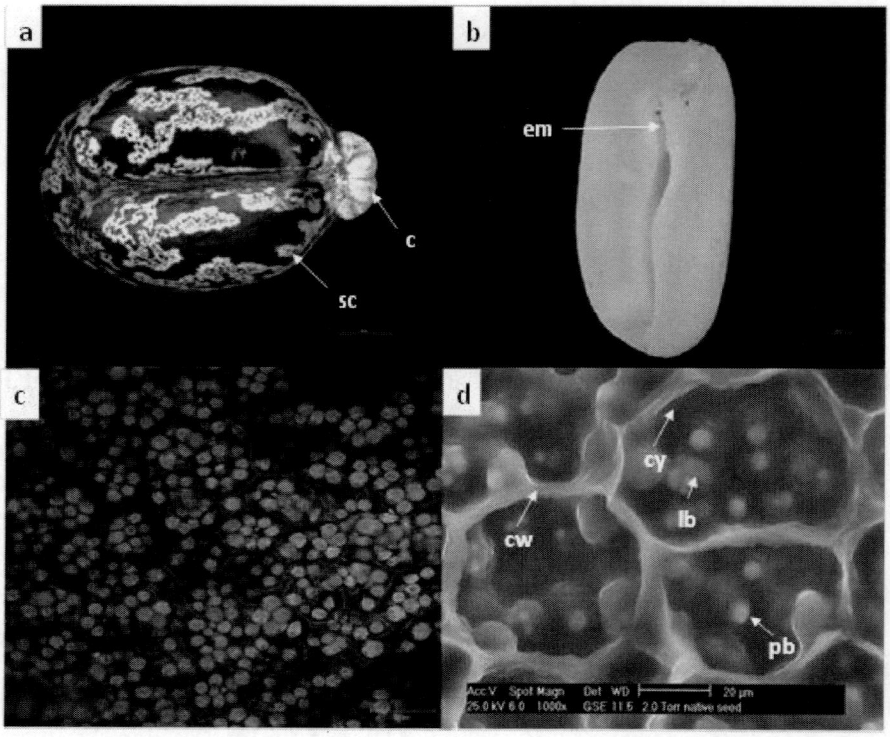

Figure 2. Stereo microscopy micrographs of *R. communis* seed, top view (a), longitudinal cut of endosperm seed (b) Confocal laser scanning microscopy micrographs show autofluorescence of endosperm cells of *R. communis* seed. Green color depicts lipid bodies and red protein bodies. (c). Environmental scanning electron microscopy micrograph of *R. communis* seed of an endosperm section (d). Where sc stands for seed coat or testa; c, caruncle; em, embryo; cw, cell wall; cy, cytoplasm; lb, lipid bodies and pb, protein bodies. (For interpretation of the references to color in this figure legend, the reader is referred to the web version of the chapter).

Table 6. Morphological features of endosperm cell and lipid bodies of *R.communis* seed

Morphological features	Endosperm cells*	Lipids bodies*
Area (μm^2)	1013.46	93.07
Roundness	0.79	0.87
Number of cell per mm^2 or lipid bodies per cell (μm^2)	570	9.57

* Results expressed as the mean value.
(Perea-Flores, et al., 2011).

Regarding to microstructure of the seed, this is covered with a shell that tends to be sclerificated; the endosperm, a spongy structure, is located inside of the seed, and it has a spongy caruncle and at the base a micropyle and a raphe are located, which are distributed over the seed (Perea-Flores, et al., 2011).

R. communis bean seeds are endospermic and its embryo is a spatulate axile. The cotyledons are thin and broad and the endosperm is the major storage tissue. The internal structure of the seed shows a copious white endosperm that consists of two cotyledons and at the center of them the embryo is located (**Figure 2b**). The embryo is about 6 mm long, consisting of the shaft of the root and a pair of flat cotyledons elongated through the center of the endosperm cell which has reached two thirds of its length. Besides, the microstructure of endosperm cells, as observed by confocal laser scanning microscopy (CLSM), show a green-red autofluorescence and clusters of lipids bodies inside of endosperm cells and proteins associated with lipids body surfaces and cell walls (**Figure 2c**). The cell wall and cellular membrane are structures composed of lipids, phospholipids, proteins and lignocellulosic compounds that exhibit autofluorescence when being excited at ultraviolet-green wavelengths. For this reason, the autofluorescence of cell wall and protein bodies (red) associated to lipid bodies (green) is easily noticed (**Figure 2c**). The endosperm cells of *R. communis* seed observed under environmental scanning electron microscopy (ESEM) are shown in the **Figure 2d**, where it can be seen that endosperm is an homogeneous tissue with a complex cytoplasmic network. The cells have a nearby isodiametric shape. Furthermore ESEM image allows observing the subcellular organelles that include protein and lipid bodies (Perea-Flores, et al., 2011).

In a previous study (Perea-Flores, et al., 2011) about microstructural characterization by means of image analysis of *R.communis* seed it was found that the endosperm cells had roundness (R) values below to 1 (0.79), these values allowed characterize the ovoid form of endosperm cells. Additionally, the number of cells per mm^2 (cell density) presented an average value of 570. Moreover, the measurements of the shape of lipids bodies obtained from CLSM (**Figure 2c**), had approximately a circular shape with R values close to 1 (0.87). This spherical shape for lipids bodies have also been reported in other seeds such as coffee, almond, *Brassica rapa L.* and peanut (Perea-Flores, et al., 2011); it is well known that the oil reserve organelles such as the lipids bodies present a spherical shape, which has proved to be very stable to storage fluid substances into the cells. The largest size and high density of lipids bodies per cell (9.57) in endosperm cells confirm its high content of lipids and high potential as oil source for several uses, in special for biodiesel production. **Table 6** summarizes the size, shape and density of endosperm cells and lipid bodies in castor bean Thereby, microstructural characterization of seeds provide relevant qualitative and quantitative information that can be helpful for its processing and utilization.

EFFECT OF *R. COMMUNIS* SEED COMPONENTS ON SOME DISEASES AND PHARMACEUTICAL USES

R. communis is an important source of compounds with pharmacological properties. The phytochemical components of the *R. communis* have been established in some studies and these include tannins, saponins, alkaloids, carbohydrates, phenols, flavonoids, sterols and resins. The presence of phytochemical constituents and pharmacological activities proved that the plant has an important capacity for the development of drugs (Jena, et al., 2012). It has been found that the ethanolic root extract of *R. communis* is effective in asthma treatment because of its antiallergic and mast cell stabilizing potential effect. Saponins has mast cell stabilizing effect and the flavonoids possess smooth muscle relaxant and bronchodilator activity; for instance the apigenin and luteolin (flavonoids) generally inhibit basophil histamine release and neutrophils beta glucuronidase release have shown antiallergic activity.

Additionally, ethanol extract of *R. communis* root resulted in anti-histaminic activity while clonidine induced catalepsy in mice. Tannins extracted from the leaves significantly increased the phagocytic function of the human leucocytes and as result produces an immune modulatory effect.

Castor oil is also used as purgative and it is considered safe and effective when used as a stimulant laxative (FDA, 2003; Ogunniyi, 2006; Scholz and DaSilva. 2008). Among its medical uses, 1.25% of castor oil emulsion has recently been shown to be effective in the clinical management of evaporative dry eye and homogenized castor oil eye drops are effective in the treatment of meibomian gland dysfunction (Morris, et al., 2011).

Methanol extracts of *R. communis* seed possess phytochemical compounds as steroids and alkaloids, which have anti-fertility activity by means of hormonal control and so it helps the inhibition of maturation of the follicle in the ovary and prevents ovulation. (Jena, et al., 2012). The Castor oil is a great skin protector, an emollient, and moisture barrier. Acting as a humectant, this oil keeps moisture close to the skin's surface due to its water-binding abilities. It also contains undecylenic acid which, due to its germicidal and disinfectant properties, is useful in treating skin disorders, particularly those which are caused due to bacterial or fungal infections. The antimicrobial activity of *R. communis* is good against dermatophytic and pathogenic bacterial strains *Streptococcus progenies*, *Staphylococcus aureus* as well as *Klebsiella pneumonia*, and *Escherichia coli*. The different solvent extracts of roots of *R. communis* possess antimicrobial activity against pathogenic microorganisms such as *Escherichia coli, Staphylococcus aureus, Pseudomonas aeruginosa, Salmonella typhimurium, Proteus vulgaris, Bacillus subtilis, Candida albicans and Aspergillus niger*. The antibacterial assay revealed that the acetone and hexane extracts of *Ricinus* possess good zone of inhibition whereas ethanolic extract having antibacterial activity only on higher concentration (Jena, et al., 2012).

R. communis seed could become one of the premier crops for health enhancing phytochemicals. An immunotoxin derived from castor bean constituents was successful in combination with chemotherapy in treating patients with AIDS-related non-Hodgkin's lymphoma. Ricin can be targeted to specific cells, such as cancer cells, by conjugating the Ricin toxin A (RTA) subunit to antibodies or growth factors that preferentially bind the unwanted cells (Embleton, et al., 1986). These immunotoxins have worked well for in vitro applications, e.g. bone marrow transplants. Although they have not worked very well in many in vivo situations, progress in this area of research shows promise for the future. The pericarp of castor bean when used at low doses to feed mice showed central nervous system effects. At lower doses, the extract improved memory consolidation.

Other example is chlorogenic acid that showed a clinically significant effect on the absorption and utilization of glucose when consumed with coffee reducing the subject's body mass and fat, additionally this acid reduced the blood pressure in people with high pressure problems. Quercetin (antioxidant, flavonoid) reduced blood pressure in hypertensive people, while the flavonoid isoquercitrin was clinically effective and safe for treatment of chronic venous. Castor oil combined with balsam of Peru (*Myroxylon balsamum L.*) and trypsin reduced edema and scabbing of wounds. A fucose rich oligo- and poly-saccharide have a significantly effect for attenuate periorbital wrinkles with a total regression of crow's feet observed in some women (Morris, et al., 2011).

These phytochemicals from *R. communis* seed have not yet been introduced into phamarceutical markets. Perhaps, the most significant medical use of compounds obtained from castor bean involves the toxic protein, ricin. Even before the mechanism of action of ricin was understood, it was shown to have anticancer properties (Morris, et al., 2004). Thereby, many countries have used castor bean to treat various health ailments for example China treats head ache, abscess, arthritis, dog bite, boil and as an expectorant, Guatemala and Ghana for head ache principally, South Africa for stomach ache, Haiti for fever, flu and asthma treatment, Trinidad for bactericide, fever, inflammation and stomach ache, Mexico for head ache, fever and colic, India for cholera and convulsion, Canada and Indochina to treat cancer, Africa, Guatemala as hair oil. (UK Cropnet 2008).

CONCLUSION

In the last years, the global demand for fuel has been increasing. The production of biodiesel from non-edible oils has become an environmentally friendly alternative as well as an energy resource. Despite of this more than 95% of biodiesel is still made from edible oils, which causes an imbalance between the utilization of energy resources and food consumption. This has motivated that *R. communis* seed has become an important crop for biodiesel production, additionally the generation of sub-products and wastes derivative of its exploitation has opened the interest on the research about genetic improvement of crops and optimization of processes associated with the use of *R. communis* seed. The research has mainly been focused on the seed detoxification and the use of its proteins and phytochemicals for applications in the food and pharmaceutical industries. In this regard various applications have been proposed to utilize the waste cake obtained from the oil extraction process applied during the biodiesel production. Thus, new applications on the biological area have been generated in recent years, as the use of proteins for the manufacture of edible films, ricin for use in cancer treatments, characterization of phytochemicals with antioxidant potential and antimicrobial among others applications traditionally known. Thereby, *R. communis* to be non-edible seed with high contents of protein, fatty acids and phytochemicals components has become an attractive crop and consequently its production is increasing, as well as research and technological development in several fields of knowledge. Thus, in this chapter is provided basic information and a brief revision about castor seed in order to visualize future trends, potential uses and applications of this crop.

REFERENCES

Akpan, UG; Jimoh, A; Mohammed, AD. Extraction, Characterization and Modification of castor seed oil. *Leonardo Journal of Sciences*, 2006, 8, 43-52.

Ali, RM; Elfeky, SS; Abbas, H. Response of salt stressed *Ricinus commnunis* L. to exogenous application of glycerol and/or aspastic acid. *Journal of Biological Sciences*, 2008, 8, 171-175.

Anjani, K. Castor genetic resources: A primary gene pool for exploitation. *Industrial Crops and Products*, 2012, 35, 1– 14.

Annongu, AA; Joseph, JK. Proximate analysis of castor seeds and cake. *Journal of Applied Science and Environmental Management*, 2008, 12, 39 – 41.

Biodiesel Technology. http://www.dovebiotech.com. Accessed in February 2013.

Bradberry, S. Ricin and abrin. *Medicine*, 2007, 35, 576-577.

Chen, GQ; He, X; Liao, LP; McKeon, TA. 2S Albumin gene expression in castor plant (Ricinus communis L.). *Journal of the American Oil Chemists' Society*, 2004, 81, 867-872.

Dubois, JL; Piccirilli, A; Magne, J; He, X. Detoxification of castor meal through reactive seed crushing. *Industrial Crops And Products*, 2013, 43, 194-199.

Duffus, C; Slaugther, C; 1985. *Capítulo 2: Formación de la semilla en Las semillas y sus usos*.(41-67), México, AGT Editor, S.A.

Embleton, MJ; Byers, VS; Lee, HM; Scannon, P; Blackball, NW; Baldwin, RW. Sensitivity and Selectivity of Ricin Toxin A Chain-Monoclonal Antibody 79IT/36. Conjugates against Human Tumor Cell Lines. *Cancer Research*, 1986, 46, 5524-5528.

Falasca, SL; Ulberich, AC; Ulberich E. Developing and agro-climatic zoning model to determine potential production areas for castor bean (*Ricinus communis* L.). *Industrial crops and products*, 2012, 40, 185– 191.

FAO, 2009. Draft Second Report on the State of the World's Plant Genetic Resources for Food and Agriculture – Final Version. Commission on Genetic Resources for Food and Agriculture, Food and Agriculture Organization of the United Nations, Rome, p. 276. www.fao.org.docrep/013/i500e._Accessed in February 2013.

FAO, 2011. Food and Agricultural commodities production. http://faostat.fao.org. Accessed in February 2013.

FDA (2003) OTC Drug review ingredient report. FDA database. FDA, Washington DC

Ferreira, P; Pereira, R; Coelho, JFJ; Silva, AFM; Gil, MH. Modification of the biopolymer castor oil with free isocyanate groups to be applied as bioadhesive. *International Journal of Biological Macromolecules*, 2007, 40, 144-152.

Greenwood, JS; Helm, M; Gietl, C. Ricinosomes and endosperm transfer cell structure in programmed cell death of the nucellus during Ricinus seed development. *Proceedings of the National Academy of Sciences of the United States of America*, 2005, 102, 2238–2243.

http://www.castoroil.in/

Industrial Castor Oil. http://www.whc-oils.com/industrial-castor-oil.html. Accessed in February 2013.

Jena, J; Gupta AK. *Ricinus Communis* Linn: A Phytofarmacological Review. *International Journal of Pharmacy and Pharmaceutical Sciences*, 2012, 4, 25-29.

Kim, S-K; Hancock, DK; Wang, L; Cole, KD; Reddy, PT. Methods to characterize Ricin for the Development of Reference Materials. *Journal of Research of the National Institute of Standars and Technology*, 2006, 111, 313-224.

Lott, JNA; Greenwood, JS; Vollmer, CM. Mineral reserves in castor beans: the dry seed. *Plant Physiology*, 1982, 69, 829-833.

Lott, JNA; West, MM; Calrk, B; Beecroft. Changes in the composition of globoids in castor bean cotyledons and endosperm during early seedling growth with and without complete mineral nutrients. *Seed Science Research*, 1995, 5, 121-125.

Madankar, CS; Pradhan, S; Naik, SN. Parametric study of reactive extraction of castor seed (*Ricinus communis* L.) for methyl ester production and its potential use as bio lubricant. *Industrial Crops and Products*, 2013, 43, 283-290.

McKeon, TA; Chen, GQ; Lin, JT. Biochemical aspects of castor oil biosynthesis. *Biochemical Society Transactions*, 2000, 28, 972-974.

Mendes, RC; Fernandez, DCdSD, Dias, PM, Berger, PG. Pre-germinative treatments in castor bean (*Ricinus communis* L.) seeds. *Revista Brasileira de Sementes*, 2009, 31(1).

Mhaske, SD; Mahatma, MK; Jha, S; Singh, P; Mahatma, L; Parekh, VB; Ahmad, T. Castor (*Ricinus communis* L.) Rc-LOX5 plays important role in wilt resistance. *Industrial Crops and Products*, 2013, 45, 20– 24.

Morris JB; Morse AS; Navas A; Soares DJ; Sofiatti V; Wang ML; Zanotto DM; Zieler H. A Review on the Challenges for Increased Production of Castor. *Agronomy Journal*, 2012, 104, 853-880.

Morris JB. Phytochemical traits in the genetic resources of castorbean. *Curr Top Plant Biol*, 2004, 5, 63–67.

Morris, JB; Wang ML; Morse SA. (2011). Ricinus. In: Kole C. (Ed), *Wild Crop Relatives: Genomic and Breeding Resources - Oilseeds*, Berlin, Heidelberg: Springer Berlin Heidelberg, Germany, 251-260.

Nielsen F. (2011) *Castor (Ricinus communis) Potential of castor for bio-fuel production*, www.fact-foundation.com/.../FACTsheet_Castor._Accessed in February 2013.

Ogunniyi, DS. Castor Oil: A vital industrial raw material., *Bioresource Technology*, 2006, 97, 1086-1091.

Pecina-Quintero, V; Anaya-López, JL; Núñez-Colína, CA; Zamarripa-Colmenero A; Montes-García, N; Solís-Bonilla, JL; Aguilar-Rangel, MR. Assessing the genetic diversity of castor bean from Chiapas, México using SSR and AFLP markers. *Industrial Crops and Products*, 2013, 41, 134– 143.

Perdomo, FA; Acosta-Osorio, AA; Herrera, G; Vasco-Leal., JF; Mosquera-Artamonov, JD; Millan-Malo, B; Rodriguez-Garcia, ME. Physicochemical characterization of seven Mexican Ricinus communis L. seeds & oil contents. *Biomass and Bioenergy*, 2013, 48, 17-24.

Plants database of the United States Department of Agriculture, USDA. http://plants.usda.gov/java/profile?symbol=rico3. Accessed in February 2013.

Perea-Flores MJ; Chanona-Pérez JJ; Garibay FV; Calderón DG; Terrés, RE; Mendoza, PJA; Herrera, BR. Microscopy techniques and image analysis for evaluation of some chemical and physical properties and morphological features for seeds of the castor oil plant (*Ricinus communis*). *Industrial Crops and Products*, 2011, 34, 1057-1065.

Pita, R; Anadón, A; Martínez-Larrañaga, MR. Ricina: una fitotoxina de uso potencial como arma. *Revista Toxicológica*, 2004, 21, 51-63.

Preview of Comprehensive Castor Oil Report. A report on castor oil & castor oil derivatives. http://www.castoroil.in/reference/report/report.html._Accessed in February 2013.

Ramanjaneyulu, AV; Vishnuvardhan RA; Madhavi, A. The impact of sowing date and irrigation regime on castor (*Ricinus communis* L.) seed yield, oil quality characteristics and fatty acid composition during posrainy season in South India. *Industrial Crops and Products*, 2013, 44, 25-31.

San Miguel, MMM. Ricino (*Ricinus communis*). *Maleza-Ricino*, 2006, 91-95.

Sathiskumar, PS; Madras, G. Synthesis, characterization, degradation of biodegradable castor oil based polyesters. *Polymer Degradation and Stability*, 2011, 96, 1695-1704.

Scholz, V; Da Silva, JN. Review Prospects and risks of the use of castor oil as a fuel. *Biomass and Bioenergy*, 2008, 32, 95-100.

Salimon, J; Noor, MDA; Nazrizawati, AT; Firdaus, MMY; Noraishah, A. Fatty Acid Composition and Physicochemical Properties of Malaysian Castor Bean *Ricinus communis* L. Seed Oil. *Journal of Saudi Chemical Society*, 2010, 39, 761–764.

Sujatha, M; Reddy, TP; Mahasi, MJ. Role of biotechnological interventions in the improvement of castor (*Ricinus communis* L.) and *Jatropha curcas* L. Biotechnology Advances, 2008, 26,424-435.

UK Cropnet (2008) http://ukcrop.net/perl/ace/grep/EthnobotDB. Accessed in February 2013.

Zapata, N; Vargas, M; Reyes, JF; Belmar, G. Quality of biodiesel and press cake obtained from Euphorbia lathyris, Brassica napus and Ricinus communis. *Industrial Crops and Products*, 2012, 38, 1– 5.

In: Seeds as Functional Foods and Nutraceuticals
Editors: R. Mora-Escobedo, J. De J. Berrios et al.
ISBN: 978-1-62808-489-4

Chapter 15

INDUSTRIAL APPLICATIONS AND POTENTIAL PHARMACEUTICAL USES OF MANGO (*MANGIFERA INDICA*) KERNEL

Nayeli Vélez Rivera[1], Stefany Cárdenas Pérez[1], José Jorge Chanona-Pérez[1*], María de Jesús Perea-Flores[2], Georgina Calderón-Domínguez[1], José Blasco Ivars[3] and Reynold Ramón Farrera-Rebollo[1]

[1]Departamento de Ingeniería Bioquímica, Escuela Nacional de Ciencias Biológicas, Instituto Politécnico Nacional, Plan de Ayala y Carpio s/n, Col Santo Tomas, D.F., México

[2]Centro de Nanociencias y Micro y Nanotecnologías, Instituto Politécnico Nacional, Unidad Profesional "Adolfo López Mateos", Luis Enrique Erro s/n, Zacatenco, D.F., México

[3]Centro de Agroingeniería, Instituto Valenciano de Investigaciones Agrarias, (IVIA). Moncada, Spain

ABSTRACT

Mango is grown across the world in tropical and warm temperate regions. Its processing is very important for exporting in many producer countries. This is primarily due to the fact that there are several limitations regarding the export of fresh fruits. Fruit processing is very advanced in many countries, in some of them it has been invented and developed various systems to enhance the fruit processing. Sale in fresh is the most common use of the mango which implies that the end products have low added value in the market, for instance, mango seed kernel is one of the by-products of food processing industry and is not commercially exploited, but are discarded, consequently there is a high wastage raw material. For this reason this chapter is aimed to made brief recompilation about Mango´s seeds, beginning since its origin and taxonomy of the

* Corresponding authcr: Email: Jorge_chanona@hotmail.com.

Mango, some important data of global production and marketing of mango´s kernel seed and its oil are also provided. During processing of mango, kernel seeds were thrown off as garbage, and thereby losing a resource for valuable seed fat. Physical and chemical properties presented important information for its contents and processing. A chemical analysis of the fresh core of the mango seed also known as the seed stone comprises 13 per cent of the fruit's total weight and around 55 to 65 per cent of the seed's weight. In fact, the kernel or the mango seed core is a significant spin-off of the mango processing industry. It is edible and has various utilities in the food industry. For instance, the mango seeds are dried and stored in India for use as meals during acute food crisis. The mango seed kernel contains substantial amount of edible fat, which is extracted and used for several purposes. Additionally the effect of the seed used as biosorbent as well as antioxidant, antibiotic and other relevant uses are also described. Finally, the future trends to improve its production and its development of new products from mango seed are also indicated.

Origin and Taxonomy

Mango (*Mangifera indica* L.) is a tropical fruit with a great demand on the world market due to consumer acceptance and high levels of production. During recent years, the mango has become more important as an export product in several producer countries, the main consumers being USA and Europe owing to the large consumer demand for fresh and processed mango products. The mango is usually found in Southern Asia, especially in Eastern India, China, Burma, Andaman Islands and Central America. Mangoes belong to the genus *Angifera*, consisting of numerous species of tropical fruiting trees in the flowering plant family Anacardiaceae. They have been cultivated in the Indian Subcontinent for thousands of years and reached East Asia between the 5th-4th century BC. By the 10th century AD, they were transported to East Africa and subsequently introduced to Brazil, West Indies and Mexico, where climate allows its appropriate growth. The origins of mango are thought to have been from a plant of Malaysia, India and Indonesia. It probably was grown in Southeast Asia before the seventh century. (Kittiphoom, 2012). The Spanish introduced the crop to its colonies in the Americas, through the traffic between the Philippines and the west coast of Mexico for the XV and XVI (Infoagro, 2006).

The original wild mangos were small fruits with scant, fibrous flesh, and it is believed that natural hybridization has taken place between *M. indica* and *M. sylvatica* Roxb in Southeast Asia. Mango is now widely cultivated as a fruit tree in frost-free tropical and warmer subtropical climates throughout the Indian subcontinent. Mango can grow in climates where the average temperature of the coldest month is greater than 15° C, it is considered that the optimum production temperature is 24 to 27 ° C and growth stops between 4 and 6 ° C (Martinez et al., 2006).

The nomenclature of Mangifera species and mango cultivars has been complicated by the widespread use of synonyms. This family consists of sixty four genera, mostly of trees or shrubs, often containing milky or acrid juice, some of which are even poisonous. The leaves are estipulate, usually alternate and simple. The inflorescence is generally an axillary or terminal panicle or spike bearing small and regular flowers. They may be unisexual (usually male) or bisexual, borne on the same or different trees. Mango possesses a very long taproot.

The fruit is usually a drupe and seed is exalbuminous and is located inside the stony endocarp. (Vasanthaiah, 2007)

Mango varieties can be classified as having either mono-embryonic or poly-embryonic seed embryos. In mono-embryonic varieties, the seed contains only one embryo that is a true sexual (zygotic) embryo. Mono-embryonic seeds are a cross between the maternal and paternal (pollen) parents. Fruit from mono embryonic seedlings will often vary from the parent trees, so propagation by grafting is used to produce true-to-type monoembryonic trees. Poly-embryonic seeds contain many embryos, most of which are asexual (nucellar) in origin and genetically identical to the maternal parent. Poly-embryonic seeds also contain a zygotic embryo that is the result of cross-pollination. The mono-embryonic seedling usually has less vigor than a nucellar seedling for use as a rootstock. In some varieties this is reversed and the zygotic seedling is the most vigorous. The occurrence of off-types in orchards is often attributed to use of zygotic seedlings. (Abdullahi, 2012)

Table 1. Mango's groups classification in Mexico

Characteristics	Mulgova Group	Varieties	Hindustani Group	Varieties
Shape	Flattened and elongated	Manila, Ataulfo, Diplomático, Panameño, Obo and Criollo. 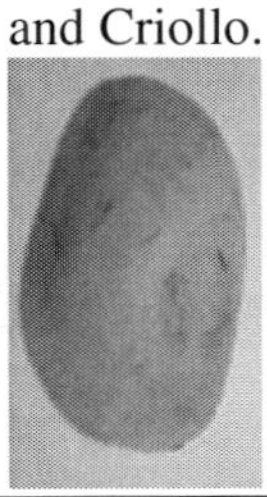	Oval or round	Haden, Keitt, Kent, Tommy Atkins, Irwin, Zill, Sensation, Oro, Vandycke and Vishis (Manzanillo). 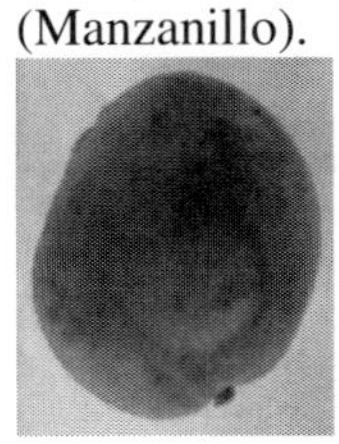
Peel	Soft and yellow		Leathery, slightly Green and reddish	
Pulp	Fibrous, aromatic, with a good flavor.		Yellow color, slightly fibrous, aromatic and with good flavor.	

NOM-129-SCFI-1998.

Table 2. Availability of mango along the year in different countries

Northern Hemisphere	Jan	Feb	Mar	Apr	May	Jun	Jul	Ago	Sep	Oct	Nov	Dec
India												
México												
Filipinas												
Puerto Rico												
Costa Rica												
Israel												
Southern Hemisphere												
Brazil												
Ecuador												
Peru												
South Africa												

(Ruiz, 2007).

The overall mango varieties in México are divided into two main groups (Mulgova and Hindustani) according to their physical characteristics. The most important varieties in Mexico are "Manila", "Ataulfo" and "Haden" (Ornelas-Paz, et al., 2008). **Table 1** shows the main characteristics of these groups.

THE GLOBAL MARKET

Mangoes are produced in over 90 countries worldwide. Asia accounts for approximately 74% of global mango production (Table 1), and the Americas account for approximately 9%. Although not a major mango producer, the United States has developed most of the popular cultivars traded on the international market. It is also the largest single-country mango importer (Evans, 2008). In terms of distribution, Mexico, Brazil, Peru, Ecuador, and Haiti supply the majority of mango imports to the North American market. India and Pakistan are the predominant suppliers to the West Asian market. The Philippines and Thailand supply most of the Southeast Asian market. The European Union mainly buys mangoes from South America and Asia. (Evans, 2008).

The FAO reported that in 2011 Mexico was ranked 5th among 58 top mango-producing countries and was also the world's number one exporter (FAOSTAT, 2011). Mexico produces more than 50 mango cultivars and the "Manila" and "Ataulfo" cultivars are the most important for national and international markets (Ornelas-Paz et al., 2008). In Brazil, most mangos are produced in the state of Minas, Gerais. The bulk of the crop is for domestic consumption. Although only 3% of the world production of mango is traded globally, it represents a noticeable increase over the quantities traded 20 years ago. In 2011 India was the biggest mango producer, with a 15.1 millions of tons, followed by China (4.5 millions of tons) and Thailand (3.2 millions of tons) as shown in **Table 3**. Other leading mango producers during 2011 period include Indonesia (5.83%), Pakistan (5.17%), Mexico (5 %), Brazil (3.4%). (http://faoststst.fao.org).

Table 3. Mango production in 2011, Int: International Dollars, MT: Million tonnes

Country	Production (Int $1000)	Production (MT)	% Production
India	9100148	15188000	41.54
China	2707863	4519380	12.36
Thailand	1963620	3277250	8.96
Indonesia	1276908	2131140	5.83
Pakistan	1131496	1888450	5.17
Mexico	1094863	1827310	5.00
Brazil	748671	1249520	3.42
Others	3883221	6481045	17.73
Total	21906790	36562095	100

FAO Stat.

India ranks first among world's mango producing countries accounting for 41.5 per cent of the total world's mango production of 21.9 million tonnes (Table 3). In India, the yields of mangoes range one million tonnes of mango seed annually. It is estimated that the mango kernel available out of these seeds is about 0.6 million tonnes per year with starch content between 60-65%. (Velan, 1995). Otherwise, it has been estimated that in India alone some 30,000 tons of oil could be extracted from 4 million tons of the total annual harvest of 7 million tons. Some mango kernel oil has been commercially extracted in India with 150 tons being exported in 1976 rising to 850 tons in 1978. (Board of Investments, 2011).

In Mexico, the agribusiness chain is directed by two-thirds to the sector as sold fresh primary and a third part of the secondary sector or processing and packaging fruits, of which 80% is directed to the production of pulps, juices and nectars. (INEGI, 2008).

Unfortunately, after eaten away the juicy fruit pulp when ripe, the hard mango seeds are all thrown off as garbage, and thereby losing a resource for valuable seed fat. Although the resource has not been fully exploited, the mango seed fat is known to have considerable nutritional and commercial importance (Bhattacharjee, 2011). In the particularly case of México, the major problem is the minimum handle of technological development for industrialization, so that the sale in fresh is the most common use which implies that the end products have low added value in the market, therefore there is a high wastage raw material (approximately 40%). One way to address this situation is to form a network of value, in where it is indispensable to know the needs of consumer to develop new marketing strategies, handle all this in order to increase the competitiveness of the product. The value networks are a new concept in agribusiness, where the economy is headed consumer needs for differentiated products. Some of the main properties that can add value to a product are the nutritional, functional properties and the total utilization of the organic industrial waste. The agribusiness is the pivot function that allows the value network, as it is the link between consumer needs and producers, input suppliers and researchers, through knowledge transfer and the processing of products adding high value and a better yield (Sumaya-Martínez, et al., 2012).

FORMER SCOPE AND FUTURE TRENDS

Mango seed kernel could be used as a potential source for functional food ingredients, antimicrobial compounds and cosmetic due to its high quality of fat and protein as well as high levels of natural antioxidants. The mango stone obtained after decortication of mango seed can be utilized as adsorbent. (Kittiphoom, 2012).

The kernel is a major by-product of the mango-processing industry. In times of food scarcity in India, the kernels are roasted or boiled and eaten. After soaking to dispel the astringency (tannins), the kernels are dried and ground to flour which is mixed with wheat or rice flour to make bread and it is also used in puddings. The fat extracted from the kernel is white, solid like cocoa butter and tallow, edible, and has been proposed as a substitute for cocoa butter in chocolate (Morton, 1987). Additionally, most parts of the mango can be fully utilized; some current uses of the whole mango are reported in Table 4.

Table 4. Most common uses of Mango and kernel

Mango section	Use
Seed kernels	After soaking and drying, the kernels are fed to poultry and cattle. Without the removal of tannins, the feeding value is low. Cuban scientists declare that the mineral levels are so low that mineral supplementation is needed if the kernel is used for poultry feed, for which purpose it is recommended mainly because it has little crude fiber.
Mango kernel decoction and powder (not tannin-free)	Used as verminfuges and as astringents in diarrhea, hemorrhages and bleeding hemorrhoids. The fat is administered in cases of stomatitis. Extracts of unripe fruits and of bark, stems and leaves have shown antibiotic activity. In some Caribbean islands the leaf decoction is taken as a remedy for diarrhea, fever, chest complaints, diabetes, hypertension and other ills. A combined decoction of mango and other leaves is taken after childbirth.
Seed fat	Having high stearic acid content, the fat is desirable for soap-making. The seed residue after fat extraction is usable for cattle feed and soil enrichment. A mango stone decorticator has been designed and successfully operated by Indian Universities.
Raw mango fat	Raw mango butter is rich in oleic acid, a mono-unsaturated omega-9 acid, and stearic acid, a saturated fatty acid. Its natural ingredients act as emollients to add moisture and soothe *dry skin* and hair. Mango butter is a common product in many commercial soaps, lotions, and even cosmetics. It may also be included in shampoos, conditioners, and hot-oil treatments for hair.
Bark	The bark possesses 16% to 20% tannin and has been employed for tanning hides. It yields a yellow dye, or, with turmeric and lime, a bright rose-pink.
Gum	A somewhat resinous, red-brown gum from the trunk is used for mending crockery in tropical Africa. In India, it is sold as a substitute for gum arabic.
Medicinal Uses	Dried mango flowers, containing 15% tannin, serve as astringents in cases of diarrhea, chronic dysentery, catarrh of the bladder and chronic urethritis resulting from gonorrhea. The bark contains mangiferine and is astringent and employed against rheumatism and diphtheria in India. The resinous gum from the trunk is applied on cracks in the skin of the feet and on scabies, and is believed helpful in cases of syphilis.

(Morton, 1987).

It has been already established by nutritional and toxicological studies that mango seed fat consumption is safe and can substitute for cocoa butter as solid nutritional fat. Otherwise, the concern for fast depletion of petroleum oil and its environmental impact has shifted interest to alternative sources of fuels, particularly biofuels, which are renewable and environmental friendly. Thus, varieties of virgin, non-edible and waste vegetable oils have been sourced for the production of biofuels. The choice of biofuel over diesel fuel includes its portability, availability, renewability, higher combustion efficiency, higher cetane number, higher biodegradability, high flash point, inherent lubricity, lower sulfur and aromatic contents (Agbede, et al., 2011).

The usefulness of the whole mango kernel, the defatted kernel flour and the oil, in comparison with mango juice, is yet to gain both cottage and industrial attraction in some African countries, particularly Congo. The underutilization could be partly due to the limited knowledge of the toxicological status of the mango kernel, the functional properties of the kernel flour and appropriate processing technology (Nzikou, et al., 2010).

Manufacturing industry has used the mango seed shell as adsorbent for the removal of methylene blue from its aqueous solution, it is potentially low cost substrates for generation of activated biosorbent, giving up to 60% dye removal. Table 5 shows the sorption capacity of mango seed shell to uptake methylene blue, from this table it was observed that the sorption capacity of mango seed shell determined is comparable enough to some of the low cost adsorbents for the uptake of methylene blue from its aqueous solutions (Kittiphoom, 2012).

Adsorption has been a successful technique for removal of heavy metals from waste water stream. There is a wide variety of adsorbent which exhibits different characteristics depending upon their chemical constituents and the synthesis techniques. Agricultural and food industry waste materials like mango kernel, which can effectively be converted into adsorbent by chemical and or thermal activation. Heavy metals rank as major environmental pollutants and its removal is today's need (Pandharipande, 2012).

Several studies have also shown that mango seed kernels contain various phenolic compounds and can be a good source of natural antioxidants. Mango seed kernel had the highest antioxidant activity and phenolic content. The mango seed could be used as a potential source for functional food ingredients natural antioxidants antimicrobial compounds, cosmetic, activated carbon and, in addition, it could be further processed into therapeutic functional food products. This suggests that the mango seeds should be further utilized rather than just discarded as waste (Kittiphoom, 2012).

Additionally, mango kernel extract has a bacteriostatic and antibacterial activity and thus can be used in the pharmaceutical frame as antibacterial agent. For instance, agents for preventing and treating acne or agents for preventing dental caries can be provided by adding kernel mango extract as an effective component.

Table 5. Low cost adsorbents and their sorption capacities for the removal of methylene blue from its aqueous solutions

Adsorbent	Sorption capacity (mg/g)
Coal	323.68
Rice husk	312.25
Cotton waste	277.77
Mango seed shell	142.85
Clay	6.3

Adapted from Kittiphoom, 2012.

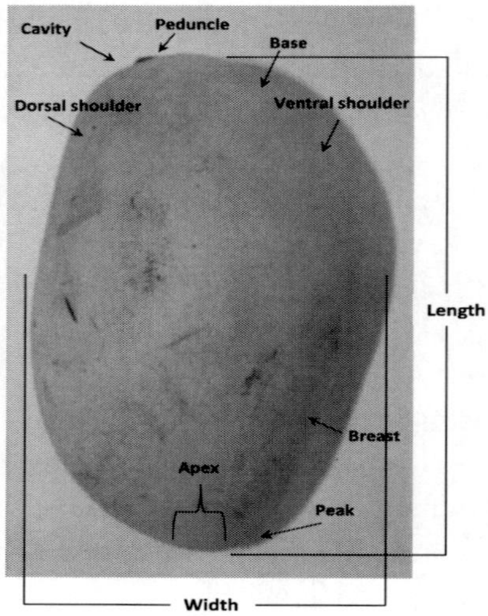

Figure 1. Morphology of Mango "*Manila*".

PHYSICAL COMPOSITION

In **Figure 1**, it is shown the morphology of the mango which has several parts: peduncle, cavity, base, dorsal shoulder, ventral shoulder, breast, and peak or apex. However, the most common measures are the length and width of mango (Guzmán et al., 1996).

The mango "Manila" is a drupe with an outer leathery skin exocarp, a fleshy mesocarp and a hard stony endocarp (pit) surrounding the large kernel seed as shown in **Figure 2**.

The peel and seed of the mango comprise 50 % of the fruit and these are just discarded after separating the flesh. The kernel makes up 75% of the seed weight and an average mango kernel contains about 8% to 15% extract potential (butter and oil). Mango kernel oil is pale yellow in color. After mango is processed, different components are obtained during mango pulp extraction as shown in Table 6. (Kittiphoom, 2012)

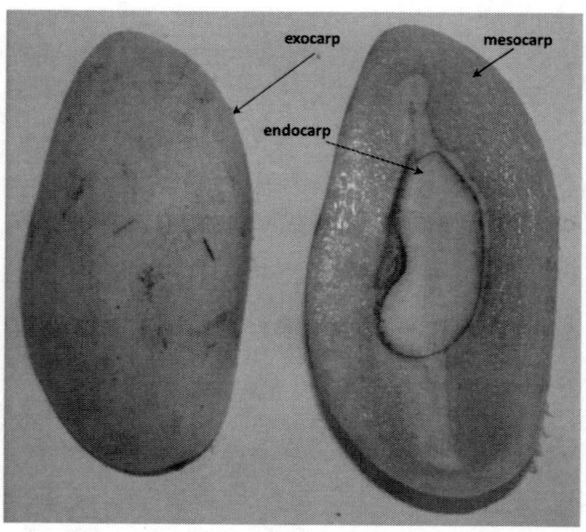

Figure 2. Mango "Manila" halved, showing exocarp, mesocarp and endocarp features.

Table 6. Different components obtained during mango pulp extraction

Component	%
Mango pulp	45-65
Peels	15-20
Pulpier waste	10-15
Stones (seed)	10-20

(Kittiphoom, 2012).

CHEMICAL COMPOSITION

Mango seed is fair in crude protein, high ether extract and low in fiber as shown in **Table 7**. Some studies reported that mango seed contained minerals such as sodium, potassium, calcium, magnesium, iron and copper Table 7. Additionally it has been reported that mango seed is rich in glutamate, and poor in methionine. Among the essential amino acids, reported that leucine has the highest value (8.40 g/100 g of protein). This is followed by arginine which has the value of 5.17 g/100 g of protein. The table 3 shows the amino acid profile of mango seed. Phenylalanine/tyrosine had the highest amino acid value while lysine has the lowest value and it is the limiting amino acid in mango seed. Moreover, mango seed is richer in vitamins like vitamin A, vitamin E, vitamin K, vitamin B1, vitamin B2, vitamin B6, vitamin B12 and vitamin C as reported in **Table 7.**

Table 7. Seed kernel chemical composition, amino acids and vitamins

Seed Kernel composition	%	Amino acid	Quantity (%)	Vitamins	Amount (mg\100 g)
Moisture	10.55-11.35	Lysine	3.13	A	15.27 (IU)
Protein	4.76-8.5	Alanine	6.4	E	1.3
Fat	6-15	Histidine	2.31	K	0.59
Starch	40-72	Cysteine	2.3	B1	0.08
Sugar	1.07	Arginine	5.17	B2	0.03
Fiber	1.17-2.6	Valine	3.8	B6	0.19
Ash	1.72-3.66	Aspartate	6.33	B12	0.12
Silica	0.41	Methionine	1.04	C	0.56
Iron	0.03	Threonine	2.04		
Calcium	0.11-0.23	Isoleucine	3.23		
Magnesium	0.34	Serine	2.93		
Phosphorus	0.21-0.66	Leucine	8.4		
Sodium	0.28	Glutamate	13		
Potassium	1.31	Tyrosine	3.17		
Sulfur	0.23	Proline	3		
Carbonate	0.09	Glycine	3.5		
Copper	0.1	Phenylalanine	4.46		

Adapted from Abdullahi, 2012.

Mango seed kernel is generally 10% of the fruit, and its fat content is about 10% which can potentially yield more than 80,000 tons of mango kernel fat per year according with reports from India. The fatty acid composition of mature mango seed fat varies with the variety of mango, but essentially it is stearate and oleate rich: C16:0: 6–10%, C18:0: 24–49%, C18:1: 33–50%, C18:2: 1–10% and C20:0: 1–3%. (Bhattacharjee, et al., 2011). Kernel fat is obtained from the seed kernels of mango (*Mangifera indica*). The fat extracted from the kernel is white, solid like cocoa butter and tallow, edible, and has been proposed as a substitute for cocoa butter in chocolate. Mango fat contains 8-12% oil. The major fatty acids are oleic (38.2 – 49.7%) and stearic (33.96 – 47.8%) as shown in table 4. Refined fat is suitable as an edible fat, and as a cocoa butter extender substitute in suppository bases and in cosmetic formulations. The contents of stone in mango fruit, fat in kernel vary with the variety of fruits. On dry basis, stone comprises 4 – 14% of fruit, kernels: 61 – 83% of stone and fat: 7 - 11 % of kernels; the kernel fat composition is reported in **Table 8** (Morton, 1987).

The mango kernel meal can be used as cattle feed, manure, and for production of starch, with good export potential. Mango kernels contain 7-11% grayish-white colored fat which is suitable for edible purpose after refining. The fat is solid, closely resembling cocoa butter in physical and chemical characteristics and is therefore used as a substitute or extender. It has a good potential as edible oil. The raw fat is used in soap making. Refined fat is suitable as an edible fat, and in cosmetics formulations (Manorama group, 2013).

Mango seed kernels were shown to be a good source of polyphenols, phytosterols as campesterol, sitosterol and tocopherols. In addition, mango seed kernel could be used as a potential source for functional food ingredients, antimicrobial compounds and cosmetic due to its high quality of fat and protein as well as high levels of natural antioxidants. The mango stone obtained after decortication of mango seed can be utilized as adsorbent (Kittiphoom, 2012).

Table 8. Mango kernel fat composition

Kernel Fat	**g/100g**
Fully saturated glycerides	14.2
Mono-oleoglycerides	24.2
Di-oleoglycerides	60.8
Tri-unsaturated glycerides	0.8
Fatty Acids:	%
Myristic	0.69
Palmitic	4.4-8.83
Stearic	33.9 - 47.8
Arachidic	2.7 - 6.74
Oleic	38.2 - 49.7
Linoleic	4.4 - 5.4
Linolenic	0.5

Adapted from Morton, 1987.

Table 9. Mango kernel flour composition (Morton, 1987)

Kernel flour	g/100 g
Protein	5.56
Fat	16.17
Carbohydrates	69.20
Ash (minerals)	0.35

Mango kernel is reported to contain some tannin which exerts anti-nutritive effects in poultry and man. Boiling has been reported to be an effective method of tannin reduction. Recently, has been reported a replaced 20% of dietary maize in broiler chickens with raw or boiled mango kernel meal and observed no significant differences in performance between the control and the boiled kernel meal diets (Abdullahi, 2012)

After soaking to dispel the astringency (tannins), the kernels are dried and ground to flour which is mixed with wheat or rice flour to make bread and it is also used in puddings. Mango flour kernel reveals high content of carbohydrates (**Table 9**). The kernel flour obtained from mango fruit could be equal to rice in food value, if tannin-free.

Mango seed has been reported to contain alkaloid (0.01 ± 0.0 mg/100g), tannins (1.03 ± 0.01 mg/100 g), phytate (1.44 ± 0.01 mg/100 g), saponin (0.04 ± 0 mg/100 g) and oxalate (1.49 ± 0.01 mg/100 g). Its trypsin inhibitor activity was found to be 18.42 ± 2.54 TIU/mg proteins. Sherry mango seed does not contain cyanide which shows that it is less toxic when compared with cassava (Abdullahi, 2012).

Mango Kernel Seed Oil as Potential Biofuel

The specific gravity of oils generally characterizes the fluidity of oils in automobile engines and the value obtained for mango kernel seed was of 0.885 which is acceptable for biodiesel production (**Table 10**). The pH of the mango (5.44) is generally low which indicate presence of a reasonable amount of free fatty acids in the oil and this is a potential characteristic of its utilization as fuel oil. The retention factor (RF) of the seed oil is 3.78 which showed appreciable degree of purity of the oil.

The iodine value (IV) indicates the level of unsaturation in oil, the values for mango seed oil was 76.40 mg Na_2SO_3 g-1, which was below 100 mg Na_2SO_3 g^{-1} thus classified it as non-drying oil suitable for lubricant industry. Moreover, this value was lesser than the lowest IV of oils obtained from peanut kernel, cottonseed and rapeseed (**Table 11**). Furthermore, the low iodine value (IV) of mango seed oil indicates their potential as non-drying oil for industry (Agbede, et al., 2012). The acid value (AV) of oil indicates the content of free acid in the oil samples and it is an important property that influences the shelf life of the oil. The AV of mango seed oils was 0.32 mg $KOHg^{-1}$, higher than the AVs of cottonseed (0.07 mg KOH g^{-1}), peanut kernel (0.20 mg KOH g^{-1}), rapeseed (1.14 mg KOH g^{-1}) (**Table 11**).

Table 10. Mango seed oil characteristics

Oil characteristics	Mango seed oil
pH	5.44
Refractive Index at 25°C	1.45
Retention factor (cm)	3.78
Congealing Temperature (0°C)	-18 to -30
Flame nature	Slightly sooty
Product yield (%)	38.5
Specific gravity at 20°C	0.885

(Agbede, et al., 2012).

High AV favors the choice of oil seeds for the production of biodiesel because most oils with low AV are mainly edible oils fit for human consumption and food processing industries. Generally, the maximum AV recommended for biodiesel is 0.5 mg KOH g^{-1} and the AV of biodiesel to be produced from these oils is expected to be close to the values of their raw oils. Peroxide Value (PV) measures the rancidity of an oil sample and 10 mg Na_2SO_3 g^{-1} has been accepted as convectional for majority of the oils. The PV of mango seed oils was 10.08 mg Na_2SO_3 g^{-1} and compared very well with PV of sunflower oil 10.7 mg Na_2SO_3 g^{-1}which has been sourced as oil seed for the production of biodiesel. Flash point (FP) of oil is the minimum temperature at which the oil becomes flammable and high value of FP suggests that the oil is less susceptible to fire related hazard in transportation, storage and utilization. The FP of seed mango (173°C) is generally lower than other seed oils shown in Table 11, which indicate that they are safer to handle. The acceptable FP of biodiesel ranges between 130-131°C so mango seed oil compare well with most seed oils listed in Table 11. Heat of combustion (HC) of oil sample indicates the energy contents of the oil when subjected to burning and related to its maximum power output (Song, 2000). The heat of combustion of the mango seed was 34 583 KJ kg^{-1}, lesser than the values obtained for similar seed oils as well as the range (37 270 - 40 480 KJ kg^{-1}) stipulated for most. This suggests that the mango seed oil can be used in engines designed for lesser tasks particularly those involving less torque. The saponification value (SV) of the mango oil was 143 mg KOH g^{-1}, which is similar to trend observed in the heat of combustion. Oils with high saponification values, like mango seed oil, are generally suitable for soap production.

Table 11. Iodine value of some plant oils used for production of biofuel

Oil	IV (mg N2SO3 g^{-1})	HC (KJ kg^{-1})	FP (°C)	AV (mg KOH g^{-1})	PV (mg N2SO3 g^{-1})	SV (mg KOH g^{-1})
Cottonseed	90-119	39468	234	0.07	64.8	207.7
Peanut kernel	80-106	39782	271	0.2	82.7	199.8
Rapeseed	94-120	39709	246	1.14	30.2	197.7
Mango seed	76.4±0.1	34583	173	0.32	10.08	143

Adapted from Agbede, et al., 2012.

Table 12. The phytochemical substances in the crude methonolic mango seed kernel extract

Phytochemicals	Result
Benedict's test (to determine reducing sugars)	Presence of glycosides
Frothing test for saponins	Presence of saponins
Borntrager's test(anthraquinone derivatives)	Absence of anthraquinone derivatives
Flavonoid test	Flavonoids present
Ferric Chloride test for tannins	Presence of tannins
Test for alkaloids	Presence of alkaloids

Adapted from Kaur, et al., 2010.

SOME PHARMACEUTICAL USES OF MANGO SEED COMPONENTS

Mango seed kernels contain various phenolic compounds and can be a source of natural phenolic compounds and natural antioxidants. Mango seed kernel meal was found to contain tannic acid, gallic acid and epicatechin (Kaur, et al., 2010). Tannins interact with salivary proteins and glycoproteins in the mouth and render the tissues astringent to. Astringency gives tannin the medicinal value in preventing diarrhea and dysentery and for controlling hemorrhage. Furthermore, tannins protect plant against dehydration, rotting, damage by animals and pathogens. When they are polymerized, insoluble protective barrier is formed which prevents microbial attack. Therefore they can be applied to wounds as protective coating. Some studied had reported that the symptoms of continued intake of tannin include gastritis as well as irritation and edema of the intestine. It has been reported that feeding 0.5% of tannic acid decreased nitrogen retention and caused 5% mortality in rats. Pure tannins extracted from mango kernels have proven inhibitory effects against various strains of bacteria including *Listeria monocitogenes*. Some research focuses on a way to recycle wood-like mango kernels, which are usually thrown away or burned. By processing the kernels for their tannins, businesses have a way to completely utilize all fruit parts and therefore increase their profit. Currently, mangos are one of the main fruits marketed globally, ranked fifth in world production among the major fruit crops (Abdullahi, 2012).

The phytochemical qualitative studies indicated the presence of an array of phytochemicals including saponins, flavonoid, glycosides, tannins and alkaloids (Table 12). Natural antioxidants are preferred in cosmetic and food processing industry (Kaur, et al., 2010).

The results of antibacterial activity studies showed that crude methanolic extract of single mango seed kernel exhibited higher antibacterial activity against methicillin resistant *Staphylococcus aureus* followed by *Escherichia coli* and *Vibrio vulnificus*. Similarly when comparing the antibacterial activity of the crude methanolic extract of mixed mango seed kernels, methicillin resistant *S. aureus* is highly sensitive followed by *E. coli* and *V. vulnificus*. Not many reports are available on the medicinal properties of mango seed kernel. Mango seed kernel is one of the by-products of food processing industry and is not commercially exploited, but is discarded as waste. The antibacterial activity of crude methanolic extracts of mango seed kernel is significantly higher compared to the commercial antibiotic chloramphenicol. However, chloramphenicol exhibited a higher bactericidal activity against *V. vulnificus* compared to the mango seed kernel extract as has been reported in a preliminary investigation on the antibacterial activity of mango (Mangifera indica L. (Kaur, et al., 2010). In some studies it have been demonstrated that dietary doses of mango seed kernel stimulated the immunity and rendered *Labeo rohita* more resistant to *Aeromonas hydrophila*. (Sahu, et al., 2007). The possible mechanism of disease control may be through microbicidal as well as activation of immune system. Further studies in this area have already been initiated using mixed mango seed kernels with different solvents, to elucidate the horizon of antibacterial activity and the structural information of the biomolecules (Kaur, et al., 2010).

CONCLUSION

During the last years it has been a significant increase in the garbage produced by food industry, for instance the waste produced by fruit processing industries which annually discarded tones of organic waste that certainly could be used as secondary product. Overall, end products have low added value in the market; therefore there is a high wastage raw material unused. Nowadays it has been an increase in the use of raw materials in order to add them an extra value and at the same time reducing the high amount of garbage wasted. Product value increases as it is transformed into products throughout the process. Adding value to the material as it is processed results in higher value to the market. This is the case of mango seeds, which effectively could be used as an alternative in the production of biodiesel; also it is already known its use as antioxidant in food since the oil extracted is edible and can be used for human consumption. Otherwise, the kernels are also used to fed poultry and cattle. Despite of this, due to its phytochemical content, some other uses have been adjudge to the kernel, so it could be used as vermifuges and as astringents in diarrhea, hemorrhages, bleeding hemorrhoids and in antimicrobial treatments since it has shown antibiotic activity. All these applications of the mango seed have enhanced high opportunities principally for developing producer countries. The utilization of sub-products and wastes has opened the interest on the research of processes associated with the use of mango seed oil. The research has been mainly focused on the chemical seed content and its phytochemicals component for applications in the food and pharmaceutical industries. In this context various applications have been proposed for the seed oil extraction process used in the biodiesel production. Thus, in this chapter it is provided basic information and a brief revision about mango seed in order to visualize future trends, potential uses and applications of the mango seed.

REFERENCES

Abdullahi I. (2012). Nutritive value of differently processed mango seed kernel (local variety (Kanbiri) and mixed samples) meal in broiler diets. *Thesis of Master of Science of Ahmadu Bello University*, Zaria, Nigeria.

Agbede, O. O., Alade, A. O., Adebayo, G. A., Salam, K. K. & Bakare, T. (2012). Evaluation of chosen fruit seeds oils as potential biofuel. *International Agrophysics*, *26*, 199-202.

Bhattacharjee, A., Ghosh, S. K., Neogi, K., Aich, A., Willard, B., Kinter, M., Sen, S. K., Ghosh, D. & Ghosh, S. (2011). Deposition of stearate-oleate rich seed fat in Mangifera indica is mediated by a Fat A type acyl-ACP thioesterase. *Phytochemistry*, *72*, 166–177.

Board of Investments (BOI). Philippine mango seed oil. Industry & Investments. http://www.boi.gov.ph/pdf/industryprofiles. Accessed in February 2013.

Evans, E. A. (2007). Recent trends in world and U.S. mango production, trade, and consumption, Institute of Food and Agricultural Science, Universit of Florida. EDIS document FE718, 1-7. Available in http://edis.ifas.ufl.edu. Accessed in February 2013.

FAO, (2011). Food and Agricultural commodities production. http://faostat.fao.org. Accessed in February 2013.

Guzmán, E. C., Alcalde, B. S., Mosqueda, V. R. & Martínez, G. A. (1996). Ecuación para estimar el volumen y dinámica de crecimiento del fruto de mango cv. Manila. *Agronomía Tropical*, *46*, 395-412.Heidelberg: Springer.

INEGI. (2008). Encuesta Industrial Mensual. Instituto Nacional de Estadística y Geografía. México. http://www.inegi.org.mx/prod_serv/contenidos/espanol/bvinegi/productos/encuestas/estable cimientos/eim/2007/2007_1.pdf. Accessed in February 2013.

Infoagro. *El cultivo del mango*. http://www.infoagro.com/frutas/frutas_tropicales/mango.htm. Accessed in February 2013.

Kaur, J., Rathinam, X., Kasi, M., Leng, K. M., Ayyalu, R., Kathiresan, S. & Subramaniam, S. (2010). Preliminary investigation on the antibacterial activity of mango (Mangifera indica L: Anacardiaceae) seed kernel. *Asian Pacific Journal of Tropical Medicine*, 707-710.

Kittiphoom, S. (2012). Utilization of Mango Seed. *International Food Research Journal*, *19*, 4, 1325-1335.

Manorama Group. Mango. http://manoramagroup.co.in/speciality_fats_mango.html. Accessed in February 2013.

Martínez, F., Tijerina, C., Arteaga, R., Vázquez, P. & Becerril, R. (2007). Determinación de zonas agroclimáticas para la producción de mango (Mangifera indica L. "Manila") en Veracruz, *México. Investigaciones Geográficas, Boletín del Instituto de Geografía, UNAM*, *63*,17-35.

Morton, J. (1987). Mango. In: Julia F. Morton, Fruits of warm climates (221–239), Miami, FL. USA, Julia F. Morton.

Norma Oficial Mexicana NOM-129-SCFI-1998, Información comercial de etiquetado de productos agrícolas Mango.

Nzikou, J. M., Kimbonguila, A.,Matos, L., Loumouamou, B., Pambou-Tobi, N. P. G., Ndangui, C. B., Abena, A. A., Silou, Th., Scher, J. & Desobry, S. (2010). Extraction and Characteristics of Seed Kernel Oil from Mango (*Mangifera indica*). *Research Journal of Environmental and Earth Sciences*, *2*, 1, 31-35.

Ornelas-Paz, J. De. J., Yahia, E. M. & Gardea, A. (2007). Identification and quantification of xanthophyll esters, carotenes and tocopherols in the fruit of seven mexican mango cultivars by liquid chromatography-APcI+-time of flight mass spectrometry. *Journal of Agricultural and Food Chemistry*, *55*, 6628-6635.

Pandharipande, S. L. & Deshmukh, A. R. (2012). Synthesis, characterization and adsorption studies for adsorbent synthesized from aegel marmelos shell for removal of cr (vi) from aqueous solution. *International Journal of Advanced Engineering Research and Studies*, *2*, 95-96.

Rizzo Pastor, P. (1998). *El Mango Mexicano en Cifras*. http://www.sica.gov.ec/agronegocios. Accessed in February 2013.

Ruíz Ramírez M. (2007). *Cultivo del mango*. http://www.agrobanco.com.pe. Accessed in February 2013.

Sahu, S., Das, B. K., Pradhan, T., Mohapatra, B. C., Mishra, B. K. & Sarangi, N. (2007). Effect of Mangifera indica kernel as a feed additive on immunity and resistance to Aeromonas hydrophila in Labeo rohita fingerlings. *Fish & Shellfish Immunology*, *23*, 109-18.

Sumaya-Martínez, M. T., Sánchez-Herrera, L., Torres-García, G. & García-Paredes, G. (2012). Value chain of mango and its byproducts based in the nutritional and functional properties. *Revista Mexicana de Agronomos*, 826-833.

Vasanthaiah, H. K. N., Ravishankar, K. V. & Mukunda, G. K. (2007). Mango. In: Kole, C., Genome Mapping and Molecular Breeding in Plants - Fruits and Nuts: Volume *4* (303-323). Berlin, Germany: Springer-Verlag Berlin Heidelberg.

Velan, M., Krishnan, M. R. V. & Lakshmanan C. M. (1995). Conversion of mango kernel starch to glucose syrups by enzymatic hydrolysis. *Bioprocess Engineering*, *12*, 323-326.

In: Seeds as Functional Foods and Nutraceuticals
Editors: R. Mora-Escobedo, J. De J. Berrios et al.
ISBN: 978-1-62808-489-4

Chapter 16

Changes in Nutraceutical and Functional Properties of Fruit Seeds due to Different Treatments

***María Elena Ramos-Cassellis*[1,2], *María Elena Sánchez-Pardo*[1*], *Rosalva Mora-Escobedo*[1] and *Eliseo Cristiani Urbina*[1]**

[1]Departamento de Ingeniería Bioquímica, Escuela Nacional de Ciencias Biológicas, Instituto Politecnico Nacional. Prol. Carpio y Plan de Ayala S/N, Col. Sto. Tomás, D.F. México

[1,2]Facultad de Ingeniería Química, Benemérita Universidad Autónoma de Puebla, Puebla, Mexico

Abstract

Dietary bioactive compounds or nutraceuticals from different functional foods can ameliorate or even prevent diseases. Nutraceuticals and functional foods constitute a great promise to improve health and prevent aging-related chronic diseases. Fruit seeds such as avocado, mango and tamarind have much higher antioxidant activity and phenolic content than the edible portions. The contribution of some the fruit seed fractions to the total antioxidant activity and phenolic content is 70 % higher than the edible portions. However it is necessary to know the effect of different treatments in this kind of fruit seeds. The total phenolic content in extracts of mango seed kernel powder products is higher when the heating temperature has been increased. The microwave treatment cleaved and liberated phenolic compounds, hence resulting in the increase of free phenolic compounds and enhancement of antioxidant capacity. Prolonged ultrasound treatment at high power levels may induce chemical decomposition of anthocyanins. Ohmic heating processes when was performed with low voltage gradients, promoted degradation of both the ascorbic acid and the total vitamin C in a manner similar to conventional heating. In most cases, the process affects nutraceuticals and functional properties of foods. Due to these facts it is necessary to cluster the recent researches in order to know which process will be more beneficial to human health.

---

* Corresponding author: Email: alimentoselena@hotmail.com.

INTRODUCTION

Before considering the changes in nutraceutical and functional properties of fruit seeds due to different treatments it is necessary to review briefly the morphological and anatomic aspects of embryo and storage tissue. By definition, the embryos of the matured seeds of the dicots possess two cotyledons, whereas there is only one in monocots (Bewley, et al., 2012). Seeds are very diverse in their shape and size (Figure 1). In the mature state they contain an embryo, the next generation of plant, surrounded by a protective structure (the seed and/or fruit coat) and, in species in which the nutritive reserves are not stored within the cotyledons, by an alternative storage tissue (endosperm, perisperm, or megagametophyte). Most seeds contain large and characteristic quantities of polymeric reserves. These reserves are vital components of human and animal diets, and their production in crops is a basis of agriculture. Seeds also contain proteins that have been implicated in the improvements in cardiovascular health of regular consumers. Some seeds like soybeans are source of several biactives compouns such as the isoflavones that act as phytoestrogens because of structural and functional similarities to estradiol (Aluko, 2012). These compounds are called nutraceuticals and may prevent some diseases.

In recent years the trend in the market for nutraceuticals has increased, and further increase in the development and consumption of these products is forecasted, particularly for products that include in their formulation functional and natural food additives as essential oils, vitamins, minerals, carotenoids, flavonoids, anthocyanins, total antioxidant capacity and phenolic content (Torres-Giner, et al., 2010), with high nutritional value, and that may contribute to prevent cardiovascular, respiratory and carcinogenic diseases, among others (Shahidi, 2009).

Figure 1. Seeds diverse in their shape and size contain an embryo, a protective structure (the seed and/or fruit coat) and the cotyledons.

1. Nutraceuticals in Seeds

1.1. Definitions of Nutraceutical and Functional Foods

Nutraceuticals are chemicals found as a natural component of foods or other ingestible forms that have been determined to be beneficial to the human body in preventing or treating one or more diseases or improving physiological performance. A good example is a group of compounds called isoflavones that are isolated from soybean seeds and packaged into pills that women can use instead of synthetic compounds during hormone replacement therapy.

Functional foods are natural or formulated, which will enhance physiological performance or prevent or treat diseases and disorders. Functional foods include those items developed for health purposes as well as for physical performance. They are consumed as part of a usual diet, but apart from supplying nutrients they can reduce the risk of chronic diseases such as cancer, hypertension, and kidney malfunction, among others **(Aluko, 2012).**

1.2. Phenolics, Flavonoids, and Antioxidant Activities in Seeds

Plant cells produce many molecules that either do or do not seem to have direct relevance to the processes of growth and development. Those that do are often referred to as primary metabolites and include amino acids, chlorophyll, nucleotides, simple carbohydrates, and membrane lipids. Meanwhile secondary metabolites cannot be directly linked to processes as photosynthesis respiration, solute transport, translocation, and nutrient assimilation (Verhoeven, et al., 1997). Many of this secondary metabolites are involved in defense operations that help protect a plant against hervibores and insects. Other secondary metabolites may be involved in plant-to-plant competition and/or attractants for pollinators and animals.

Phenolic compounds are secondary metabolites, ubiquitous in the plant kingdom and impacting on human health, decreasing the risk of several diseases by reducing oxidative phenomena. They are important in fruits and vegetables, to which they contribute color and flavor, also being involved in astringent and bitter tastes (Heim, et al., 2002).

The total polyphenol content and antioxidant capacity vary considerably from one kind of fruit to another. Antioxidant activity and phenolic content of both edible portions and seeds of avocado, jackfruit, longan, mango and tamarind can be measured by the use of the pre-formed radical monocation of 2,2'-azinobis-(3-ethylbenzothiazoline-6-sulfonic acid (ABTS*+); ferric reducing antioxidant power (FRAP) index.

Antioxidants are substances that delay or inhibit the oxidation of foods, and are therefore of great interest to food scientists. Antioxidants are naturally present in seeds. Therefore, additional quantities are added to control oxidation, increase shelf life, and improve overall quality. The mechanism by which this occurs is termed free radical termination, and is accomplished through the donation of an electron or hydrogen atom (Haworth, 2003).

Inglett & Chen, (2011) investigated contents of phenolics, flavonoids, and antioxidant activities in skin, pulp, and seeds of Miracle Fruit, these fruits are a good source not only for flavor and color, but also antioxidants for functional food applications. They said that the free phenolic content in skin was almost 3 times of that in pulp and 4 times of that in seeds. The antioxidant activities in skin and pulp were comparable, and were significantly higher than that in seeds.

Maisuthisakul & Gordon, (2009) studied the antioxidant and tyrosinase inhibitory properties of extracts of mango seed kernel (*Mangifera indica* L.), which is normally discarded when the fruit is processed. The researchers said that the extract from mango seed kernel had high total phenol content, radical scavenging, metal-chelating and tyrosinase inhibitory activities. This extract may be suitable for use in food, cosmetic, nutraceutical and pharmaceutical applications. Besides the pulp and the peel, mango seed kernels are equally rich in polyphenols with potent antioxidative activity, but ironically the seeds are always discarded as waste during processing and consumption of the mango fruit. The extracts did not cause acute irritation of rabbit skins. The content of phenolic compounds was slightly lower than those found in extracts of mango seed kernel extracted with methanol and water (50:50) by Soong & Barlow (2004) which were expressed as gallic acid equivalents.

Soong & Barlow (2004) assayed the antioxidant activity of a variety of fruit seeds, namely, mango, jackfruit, longan, avocado, and tamarind and found that the antioxidant activity of the mango seed kernel was the highest, a fact attributed to its high polyphenolic content. Natural antioxidants are known to be safe and have potential nutritional and therapeutic value. Taking into account these results the mango seed kernel could utilized in the industry as a functional food ingredient.

Ellagic acid has previously been reported in mango seeds. Gallic acid and ellagic acid exist either in the free form or bound as gallo-tannins and ellagitannins, respectively. Hydrolysable tannins are easily hydrolyzed *in vivo* by the action of acid and/or enzymes, releasing gallic acid and ellagic acid units. Gallic acid is known to have anti-inflammatory, antimutagenic, anticancer, and antioxidant activity (Masibo & He, 2008).

Ahmed, et al., (2007) identified and quantified various polyphenolic compounds in the mango seed kernel: tannin 20.7 mg/100 g, gallic acid 6.0 mg/ 100 g, coumarin 12.6 mg/100 g, caffeic acid 7.7 mg/100 g, vanillin 20.2 mg/100 g, mangiferin 4.2 mg/100 g, ferulic acid 10.4 mg/100 g, cinnamic acid 11.2 mg/100 g, and unknown compounds 7.1 mg/100 g. The total polyphenolic content of the mango seed kernel extract was estimated to be 112 mg (GAE)/100 g.

Dietary intake of ellagitannins (tannins) is largely explained by the consumption of strawberries, raspberries and blackberries. A fraction of the ellagitannins ingested is hydrolysed in the gut and the resulting ellagic acid absorbed and metabolized. It is largely contained in the seeds, 88% (Clifford & Scalbert, 2000).

The seeds have another nutraceutical compounds, the combination of both mango seed extract and oil have higher antioxidant capacity than each one alone. It has been suggested that mango seed oil can be used as a natural food supplement due to its fatty acid pattern rich in saturated fatty acids and monounsaturated oleic acid besides the presence of different sterol fractions as D-avenasterol, fructosterol, phytosterols, and squalene (Masibo & He, 2008).

Ali & Yong-Lee, (2007) said that extracting solvent significantly affected total phenolics and antioxidant contents extracted from date seeds, they observed that acetone (50%) was the most efficient solvent for phenolic extraction; whereas, butanone was the most efficient solvent for purifying phenolic extract. Although, they indicated acetone is regarded as a low toxicity solvent, safety concerns associated with the use of organic solvents for extractions include solvent residues in the product, handling, and recovery of waste solvents; therefore, safer and more cost efficient techniques to separate phenolics from date seeds are required.

The chia fiber-rich fraction exhibited high antioxidant activity (488.8 mmol TE/g FRF), which confirms the presence of polyphenols. The physicochemical properties described here indicate the chia fiber-rich fraction to be a potential ingredient in health and diet food products such as powders, nutrition bars, breads and cookies (Vázquez-Ovando, *et al.*, 2009)

Chia seeds provide numerous health benefits and they are also a potential source of biologically-active (bioactive) peptides. Enzymatic hydrolysis is natural and safe, and effectively produces bioactive peptides from a variety of protein sources, including chia seeds (Segura-Campos, *et al.*, 2013).

1.3. Carotenoids and Dietary Fibre

Carotenoids found in the human diet are primarily derived from plants and are found in roots, leaves, shoots, seeds, fruit, and flowers. More than 600 carotenoid compounds have been characterized; 50 of these are consumed in the human diet. Approximately 12 carotenoids account for most of the dietary intake, and they are found in measurable concentrations in human blood and tissues; the most common are lycopene, lutein, α-carotene, β-carotene, β-cryptoxanthin, and zeaxanthin. Due to the presence of both carotenoids and polyphenols, mangos represent a rich source of antioxidants most of which have been shown to be of nutraceutical importance with respect to disease prevention. The dietary fibre and phenolic concentrates obtained from this process could be used as natural additives for fortified food products to enhance their functional properties (Voutilainen, et al., 2006).

Due to the presence of both carotenoids and polyphenols, mangos seeds represent a rich source of antioxidants most of which have been shown to be of nutraceutical importance with respect to disease prevention. The dietary fibre and phenolic concentrates could be used as natural additives for fortified food products to enhance their functional properties (Ali & Yong-Lee, 2008).

2. Effect of Different Treatments on Nutraceutical Components

2.1. Effect of Heat Treatment on Polyphenolic Content and Antioxidant Activity

Chamorro, et al., (2012) studied the effect of heat treatment on polyphenolic content and antioxidant activity in grape pomace and grape seed extract, which were heat treated in a furnace or autoclaved at 100 °C for 15, 30 and 60 min. In general, this treatment did not affect the total extractable polyphenols content, tannin content, procyanidin components and the antioxidant activity, neither the antioxidant activity.

The citric fruits seeds are one of the principal residues in the juice industry and their utilization can decrease significantly the problems of their final disposal. Hernández-Montoya, et al., (2009) studied the thermal degradation of three Mexican citrus seeds: orange (*Citrus sinensis*), lemon (*Citrus limon*) and grapefruit (*Citrus paradisi)* in nitrogen atmosphere. The two components (embryo and husk) of the seeds were characterized separately. The results showed that the thermal effects are very similar between the three embryos and the three husks.

Trugo, et al., (2000) evaluated the effect of heat treatment of germinated soybean, lupin, and black bean on chemical composition and protein utilization in rats. Heat treatment caused complete inactivation of trypsin inhibitors whereas it did not affect phytic acid levels. They concluded that the use of germinated heat-treated soybean, lupin, and black bean on their own and/or as food ingredients is nutritionally advantageous.

Soong & Barlow, (2004) found that the total polyphenols in extracts of mango kernel powder products increased after heating and they attributed this to the formation of phenolic substances and maillard type complexes which have antioxidative activity under milder heating temperature. This increase is correlated with the augmentation in antioxidant activity of heated mango kernel powder. Therefore, they suggested that mango kernel powder can be used in the functional food industry due to their strong antioxidative potential and associated stability to elevated temperatures of up to 160°C but recommended further studies on related toxicity.

Threlfall, et al., (2005) studied the effects of hot and cold pressing on yield, quality components, and nutraceutical content of juice, grapes, and press fractions (seeds and skins) from Black Beauty (*Vitis rotundifolia*) and Sunbelt (*Vitis labruscana* L.) grape cultivars, obtaining that seeds and skins had higher levels of phenolics and oxygen radical absorbance capacity than the whole grapes and juice. The dried seeds had more phenolics and less anthocyanins than the skins. The highest total phenolic level (95338 mg/kg) equivalents and oxygen radical absorbance capacity level (1100 μmol trolox/g) was in Black Beauty seeds from nonheated must.

Ahmed, et al., (2007) found that the antimicrobial activity of the mango kernel extracts was stable against heat (121°C, 15 min), freezing (–20°C, 16 h) and pH treatment (pH 3–9), conditions normally used in food processing. Zein et al. (2005) published that soaking and boiling treatments had a great impact in reducing of the anti-nutritional factors in mango kernel. The kernel flour after mixing with wheat or maize flour is used in making chapatti in India where also, mango kernel is traditionally roasted and eaten by the tribal people.

Carpenter, et al., (2007) incorporated grape seed extract (GSE) and bearberry (BB) to cooked pork patties which sensory properties were unaffected by the addition. Antioxidant activity of GSE and BB was observed in cooked pork patties demonstrating the thermal stability of GSE and BB. Results obtained demonstrate the potential for using health promoting nutraceutical in meat and meat products.

Siddhurajub & Beckera, (2007) studied two varieties of cowpea (*Vigna unguiculata*) seeds: raw, dry heated and hydrothermal treated samples were extracted with 70% acetone and the extracts were freeze-dried. The unprocessed light brown seeds (LB) contained significantly higher level of total phenolics and tannins than the dark brown seeds (DB). The superoxide anion radical scavenging activity was found to be significantly higher in the raw and dry heated seed extracts than the hydrothermally processed seed samples of the respective varieties. Dry heated samples of LB and DB showed the highest hydroxyl radical scavenging activity of 83.6% and 68.2%, respectively.

In Nigeria, mango kernel was processed into an edible flour, Arogba, (1999) worked with fresh mango (*Mangifera indica*) kernel (Figure 2) which was processed by soaking, sulphiting, blanching, drying and grinding into a powder. Processed mango kernel flour was incorporated into four grades of biscuits containing 50±100%. The 50:50 biscuits grades were preferred by panelists for colour, texture, flavour and overall acceptability. Other grades were favored only for the brown coloration, probably due to the contribution of the residual tannin in processed mango kernel flour. The preferred biscuit had optimum protein supplementation and an increased 6.5% energy content compared with whole-wheat biscuit.

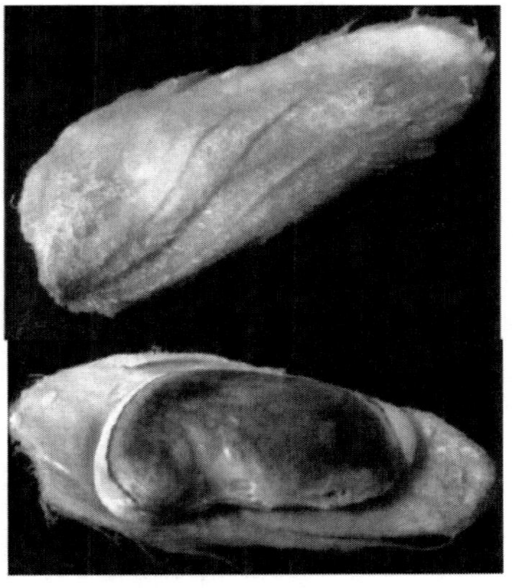

Figure 2. Mango kernel is traditionally roasted and eaten by the people in Nigeria.

2.2. Microwave Treatment

Microwave heating is an emerging technology in food processing, which bases its preservation properties in thermal reduction of the microbial counts in foods. Industrial applications of microwave heating are found for many heat treatment operations in the food-processing industries.

Results indicated that, in general, the microwave treatment cleaved and liberated phenolic compounds, hence resulting in the increase of free phenolic compounds and enhancement of antioxidant capacity of the citrus pomace extracts (Hu, et al., 1994) however, microwave treatment, due to its simplicity and rapidity could be more effective in releasing antioxidant compounds from agricultural byproducts, such as citrus pomace. A proper and reasonable microwave treatment should be carefully considered to enhance the antioxidant capacity of citrus. Pomace is the residue remaining when fruits are processed for juice, wine, or other products. Many studies have reported that fruit pomaces, although not only seeds, are rich sources of phenolic compounds and these byproducts, obtained from the juice and wine industry, might be useful raw materials for creating new value-added products (Zhou et al., 2009).

Hayat, et al, (2010) indicated that, in the free phenolic acid fraction, the sum of the individual phenolic acid contents increased significantly ($P < 0.05$) with the microwave power. Five phenolic acids, namely, gallic acid, p-hydroxybenzoic acid, vanillic acid, p-coumaric acid and ferulic acid, were identified and quantified from pomaces of two citrus mandarin cultivars. The microwave treatment of citrus pomace cleaved and liberated phenolic compounds, hence resulting in the increase of free phenolic compounds and enhancement of antioxidant capacity of the extracts. Microwave treatment, due to its simplicity and rapidity, could be more effective in releasing antioxidant compounds from agricultural by products, such as citrus pomace, on an industrial scale.

Rapeseed was treated with microwaves in order to investigate the possibility of enhancing oil extraction yield, oxidative stability and nutraceutical content, this seed was pretreated with microwaves for two different times (2 min and 4 min) and oil was extracted with a press, Azadmard-Damirchia et al., (2010) showed that microwave pretreatment of rapeseed can increase the oil extraction yield (by 10%), phytosterols (by 15%) and tocopherols (by 55%) of the oil extracted by press. Oil extracted from untreated rapeseed by press had the lowest oxidative stability (1 h); this was increased to 8 h by pretreatment of rapeseed with microwaves. They concluded that treat rapeseed with microwaves it gives a relatively good recovery of oil, with a high amount of nutraceutical.

2.3. Radiofrequency Electromagnetic Fields (RF-EMFs)

Although assessment of the RF-EMFs impact on plants is of great importance because plants have an important role in the living world as main primary producers of organic compounds and oxygen, only a few investigations of RF-EMFs effects on plants have been reported. Exposure to RF-EMFs induced a wide range of mitotic and chromatin abnormalities in the root tips of *Allium cepa* as compared with the control. At 400 MHz, only exposure to field of higher strengths and to modulate fields significantly increased the percentage of cells with aberrations while at 900 MHz all investigated exposure treatments increased mitotic and chromatin disturbances. Results show that exposure to RF-EMFs at both frequencies did not significantly change the seed-germination rate and early root growth in comparison with the control, indicating that exposure to RF-EMFs under the test conditions did not have an inhibitory effect on *Allium* (Tkalec, et al., 2009).

2.4. Ultrasound Processing

Natural antioxidants are known to be safe and have potential nutritional and therapeutic value. Ultrasonic radiation has been known for many years to be capable of disrupting, and therefore inactivating, vegetative cells. Recent research has shown that ultrasonic radiation is synergistic with mild heat, particularly if the hydrostatic pressure is slightly raised also (Gould, 1995).

The color of grape juice is a mixture of various colors depending upon the maturity stage of the grape berry and cultivar. It is mainly influenced by the presence of various anthocyanins. For example, pelargonidins are primarily responsible for the orange salmon, pink and red color of the fruit, cyanidins are primarily responsible for magenta and crimson colors, while purple, mauve, and blue colors are due to the presence of delphinidins. Combinations of various color parameters such as L*a*b*, L*a*/b*, L*/ a*b*, L*b*/a* values should be considered as the physical parameters to describe the visual color change of grape juice. Various color combinations have been reported to represent color degradation during combined thermal and high pressure processing of strawberry juice and tomato puree and ozonation of strawberry juice. Thereby the factors considered in this study caused a specific effect on the degradation of color (Rodrigo, et al., 2007). Higher retention of anthocyanins was observed in sonicated grape juice compared to sonicated strawberry (Tiwari, et al., 2008) or blackberry juice (Tiwari, et al., 2010). However prolonged treatment at high power levels may induce chemical decomposition of anthocyanins.

2.5. Effect of Space Flight Factors on Alfalfa Seeds

To explore the effect of space flight factors on the early development of alfalfa seedling, dry seeds were placed on board a satellite for a 15-day flight. A 6.2% increase in germinate potential and an 80% decrease in the number of hard seed in flight seeds. Meanwhile, negative effects included a decrease of 3.0 and 33.2% in the index of germination and vigor of flight seeds, respectively, which may be partly due to the inhibition of cell mitotic (26% less than ground control) and root growth (29.0% less than ground control) after the space flight. Moreover, the DNA and Ca^{2+} content of alfalfa seeds increased after the space flight, while the reserve energy content of alfalfa seeds, such as saccharine and fatty acid, decreased after the space flight. Conclusively, space flight factors accelerate the germination process of alfalfa seeds but restrain the root from growing due to chromosomal damage and abnormal mitosis induced by cosmic radiation (Ren, et al., 2010).

2.6. Effect of Different Chemical and Physical Treatments

Kimura, H., et al. (2006) studied natural seeds of Japanese horse chestnut (*Aesculus turbinata* Blume) these natural seeds were treated with wooden ashes to remove harshness, and they informed that these seeds contain large amounts of mixed triterpenoidal saponins called escins, which have several biological activities including anti-inflammatory action and inhibitory effects on the absorption of ethanol and glucose; including deacetylescins and desacylescins from edible seeds as novel sources for nutraceutical foods with anti-obese effects.

The effect of glucuronosylation on the color stability of anthocyanins was investigated using glucuronosylated anthocyanins isolated from the flower petals of the red daisy (*Bellis perennis*) or obtained by enzymatic *in vitro* synthesis using heterologously expressed red daisy glucuronosyltransferase BpUGT94B1. Color stability toward light and heat stress was assessed by monitoring CIELAB color coordinates and stability at pH 7.0 by As. (550). A similar increase in color stability was not observed following heat treatment. Glucuronosylation of approximately 50% of total anthocyanins proceeded fast and resulted in increased color stability in response to both heat and light. The data show that glucuronosylation may be used to stabilize industrially used extracts of natural colorants (Osmani, et al., 2009).

A germinated barley foodstuff (GBF) containing glutamine-rich protein and hemicellulose-rich fiber was made from brewer's spent grain, by physical isolation. The patients with mild to moderate, active ulcerative colitis, who had been unresponsive or intolerant of standard treatment received 20–30 g GBF, feeding daily. At 4 weeks, this treatment resulted in a significant clinical and endoscopic improvement. These results indicate that GBF feeding is a potentially new, attractive prebiotic treatment in patients with ulcerative colitis (Bomba et al., 2002).

Researchers report on the terrestrial growth of wheat seeds after exposure to microgravity and cosmic radiation in space. Some groups of experimental and control seeds were treated with cysteine before flight. The germination of space-flown seeds was not different from ground controls. Cysteine promoted growth in both experimental and control groups of seeds. Roots tips of seedlings grown from experimental seeds exhibited aberrant cells; preflight treatment with cysteine decreased the number of aberrant cells (Gu & Shen, 1989).

2.7. Effect of Abiotic Stresses

Abiotic stresses such as wounding and phytohormones may lead to an increased phenolic concentration with health-promoting properties when applied to tissues. When wounding is applied to purple-flesh potato or orange carrot tissues, there is an increase in phenolic content and antioxidant capacity after 48. In sliced potatoes the increase in total phenolic content is approximately 76.7%, with a corresponding increase of antioxidant capacity of about 95.5%. For sliced carrots the increase in phenolic content is about 35.7%, whereas the antioxidant capacity increases by 16.6%. The phenolic compounds induced in both tissues differ in their specific antioxidant activities (defined as the ratio between antioxidant capacity and phenolic content of tissue (Cisneros-Zevallos, 2003).

Pagamas & Nawata, (2008) investigated the sensitivity of developmental stages to high temperature (heat stress) in chili pepper (*Capsicum annuum* L. var. Shishito). Plants were subjected to heat stress (38/30°C-8°C day/night) immediately after anthesis for 5 or 10 days, or from 10 to 30 days after anthesis (DAA), from 30 DAA until harvest of the seeds, or immediately after anthesis until harvest of the seeds. Control plants were grown at 30/22°C-8°C (day/night). Exposure to heat stress affected fruit growth, seed yield, and seed quality in chili pepper. The early stage of seed development from anthesis until 10 DAA was sensitive to high temperature, which affected fruit length, fruit weight and seed set. High temperatures from 10 DAA until 30 DAA inhibited carbohydrate accumulation and adversely affected seed germinability and vigor.

2.8. The Ohmic Heating Processes

A comparative evaluation of the conventional and the ohmic heating processes showed that ohmic heating, when performed with low voltage gradients, promoted degradation of both the ascorbic acid and the total vitamin C in a manner similar to conventional heating. However, high voltage gradients induced greater ascorbic acid degradation because of electrochemical reactions. These reactions increased when high electric fields were applied, producing compounds that catalyzed the degradation pathways of ascorbic acid in the presence of oxygen. Electric field and metal corrosion when stainless Steel electrodes were used, did not affect the final concentration of ascorbic acid (Domeneghini, et al., 2012).

2.9. Pulsed Electric Fields (PEF) Processing

The Pulsed Electric Fields (PEF) system is composed by a high voltaje charging unit followed by a voltage multiplier, 50 Ω coaxial cables, an air spark gap switch triggered by a trigger generator and a treatment cell. PEF in principle preserves nutritional, functional and sensory characteristics of foods since the treatment does not require high temperatures. It has been observed that PEF treatment causes the electroporation of the cell membrane and consequently the micro-organism destruction (Cortesea, et al., 2011).

PEF-processing may allow obtaining safe and shelf stable plant-based products with increased contents in health-related phytochemicals and other quality-related compounds. In general, temperature during processing and storage is an important factor affecting the bioactive content of the processed product. However, not only temperature but other factors are involved in quality losses. Therefore, individual specific approaches must be used for successfully scaling-up PEF-processing for fruit and vegetable juices at an industrial level (Odriozola-Serrano, et al., 2011)

2.10. Ultraviolet Light-C (UV-C) Radiation Treatments of Fresh-Cut Fruits

UV irradiation processing of fresh-cut fruits leads to increase in antioxidants, polyphenols, and flavonoids. Hence, apart from the application of UV for microbial safety at industrial levels, this novel technology can also be exploited for enhancement of health promoting compounds for benefit of consumers (Alothman, et al., 2009).

Blueberries (*Vaccinium corymbosum, cvs.* Collins, Bluecrop) were exposed to ultraviolet light-C (UV-C) radiation treatments from 0 to 4 kJ/m2 prior to 7 days storage at 5 °C plus 2 days at 20 °C, 90% RH. Light treatment did not affect its weight and firmness. Antioxidants as measured by total anthocyanin, total phenolics, and ferric reducing antioxidant power (FRAP) were higher in 'Collins' fruit given 0 or 1 kJ/m2 UV-C compared to unstored fruit. In 'Bluecrop', total anthocyanin content and FRAP values increased with treatment intensity, with highest values seen in fruit given 2 or 4 kJ/m2 UV-C, but no clear treatment effects were seen in total phenolic content. The postharvest application of UV-C radiation may enhance antioxidant levels. Stimulation of antioxidants by UV-C radiation appears to be dependent on cultivar, and the individual flavonoid changes in blueberries following UV-C treatment need to be explored (Perkins-Veazie, et al., 2008).

Erkan, et al., (2008) researched the changes in antioxidant capacity, enzyme activity and decay development in strawberry fruit (*Fragaria x ananassa* Duch.) illuminated with different UV-C dosages. Three UV-C illumination durations and dosages, 1, 5 and 10 min, (0.43, 2.15 and 4.30 kJm−2) tested promoted the antioxidant capacity and enzyme activities and significantly reduced the severity of decay during storage at 10 ◦C compared to the control. UV-C illumination for 5 and 10 min showed the best results for enhancing antioxidant capacity. All UV-C dosages increased the phenolic content of strawberry fruits as well. Total anthocyanin content increased during storage in all treatments.

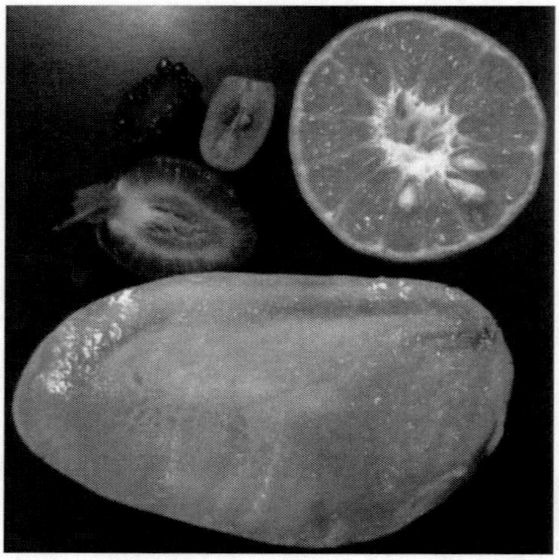

Figure 3. Fruits: grapefruit, mango, citrus, strawberries.

Nowadays there is not information about the effect of ultraviolet (UV-C) treatment on total phenol, flavonoid, and vitamin C content in seeds, only in fresh-cut fruits, as grapefruit, mango, citrus, and others (Figure 3).

Conclusion

The fruit seeds such as mango, grape, strawberries, raspberries, blackberries, among others have antioxidant activity and phenolic content as the edible portions. Nutraceuticals and functional seeds improve health and prevent aging-related chronic diseases. The effect of different treatments in this kind of foods increases nutraceutical and functional properties of seeds.

References

Ahmed, A., Saeid, D., Eman, A. & Reham, E. (2007). Egyptian mango by-product 1. Compositional quality of mango seed kernel. *Food Chemistry*, *103*, 1141–52.

Ali, A. M., Yong-Lee, Ch. (2008). Optimization of phenolics and dietary fibre extraction from date seeds. *Food Chemistry*, *108*, 977–985

Alothman, M., Bhat, R. & Karim, A. A. (2009). UV radiation-induced changes of antioxidant capacity of fresh-cut tropical fruits. *Innovative Food Science and Emerging Technologies*, *10*, 512–516.

Aluko, R. E. (2012). *Functional Foods and nutraceuticals*. ISSN 1572-0330 Springer. New York. 3-5.

Arogba, S. S. (1999). The performance of processed mango (*Mangifera indica*) kernel flour in a model food system. *Bioresource Technology*, *70*, 277-281.

Azadmard-Damirchia, S., Habibi-Nodeha, F., Hesaria, J., Nematib, M. & Fathi Achachloueic, B. (2010). Effect of pretreatment with microwaves on oxidative stability and nutraceuticals content of oil from rapeseed. *Food Chemistry*, *121*(4), 1211–1215.

Bamba, T., Kanauchi, O., Andoh, A. & Fujiyama, Y. (2002). A new prebiotic from germinated barley for nutraceutical treatment of ulcerative colitis. Journal of *Gastroenterology and Hepatology*, *17*(8), 818–824.

Bewley, J. D., Bradford, K. J., Hilhorts H. W. M. & Nanogaki, H. (2012). Development and Maduration Seeds. *Physiology of Development Germination and Domancias*. Chapter 2. Third Edition. ISBN 978-1-4614-4692-7. Springer. New York. Heidelberg Dordrecht London.

Carpenter, R., O'Grady, M. N., O'Callaghan, Y. C., O'Brien, N. M. & Kerry, J. P. (2007). Evaluation of the antioxidant potential of grape seed and bearberry extracts in raw and cooked pork. *Meat Science*, *76*(4), 604–610.

Chamorro, S., Goñi, I., Viveros, A., Hervert-Hernández, D. & Brenes, A. (2012). Changes in polyphenolic content and antioxidant activity after thermal treatments of grape seed extract and grape pomace. *European Food Research Technology*, *234*,147–155.

Cisneros-Zevallos, L. (2003). The use of controlled postharvest abiotic stresses as a tool for enhancing the nutraceutical content and adding-value of fresh fruits and vegetables. *Journal of Food Science*, *68*, Nr. 5.

Clifford, M. N., Scalbert, A. (2000). Ellagitannins-nature, occurrence and dietary burden. *Journal of the Science of Food and Agriculture*, *80(7)*, 1118–1125.

Cortesea, P., Dellacasaa, G., Gemmea, R., Bonetta, S., Bonetta, Si., Carrarob, E., Motta, F., Paganoni, M. & Pizzichemi, M. (2011). A pulsed electric field (PEF) bench static system to study bacteria inactivation. *Nuclear Physics B*, *Proceedings Supplements*, (215) *1*, 162–164.

Domeneghini Mercali, G., Pez-Jaeschke, D., Tessaro, I. C. & Damasceno Ferreira M. L. (2012). Study of vitamin C degradation in acerola pulp during ohmic and conventional heat treatment. *LWT - Food Science and Technology*, *47*, 91-95

Erkan, M., Wang, S. Y. & Wang, Ch. Y. (2008). Effect of UV treatment on antioxidant capacity, antioxidant enzyme activity and decay in strawberry fruit. *Postharvest Biology and Technology*, *48*, 163–171.

Gould, G. W. (1995). Biodeterioration of foods and an overview of preservation in the food and dairy industries. *International Biodeterioration & Biodegradation*, *36*(3–4), 267-277.

Gu, R. Q. & Shen H. M. (1989). Effects of space flight on the growth and some cytological characteristics of wheat seedlings. *Acta Phytophysiol Sinica*, *15*(4), 403-7.

Haworth, J. E. (2003). Natural Antioxidants, Review. *Ingredient Technology. Proceedings of the 56th American Meat Science Association*. Reciprocal Meat Conference. Columbia, Missouri. 95-98. www.meatscience.org

Hayat, K., Zhang, X., Farooq, U., Abbas, S., Xia, S., Jia, Ch., Zhong, F. & Zhang, J. (2010). Effect of microwave treatment on phenolic content and antioxidant activity of citrus mandarin pomace. *Food Chemistry*, *123*(2), 423–429.

Heim, K. E., Tagliaferro, A. R. & Bobilga, D. J. (2002). Flavonoid antioxidants: chemistry, metabolism and structure-activity relationship. *Journal of Nutritional Biochemistry*, *13*, 572-584.

Hernández-Montoya, V., Montes-Morán, M. A. & Elizalde-González, M. P. (2009). Study of the thermal degradation of citrus seeds. *Biomass and Bioenergy*, *33*,1295–1299

Hu, Y. Y., Xu, Y. & Pan, J. Z. (1994). Effect of microwave treatment ongermination of rice seeds. - *Plant Physiology Commun*, *30*, 414-416.

Inglett, G. E. & Chen, D. (2011). Contents of phenolic and flavonoids and antioxidant activities in skin, pulp and seeds or miracle fruit. *Journal of Food Science*, *76*(Nr. 3), 479-482.

Kimura, H., Ogawab, S., Jisaka, M., Kimura, Y., Katsube, T. & Yokota, K. (2006) Identification of novel saponins from edible seeds of Japanese horse chestnut (*Aesculus turbinata* BLUME) after treatment with wooden ashes and their nutraceutical activity. *Journal of Pharmaceutical and Biomedical Analysis*, *41*(5), 1657–1665.

Maisuthisakul, P. & Gordon, M. (2009). Antioxidant and tyrosinase inhibitory activity of mango seed kernel by product. *Food Chemistry*, *117*, 332–341

Masibo M. & He, Q. (2008). Major mango polyphenols and their potential significance to human health. *Comprehensive Reviews in Food Science and Food Safety*, *7*(4), 309–319.

Odriozola-Serrano, I., Aguiló-Aguayo, I., Soliva-Fortuny, R. & Martín-Belloso, O. (2011). Pulsed electric fields processing effects on quality and health-related constituents of plant-based foods. *Trends in Food Science and Technology*, Available online 25 October 2011, 1-10.

Osmani, S. A., Hansen, E. H., Malien-Aubert, C., Olsen, C. E., Bak, S. & Møller, B. L. (2009). Effect of glucuronosylation on anthocyanin color stability. *Journal Agriculture of Food Chemistry*, *22*, *57(8)*, 3149-55.

Pagamas, P. & Nawata, E. (2008). Sensitive stages of fruit and seed development of chili pepper (Capsicum annuum L. var. Shishito) exposed to high-temperature stress. *Scientia Horticulturae*, *117*, 21–25

Perkins-Veazie, P., Collins, J. K. & Howard, L. (2008). Blueberry fruit response to postharvest application of ultraviolet radiation. *Postharvest Biology and Technology*, *47*, 280–285.

Ren, W., Zhang, Y., Deng, B., Guo, H., Cheng, L., & Liu, Y. (2010). Effect of space flight factors on alfalfa seeds. *African Journal of Biotechnology*, *9*(43), 7273-7279.

Rodrigo, D., van Loey, A. & Hendrickx, M. (2007). Combined thermal and high pressure color degradation of tomato puree and strawberry juice. *Journal of Food Engeeniering*, *79*, 553–560.

Rukmini,C. & Vijayaraghavan, M. (1984). Nutritional and toxicological evaluation of mango kernel oil. *JAOCS*, *61*(4), 789-792

Shahidi, F. (2009). Nutraceuticals and functional foods: Whole versus processed foods. *Trends in Food Science and Technology*, *20*, 376–387.

Segura-Campos, M. R., Salazar-Vega, I. M., Chel-Guerrero, L. A. & Betancur-Ancona, D. A. (2013). Biological potential of chia (Salvia hispanica L.) protein hydrolysates and their incorporation into functional foods. *LWT - Food Science and Technology*, *50*, 723-731.

Siddhurajub, P. & Beckera, K. (2007). The antioxidant and free radical scavenging activities of processed cowpea (*Vigna unguiculata* (L.) Walp.) seed extracts. *Food Chemistry*, *101*(1), 10–19.

Soong, Y. & Barlow, J. (2004). Antioxidant activity and phenolic content of selected fruit seeds. *Food Chemistry*, *88*, 411–7.

Threlfall, R., Morris, J. R., Howard, L.R., Brownmiller, C. R. & Walker, T. L. (2005). Pressing effects on yield, quality, and nutraceutical content of juice, seeds, and skins from black beauty and sunbelt grapes. *Journal of Food Science*, *70*(3), S167–S171.

Tiwari, B. K., Patras, A., Brunton, C. & O'Donnell, C. P. (2010). Effect of ultrasound processing on anthocyanins and color of red grape juice. *Ultrasonics Sonochemistry*, *17*, 598–604.

Tiwari, B.K., O'Donnell, C.P., Patras, A. & Cullen, P. J. (2008). Anthocyanin and ascorbic acid degradation in sonicated strawberry juice. *Journal Agriculture of Food Chemistry*, *56*, 10071–10077.

Tkalec, M., Malarić, K., Pavlica, M., Pevalelek-Kozlina, B. & Vidaković-Cifrek, Ž. (2009). Effects of radiofrequency electromagnetic fields on seed germination and root meristematic cells of *Allium* cepa L. *Mutation Research Genetic Toxicology and Environmental Mutagenesis*, *672*(2), 76-81

Torres-Giner, S., Martinez-Abad, A., Ocio, M. J. & Lagaron, J. M. (2010). Stabilization of a nutraceutical Omega-3 fatty acid by encapsulation in ultrathin electrosprayed zein prolamine. *Journal of Food Science*, *75*(6), N69–N79.

Trugo, L. C. C., Donangelo, M., Trugo, M. F. N. & Bach-Knudsen, K. E. (2000). Effect of heat treatment on nutritional quality of germinated legume seeds. *Journal Agriculture of Food Chemistry*, *48*(6), 2082–2086.

Vázquez-Ovando, A., Rosado-Rubio, G., Chel-Guerrero, L., & Betancur-Ancona, D. (2009). Physicochemical properties of a fibrous fraction from chia (*Salvia hispanica* L.). *LWT - Food Science and Technology*, *42*,168–173.

Verhoeven, D., Assen, N., Goldbohm, R.A., Dorant, E., Vantveer, P., Hermus, R.J.J., & Vandenbrandt, P.A. (1997). Vitamins C and E, retinol, beta-carotene and dietary fibre in relation to breast cancer risk: a prospective cohort study. *British Journal Cancer, 75,*149-155.

Voutilainen, S., Nurmi, T., Mursu, J. & Rissanen, T.H. (2006). Carotenoids and cardiovascular health. *American Journal Clinical of Nutrition*, 83,1265–71.

Zhou, S. H., Fang, Z. X., Lü, Y., Chen, J. C., Liu, D. H., & Ye, X. Q. (2009). Phenolics and antioxidant properties of bayberry (*Myrica rubra* Sieb. et Zucc.) pomace. *Food Chemistry*, *112*, 394–399.

INDEX

B

C

E

F

G

H

I

J

K

L

M

N

T

U

V

W

X

Y

Z